Plasma Physics by Laser and Applications 2013 Conference

(PPLA 2013)

Journal of Physics: Conference Series Volume 508

Lecce, Italy
2 – 4 October 2013

ISBN: 978-1-63439-288-4
ISSN: 1742-6588

Printed from e-media with permission by:

Curran Associates, Inc.
57 Morehouse Lane
Red Hook, NY 12571

Some format issues inherent in the e-media version may also appear in this print version.

Printed by Curran Associates, Inc. (2014)

For permission requests, please contact the Institute of Physics
at the address below.

Institute of Physics
Dirac House, Temple Back
Bristol BS1 6BE UK

Phone: 44 1 17 929 7481
Fax: 44 1 17 920 0979

techtracking@iop.org

Additional copies of this publication are available from:

Curran Associates, Inc.
57 Morehouse Lane
Red Hook, NY 12571 USA
Phone: 845-758-0400
Fax: 845-758-2634
Email: curran@proceedings.com
Web: www.proceedings.com

Plasma Physics by Laser and Applications 2013 Conference

(PPLA 2013)

Journal of Physics: Conference Series Volume 508

Lecce, Italy
2 – 4 October 2013

ISBN: 978-1-63439-288-4
ISSN: 1742-6588

IOP Institute of Physics

TABLE OF CONTENTS

PAPERS

THE PARTICLE LASER-PLASMA ACCELERATION IN ITALY 1
D Giulietti

TNSA ION ACCELERATION AT 10^{16} W/CM^2 SUB-NANOSECOND LASER INTENSITY 7
L Torrisi, M Cutroneo, L Calcagno, M Rosinski, J Ullschmied

QUASI-MONOENERGETIC PROTON BEAMS BY LASER-PLASMA X-RAYS 14
D Delle Side, J Krása, V Nassisi, L Velardi

DISTANCE-OF-FLIGHT SPECTRA OF CHARGE DENSITY OF IONS GENERATED WITH A LOW LASER INTENSITY 21
J Krása, L Velardi, A Lorusso, D Delle Side, V Nassisi

ON THE EMITTANCE DEPENDENCE ON ANODE MORPHOLOGY OF LASER INDUCED ION BEAMS 26
L Velardi, D Delle Side, V Nassisi

THE LICPA ACCELERATOR OF DENSE PLASMA AND ION BEAMS 31
J Badziak, S Jablonski, T Pisarczyk, T Chodukowski, P Parys, P Ra̧czka, M Rosinski, E Krousky, J Ullschmied, R Liska, M Kucharik, L Torrisi

BASIC FEATURES OF ELECTROMAGNETIC PULSE GENERATED IN A LASER-TARGET CHAMBER AT 3-TW LASER FACILITY PALS 39
M De Marco, M Pfeifer, E Krousky, J Krasa, J Cikhardt, D Klir, V Nassisi

OPTICAL EMISSION FROM AL TARGET IRRADIATED BY FLASH 43
M Stránský, K Rohlena

MEV ION BEAMS GENERATED BY INTENSE PULSED LASER MONITORED BY SILICON CARBIDE DETECTORS 48
L Calcagno, P Musumeci, M Cutroneo, L Torrisi, F La Via, J Ullschmied

ELIMED, MEDICAL AND MULTIDISCIPLINARY APPLICATIONS AT ELI-BEAMLINES 54
F Schillaci, A Anzalone, G A P Cirrone, M Carpinelli, G Cuttone, M Cutroneo, C De Martinis, D Giove, G Korn, M Maggiore, L Manti, D Margarone, A Musumarra, F M Perozziello, I Petrovic, P Pisciotta, M Renis, A Ristic-Fira, F Romano, F P Romano, G Schettino, V Scuderi, L Torrisi, A Tramontana, S Tudisco

HIGH ENERGY X-RAY EMISSION DRIVEN BY HIGH VOLTAGE CIRCUIT SYSTEM 61
M Di Paolo Emilio, L Palladino

CHARACTERIZATION OF THIN FILMS FOR TNSA LASER IRRADIATION 68
M Cutroneo, L Torrisi, L Calcagno, A Torrisi

A PROTON SOURCE VIA LASER ABLATION OF HYDROGENATED TARGETS 75
D Delle Side, V Nassisi, L Velardi

INTRAOCULAR LENS EMPLOYED FOR CATARACT SURGERY 82
A M Roszkowska, L Torrisi

RESEARCH ON LASER-REMOVAL OF A DEUTERIUM DEPOSIT FROM A GRAPHITE SAMPLE 89
M Kubkowska, E Skladnik-Sadowska, K Malinowski, M J Sadowski, M Rosinski, P Gasior

HIGH HARMONIC GENERATION FROM ATOM CLUSTERS 95
M Aladi, I B Földes

HALL MAGNETOHYDRODYNAMIC MODEL OF THE QUASI-SINGLE HELICITY SELF-ORGANIZATION OF RFP 99
A Gurin

LASER ABLATION CHARACTERIZATION IN LABORATORI NAZIONALI DI LEGNARO 104
D Scarpa, P Nicolosi, A Franci, A Tomaselli, M Manzolaro, S Corradetti, J Vasquez, M Rossignoli, M Calderolla, A Monetti, A Andrighetto, G Prete

HIGH HARMONICS GENERATION AT THE INTERACTION OF AN INTENSE FS LASER PULSE WITH AN OVERDENSE PLASMA LAYER 108
A Mihailescu, V Stancalie, V Pais

THOMSON PARABOLA SPECTROMETRY OF LASER GENERATED PLASMA AT PALS LABORATORY 113
M Cutroneo, L Torrisi, S Cavallaro, L Ando', A Velyhan

A STUDY OF THE SECONDARY ELECTRONS YIELD PRODUCED BY LASER INDUCED IONS ON FARADAY CUPS AS A FUNCTION OF BEAM ANGLE 118
V Nassisi, D Delle Side, L Velardi, G Buccolieri, A De Benedittis, F Paladini

HIGH MAGNETIC PULSE BY SOLENOIDS FOR INTENSE ION BEAM TRANSPORT 123
E Giuffreda, L Velardi, D Delle Side, F Paladini, V Nassisi
NEUTRON FLUENCES OF THE D-D FUSION REACTION AT 10^{16} W/CM^2 LASER-TARGET INTERACTIONS 129
S Cavallaro, L Torrisi, M Cutroneo, J Krása, J Ullschmied
POTENTIAL OF A LASER-DRIVEN SOURCE OF CHARACTERISTIC γ-RAYS FOR FAST DETECTION AND IDENTIFICATION OF NUCLEAR MATERIALS 133
F Belloni
LASER ABLATION COUPLED TO MASS QUADRUPOLE SPECTROMETRY FOR ANALYSIS IN THE CULTURAL HERITAGE 138
A Torrisi, M Cutroneo, E D Castrizio, L Torrisi
ROOT CANALS DECONTAMINATION BY COHERENT PHOTONS INITIATED PHOTOACUSTIC STREAMING (PIPS) OF IRRIGANTS: AN EX-VIVO STUDY 143
E Pedullà, C Genovese, C Scolaro, M Cutroneo, G Tempera, E Rapisarda, L Torrisi
ND:YAG LASER IRRADIATION OF SINGLE LAP JOINTS MADE BY POLYETHYLENE AND POLYETHYLENE DOPED BY CARBON NANOMATERIALS 149
A M Visco, V Brancato, M Cutroneo, L Torrisi
UV-VIS-NIR SPECTRAL OPTICAL PROPERTIES OF SILVER IODIDE BORATE GLASSES 154
C Crupi, G Di Marco, L Torrisi, C Branca, G Carini, U Wanderlingh, G D'Angelo
AN ANTIBACTERIAL COATING OBTAINED THROUGH IMPLANTATION OF TITANIUM IONS 159
D Delle Side, V Nassisi, E Giuffreda, L Velardi, P Alifano, A Talà, S M Tredici
WETTING ABILITY MODIFICATIONS IN BIOCOMPATIBLE POLYMERS INDUCED BY PULSED LASERS 164
C Scolaro, L Torrisi, M Cutroneo, L Velardi
A PULSED MAGNETIC STRESS APPLIED TO DROSOPHILA MELANOGASTER FLIES 169
D Delle Side, M P Bozzetti, A Friscini, E Giuffreda, V Nassisi, V Specchia, L Velardi
ANALYSIS OF SELECTIVE LASER CLEANING OF PATINA ON BRONZE COINS 174
G Buccolieri, V Nassisi, L Torrisi, A Buccolieri, A Castellano, M Di Giulio, E Giuffreda, D Delle Side, L Velardi
RESEARCH ACTIVITY IN THE LABORATORY FOR INERTIAL CONFINEMENT FUSION IN ENEA – FRASCATI 180
R De Angelis, P Andreoli, F Consoli, G Cristofari, G Di Giorgio
Author Index

Journal of Physics: Conference Series **508** (2014) 011001 doi:10.1088/1742-6596/508/1/011001

Preface

The "Plasma Physics by Laser and Applications" Conference (PPLA 2013) is a biennial meeting in which the National teams involved in Laser-Plasma Interaction at high intensities communicate their late results comparing with the colleagues from the most important European Laser Facilities. The sixth appointment has been organized in Lecce, Italy, from 2 to 4 October 2013 at the Rector Palace of the University of Salento.

Surprising results obtained by laser-matter interaction at high intensities, as well as, non-equilibrium plasma generation, laser-plasma acceleration and related secondary sources, diagnostic methodologies and applications based on lasers and plasma pulses have transferred to researchers the enthusiasm to perform experiments *ad maiora*. The plasma generated by powerful laser pulses produces high kinetic particles and energetic photons that may be employed in different fields, from medicine to microelectronics, from engineering to nuclear fusion, from chemistry to environment. A relevant interest concerns the understanding of the fundamental physical phenomena, the employed lasers, plasma diagnostics and their consequent applications.

For this reason we need continuous updates, meetings and expertise exchanges in this field in order to follow the evolution and disclose information, that has been done this year in Lecce, discussing and comparing the experiences gained in various international laboratories.

The conference duration, although limited to just 3 days, permitted to highlight important aspects of the research in the aforementioned fields, giving discussion opportunities about the activities of researchers of high international prestige. The program consisted of 10 invited talks, 17 oral talks and 17 poster contributions for a total of 44 communications. The presented themes covered different areas and, far from being exhaustive gave updates, stimulating useful scientific discussions.

The **Organizers** belong to three Italian Universities, Prof. V. Nassisi of Salento University, Prof. D. Giulietti of Pisa University and Prof. L. Torrisi of Messina University.

The **Scientific Committee** was constituted by colleagues coming from different European laboratories: Dr. F. Belloni from European Commission, Bruxell, Belgium; Prof. M. Borghesi from the Queen's University of Belfast, United Kingdom; Prof. L. Calcagno from Catania University, Italy; Prof. D. Giulietti from Pisa University, Italy; Dr. J. Krása from Academy of Science of Czech Republic, Prague; Prof. V. Malka from Laboratoire d'Optique Appliquée, Palaiseau, France; Prof. V. Nassisi from Salento University, Italy; Prof. L. Palladino from L'Aquila University, Italy; Prof. L. Torrisi from Messina University, Italy; Prof. Ullschmied from Academy of Science of Czech Republic, Prague; Prof. J. Wolowski from Institute of Plasma Physics and Laser Microfusion of Warsaw, Poland and Dr. J. Badziak from Institute of Plasma Physics and Laser Microfusion of Warsaw, Poland.

The **Local Organizing** team was composed by: Dr. G. Buccolieri, Dr. D Delle Side, Dr. F. Paladini and Dr. L Velardi from Salento University and Dr. M. Cutroneo from Messina University. The **Scientific secretariat** was coordinated by Dr. D. Dell'Anna from Salento University.

The **Topics** discussed in the conference were:

- Laser-Matter interactions;
- Laser ion sources;
- Electron beam generation;
- Physics of non-equilibrium plasmas;
- Theoretical models in plasmas;
- Photons and particles emission from pulsed plasmas;
- Ion acceleration from plasma;
- Fs laser pulses;
- Pulsed laser deposition;
- Applications of laser beams and pulsed plasmas;
- Techniques of characterization of plasmas.

The colleagues attending the conference were about 80. **The Chairmen and Presidents** of the different Conference sessions were: Prof. V. Nassisi, Prof. D. Giulietti, Prof. L. Torrisi, Prof. M. Borghesi, Dr. K. Rohlena (ASCR of Prague, Czech Republic), Prof. D. Neely (RAL, Oxon, UK), Dr. J. Ullschmied (ASCR, Prague, Czech Republic), Prof. S. Ratynskaia (Royal Institute of Technology, Stockholm, Sweden), Dr. J. Krása, Dr. J. Badziak.

The award Leos Laska, a Czech colleague which gave in its country relevant contributions to development of the experimental activities in these research fields, has been proposed in memory to his work and to stimulate the interest of young researchers in this important sector. The Scientific Committee conferred the prize to Dr. Mariapompea Cutroneo, PhD in Physics of Messina University, for her activity in the field of new methodologies related to the ion acceleration in laser-generated plasma.

The widespread success of the event suggests we will meet again, next 2015, in another South Italy venue, as wonderful and welcoming as Lecce was.

Vincenzo Nassisi, Danilo Giulietti, Lorenzo Torrisi and Domenico Delle Side

The **Sponsors** of the conference, besides the Salento, Messina and Pisa Universities, were:

Dipartimento di Matematica e Fisica *"Ennio De Giorgi"*

Istituto Nazionale di Fisica Nuclare

Banca Monte dei Paschi di Siena

Teledyne Lecroy

Rhode & Schwarz

Crisel Instruments

Journal of Physics: Conference Series **508** (2014) 011002 doi:10.1088/1742-6596/508/1/011002

Ariator Plast

Flying shoes

Casciaro gardens

P & P socks

Centro Diagnostico Città di Casarano

Acli-Racale

We gratefully acknowledge their support.

Conference photographs

Plenary photo, during the second day of the works.

Journal of Physics: Conference Series **508** (2014) 011003 doi:10.1088/1742-6596/508/1/011003

Dr M Cutroneo (left) being awarded with the Leos Laska prize by Professor K Rohlena (right).

A moment of banqueting during the conference works.

INFN

Dipartimento di Matematica e Fisica
"Ennio De Giorgi"

UNIVERSITÀ DI PISA

PPLA13

Plasma Physics by Laser and Applications 2013

Lecce, October 2nd- 4th, 2013

http://ppla13.dmf.unisalento.it

Topics:

Laser-Matter interaction
Laser ion-sources
Electron beam generation
Physics of non-equilibrium plasmas
Theoretical models in plasmas
Photons and particles emission from pulsed plasmas
Ion acceleration from plasma
Fs laser pulses
Pulsed Laser Deposition
Applications of laser beams and pulsed plasmas
Techniques of characterization of plasmas

Scientific Committee

- Dr. F. Belloni, European Commission, Bruxelles, Belgium
- Prof. M. Borghesi, Queen's University of Belfast, UK
- Prof. L. Calcagno, University of Catania, Italy
- Prof. D. Giulietti, University of Pisa, Italy
- Dr. J. Krasa, ASCR, Prague, Czech Republic
- Prof. V. Malka, LOA, Palaiseau, France
- Prof. V. Nassisi, University of Salento, Italy
- Prof. L. Palladino, University of L'Aquila, Italy
- Prof. L. Torrisi, University of Messina, Italy
- Dr. J. Ullschmied, ASCR, Prague Czech Republic
- Prof. J. Wolowski, IPPLM, Warsaw, Poland

Invited Speakers

- Prof. M. Borghesi, QUB, Belfast, UK
- Prof. V. Malka, LOA, Palaiseau, FR
- Prof. J. Wolowski, IPPLM, Warsaw, PL
- Dr. J. Krasa, ASCR, Prague, CZ
- Prof. J. Badziak, IPPLM, Warsaw, PL
- Prof. D. Neely, CLF-RAL, Didcot, UK
- Prof. S. Ratynskaia, RIT, Stockholm, SE
- Prof. G. Foti, INFN-LNS, Catania, IT
- Dr. P. Cirrone, INFN-LNS, Catania, IT
- Dr. M. Galimberti, CLF-RAL, Didcot, UK

Organizing Committee

- Prof. D. Giulietti, University of Pisa, Italy
- Prof V. Nassisi, University of Salento, Italy
- Prof. L. Torrisi, University of Messina, Italy

Local Organizing Team

- Dr. G. Buccolieri, University of Salento, Italy
- Dr.ssa M. Cutroneo, University of Messina, Italy
- Dr. D. Delle Side, University of Salento, Italy
- Dr F. Paladini, University of Salento, Italy
- Dr L. Velardi, University of Salento, Italy

Scientific Secretariat

Dr.ssa D. Dell'Anna, University of Salento, Italy daniela.dellanna@unisalento.it

Peer review statement

All papers published in this volume of *Journal of Physics: Conference Series* have been peer reviewed through processes administered by the proceedings Editors. Reviews were conducted by expert referees to the professional and scientific standards expected of a proceedings journal published by IOP Publishing.

The particle laser-plasma acceleration in Italy

D Giulietti[1]
Physics Department, University of Pisa and INFN, Italy

E-mail: danilo.giulietti@df.unipi.it

Abstract. In this paper the beginning activity on Laser Plasma Acceleration in Italy and the recent results obtained in the frame of the INFN Strategic Project PLASMONX will be presented. The project, involving eight groups and two National Laboratory disseminated along Italy, is aimed at the development of an innovative, high-gradient acceleration with super-intense and ultra-short laser pulses, and a tuneable, hard X/γ-ray source, based upon Thomson scattering of optical photons by energetic electrons. Both experiments require very high power, ultra-short laser pulses in combination with very bright and short electron bunches generated either by conventional acceleration (LINAC) or by laser-driven acceleration in plasmas.

1. Introduction

In a seminal paper [1] Tajima and Dawson proposed in 1979 to accelerate electrons by the very strong longitudinal electric fields associated with plasma waves generated, for example, by the beating wave process or by a wake-field of intense laser pulses. The latter requires a suitable matching between the plasma density and the laser pulse duration. However, even when matching is not satisfied, self-modulation of the pulse can provide useful conditions for acceleration. In fact, the **S**elf-**M**odulated **L**aser **W**ake-field (SMLW) has been demonstrated as a scheme suitable for the acceleration of electrons trapped from the plasma itself [2, 3]. More recently additional mechanisms have been proposed as the "forced wake-field" [4] or the "bubble" regimes [5]. All these acceleration physical processes became experimentally accessible with the advent of the powerful CPA (**C**hirped **P**ulse **A**mplification) [6] pulses. In fact several groups active in laser-matter interaction at high intensities by the early 90's of the last century become involved in Laser Plasma Acceleration experiments at the major European and Nord America laser facilities. At that time in Italy the CNR-University Pisa laser group, lead by Antonio and Danilo Giulietti, carried out a successful experimental campaign at Laboratoire d'Optique Appliquée (LOA, École Polytechnique) [7, 8, 9, 10] that induced interest in the National Institute for Nuclear Physics (INFN). After a series of seminars at Laboratori Nazionali di Frascati (LNF) given by Danilo Giulietti on the new acceleration techniques and the results obtained by his group at LOA, INFN launched a Strategic Project: **PLASMONX** (**PLAS**ma acceleration and **MON**ochromatic **X**-ray production) [11]. The facility was built in close interaction with the SPARC (**S**orgente **P**ulsata **A**uto-amplificata di **R**adiazione **C**oerente) project [12] based on an advanced 150 MeV LINAC at LNF. The main purpose of the facility consists in R&D activity aimed at the following objectives: 1) demonstration of high-gradient acceleration of electrons injected into electron plasma waves excited by ultra-short, high power laser pulses; 2) development of a monochromatic and tuneable X-ray source in the 20-1000 keV range, based upon Thomson Scattering of laser pulses by

[1] To whom any correspondence should be addressed.

relativistic electrons; 3) development of secondary sources of particles (protons, ions, positrons, muons, ...) and energetic radiation (Bremsstahlung, betatron, FEL, ...). All these lines require very bright and short electron bursts together with very high power laser pulses of sub-100fs duration. Studies at the forefront of research in this field consistently indicate that power levels and the focused intensities for a laser system appropriate for these studies are >10 TW and $>10^{19}$W/cm^2, respectively. For this reason the PLASMONX facility has been endowed by a Ti:Sapphire laser delivering pulses up to 8J in 20fs at 10Hz, that once focused on target can exceed intensities of 10^{21} W/cm^2.

2. Exploding-foil technique at ultra-relativistic intensities

At present CPA technique can produce the most powerful laser pulses due to their shortness, however, such pulses are accompanied by a low intensity ''pedestal'' of nanosecond duration arising from **A**mplified **S**pontaneous **E**mission (ASE) in the amplifier chain. This pedestal results in a precursor target irradiation which can affect the dynamics of ultra-short laser interactions with matter. A special feature of the experiments we performed in the early 2000's at LOA [8, 9, 10] where we used the ASE pedestal in order to produce, via the exploding-foil technique, a pre-formed plasma, providing favourable conditions for electron acceleration with a 35 fs pulse. This technique, already tested and validated in previous experimental works, provides a novel method to investigate acceleration schemes in conditions different from the ones using gas-jet targets. In fact, our technique allows interactions to be achieved with preformed plasmas of smaller size (typically a few tens of μm scale length) compared with the size of typical gas-jet plasmas (mm length) used in acceleration experiments. The laser-target configuration is basically the same already used in previous experiments. The linearly *p*-polarized beam of the Ti:Sapphire laser of the LOA (Salle Jaune), 1 J in 35 fs at 815 nm wavelength, was focussed by an off-axis F/5 parabola on 1.0 μm thick, 500 μm wide plastic foil target, at an angle of incidence of 20 deg. The laser intensity distribution in the focal spot was Gaussian, with 50% of the total laser energy within a circle of ≈ 4 μm diameter resulting in a laser intensity on target of $I_L \approx 8\ 10^{19}$ W/cm^2. The CPA/ASE intensity contrast ratio was of the order of 10^6. Part of the main beam (100 mJ), frequency doubled using a 2 mm thick KDP crystal and timed within a fraction of a picosecond with respect to the main beam, was used as an optical probe beam parallel to the thin foil surface. In the experiment we were able to produce for the first time interferograms in the Nomarski configuration with a femtosecond probe pulse. Energetic electrons emitted forward were the main subject of the experimental investigation. Three distinct diagnostics were used in order to measure (a) the number of electrons, (b) their angular distribution, and (c) their energy spectrum. Further diagnostics included optical imaging and spectroscopy of the transmitted laser light and detection by scintillator detectors of the gamma-rays emitted by bremsstrahlung of the energetic electrons. The plasma density was measured using interferograms taken at different probing times before and after the arrival of the main pulse in the plasma. The longitudinal plasma density profile (along the laser path) as obtained from a typical interferogram taken 50 ps before the arrival of the main pulse shows a well defined density profile, having an exponential decrease with a scale length typically of a few tens of micrometers starting from a top density that can be reasonably inferred to be approximately $4\text{x}10^{19}$ el/cm^3. The number of electrons emitted forward (i.e., in the direction of the laser pulse) was measured with a calibrated coil-based charge detector having an acceptance cone of aperture $\theta_{coll} \approx 7$ deg. The detector was placed beyond the target and after a 100 μm thick quartz plate (necessary for optical measurements), which substantially cuts-off electrons of energy below 0.2 MeV. The charge collected in this condition was about 0.2 nC per shot, corresponding to approximately 10^9 electrons. The electron angular distribution is a crucial measurement which was performed using a detector based on radiochromic films (MD55–GAFCHROMICth). After calibration, radiochromic films straightforwardly provide the angular distribution of impinging electrons. The detector basically consists of a stack of radiochromic films some of which were separated by aluminum plates of suitable thickness. The detector was placed 2.6 cm behind the target, with the center of the films aligned with the laser propagation axis. Typical radiochromic film results obtained after a single exposure to the electrons generated by focusing the CPA pulse at an intensity of $\approx 8\text{x}10^{19}$ W/cm^2 in the plasma,, show a small central spot aligned with the laser propagation axis. This feature is visible on all seven layers of the stack while other features surrounding this bright spot are only visible on the first layer. These results suggest that besides an intense electron flux of lower energy in a cone of $\theta_{ring}\approx 18$ deg aperture, there is a bunch of very collimated, high energy electrons accelerated along the laser propagation axis. This electron bunch is confined in a solid angle $\approx 10^{-3}$ sr. The electron energy spectrum was obtained with a spectrometer based on an electro-magnet coupled with a set of four photodiodes "surface barrier detectors". The entrance axis of the spectrometer was carefully aligned on the laser propagation axis in order to analyse the spectrum of the electrons in the central bright beam. A typical spectrum is shown in Fig. 1. According to this plot, the electron beam consists of a sizeable number of electrons with energies from a few MeV up to 40 MeV. The experimental results show that the accelerating field in the plasma could approach 1TV/m.

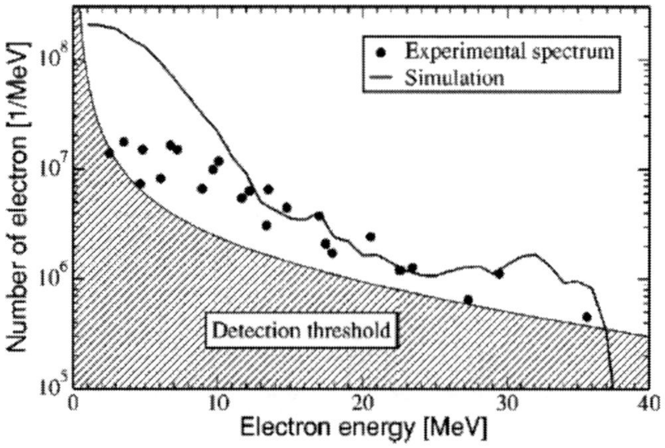

Figure 1. Spectrum of the electrons in the collimated beam obtained by magnetic spectrometer. Black dots represent the experimental data. Also shown is the detection threshold of the electron spectrometer. The solid line is the spectrum of the electrons in the collimated beam (within 3 deg around the axis) as obtained from the PIC simulation.

3. First PLASMONX experimental activity

During the construction of the PLASMONX laboratory and the assembly of the 300TW Ti: Sapphire laser by Amplitude Technologies at LNF the Italian experimental activity on LPA (**L**aser-**P**lasma **A**cceleration) was developed at CNR-Pisa laboratory an other European facility. The Pisa group devoted its main interest to the study of the propagation of the intense laser pulses in plasmas, taking up the pioneering experiment on laser guiding was performed in the late 1980s [13]. In that experiment two collinear Q-switching Nd-laser pulses, separated each other of few nanoseconds, were focused on an under-dense plasma. While the first pulse was strongly diffracted as a consequence of the suffered self-focusing effects, the second one propagated almost un-attenuated for several Rayleigh lengths in a low density channel formed by the first one. Adapting such studies to the demands of the Laser-Plasma Acceleration techniques the same group conceived an experiment oriented to find the most suitable conditions to improve the electron acceleration length, producing hollow cylindrical plasmas for guiding intense laser pulses [14]. To do so, an experimental arrangement to probe the laser-gas-jet interaction with a fast interferometric technique was setup at INO-CNR (Istituto Nazionale d'Ottica of the Consiglio Nazionale delle Ricerche) in Pisa (Italy). The laser system employed to study the plasma evolution driven by an ASE-like laser pulse is a 3.3 GW

Nd:YLF delivering two beams with a maximum energy of 5 J per beam at a wavelength of 1053 nm and duration of 3 ns FWHM. The system can provide intensities up to 10^{15} W·cm^{-2} on a target with a high temporal and spatial quality. The production of quasi-cylindrical plasmas extending over several millimeters and with a hollow density channel on their axis was demonstrated. The mean on-axis electron plasma density was ≈4x10^{18} cm^{-3}. As a speculation, it can be noted that these values of electron density are quasi-resonant in the LWFA (Laser Wake-Field Acceleration) regime with laser pulses of duration τ≈ 30 fs FWHM. For what concerns the transverse density profile, the density minimum has a value in the range of 60 to 70% with respect to the channel walls, while its radius r_{ch} at half of the depth is of the order of 30 to 50 μm. These values are suitable to optically guide a driving pulse. The density distribution can be described as

$$n(r) = n_0 + \Delta n \cdot \left(\frac{r}{r_{ch}}\right)^2 \tag{1}$$

the values for n_0 and $\Delta n(r_{ch})^{-2}$ being 4.54x10^{18} cm^{-3} and 3.74x10^{22} cm^{-5}, respectively. With reference to equation

$$w_0 = \frac{r_{ch}^2}{(\pi r_e \Delta n)^{1/4}}, \tag{2}$$

which relates the plasma channel density profile with the laser spot size to match the optical guiding condition [14], the obtained data are consistent with w_0 = 23 μm. It means that pulses focused in spots smaller than w_0 can be refractive-guided in such plasma channels. Plasmas similar to that produced in this experiment are found to be suitable for refractive laser-guiding over many Rayleigh ranges [15, 16], opening to an efficient use in high gain, long-scale acceleration experiments.

The role played by the laser guiding in LPA has been evidenced also in a recent experiment performed in the PLASMONX [17] laboratory at LNF-INFN, where 30 TW, 30 fs pulses from a Ti:Sa laser were focused on a supersonic gas-jet at an intensity of I ≈6x10^{18} Wcm^{-2}. The plasma electron density of the ionized N_2 at ≈7–12 bar backing pressure was n_e≈10^{19} cm^{-3}. Fig 2a shows the image of the electron bunch detected by a LANEX screen after dispersion in the 0.9 T magnetic spectrometer, while Fig. 2b shows the corresponding lineout together with a calibrated energy scale. During the experiment, mainly devoted to individuate the best conditions for LPA, the magnetic spectrometer was operated without the usual entrance slit. So, while the vertical dimension of the electron bunch on the LANEX depends on its divergence (μ5 mrad), the horizontal one depends also on its energy spread. The two dimensions being more or less the same (spot almost circular) it evidences that the electron bunch energy spread is less than the one you can deduce from the LANEX image, i.e. $\Delta E/E < 10\%$. In the experiment the activated Thomson scattering (TS) diagnostic evidenced the extension of acceleration length (≈2 mm), more than 4 times the Rayleigh length of the employed focusing optics, due to the onset of the relativistic self-focusing, whose threshold power in our experimental conditions was

$$P_c = \frac{2m^2e^5}{e^2}\frac{n_c}{n_e} \approx 17\frac{n_c}{n_e}GW \approx 3TW. \tag{3}$$

The Thomson scattering diagnostics together with the direct imaging of the electron bunch on LANEX screen also shows the formation of two acceleration "channels", due to the splitting of the main laser beam, most probably a consequence of aberrations on the main laser beam.

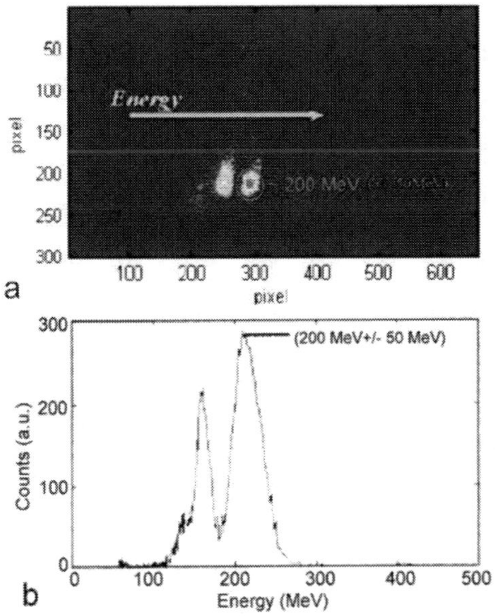

Figure 2. Electron bunch detected on LANEX after deflection in the magnetic spectrometer (a) and corresponding horizontal lineout (b). The uncertainty in the energy measurement is ± 50 MeV.

More recently [18], at LNF electron bunches (≈150pC) were produced in a laser-plasma acceleration experiment using a 10 mm helium gas-jet with backing pressures of 5, 8 and 15 bars. The pulse of Ti:Sapphire laser (duration of τ<30 fs) was focused with an off-axis parabola F/10 in a 15 μm diameter (waist) spot. The intensity was I_0=2x10^{19} W/cm^2, which corresponds to a normalized vector potential a_0=3.1 (see Fig. 3).

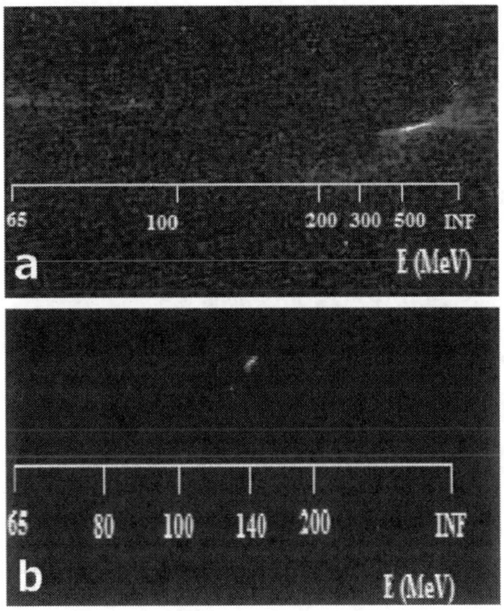

Figure 3. a) Electron spectrum obtained at 8 bars, showing a cutoff energy above 500MeV. b) Electron spectrum obtained showing the lowest energy spread (4.6 %), obtained again at 8 bars.

4. Conclusions and perspectives

The experimental and theoretical activity in LPA has been in the last fifteen years exceptionally intense and very rich of impressive results. In the next future we expect even more relevant results, as those pursued by the **B**erkeley **L**ab **L**aser **A**ccelerator (**BELLA**) project, under realization at Lawrence

Berkeley National Laboratory by the Wim Leemans group [19]. At present, several regimes for LPA have been successfully investigated, aimed to optimize the maximum energy, the energy spread, the charge and divergence of the accelerated bunches. The extension of the acceleration length, largely overcoming the Rayleigh length, is tempted following different experimental strategies; among them the control of the propagation of the intense laser pulse in the gases or plasmas confined in different ways (cells, capillaries,...). Most probably the next challenge will be the staged injection, in which in a first cell a bunch of energetic and quite monochromatic electrons is created by ionization injection [20] and further accelerated in a second stage, setup to work in the external-injection regime.
Also Italy is intensifying the engagement in this field[21, 22]. In fact the National Institute of Nuclear Physics (INFN) has recently launched the Strategic Project **PLASMONX** in which LPA techniques will be exploited using a 300TW Ti:Sapphire laser, synchronized with an advanced 150 MeV LINAC. First encouraging results are producing at LNF where the INFN, operating the most powerful Italian laser, is greatly contributing to the growth of a National community involved also in the larger European Projects **ELI** (**E**xtreme **L**ight **I**nfrastructure) and **HiPER** (**Hi**gh **P**ower laser **E**nergy **R**esearch facility) [11] since their Preparatory Phase.

Acknowledgements
The author acknowledges the contribution of the large Italian community participating to the PLASMONX Project and the Preparatory Phase of ELI and HiPER.

References
[1] Tajima T, Dawson J, Laser electron accelerator, 1979, Phys Rev Lett, **43**, 267–270
[2] Modena A, Dangor A E, Najmudin Z *et al.*, 1995, Nature (London), **377**, 606
[3] Gordon D, Tzeng K C, Clayton C E *et al.*, (1998), Phys. Rev. Lett, **80**, 2133
[4] Malka V, Fritzler S, Lefebvre E et al. 2002, *Science* **298** 1596-1600
[5] Pukhov A and Meyer-ter-Vehn J, 2002 *Appl. Phys. B* **74** 355
[6] Strickland D and Mourou G, 1985, *Opt. Commun.* **56**, 216
[7] Giulietti D, Gizzi L, Giulietti A, Macchi M et al., 1997 *Phys. Rev. Lett.* **79** 3194
[8] Galimberti M, Giulietti A, Giulietti D, Gizzi L A, Balcou P, A. Rousse and Rousseau J P, 2001 *Laser Part. Beams.* **19** 47
[9] Giulietti D, Galimberti M, Giulietti A, Gizzi L A, Borghesi M, Balcou P, Rousse A and Rousseau J P 2001 *Phys. Rev. E* **64** 015402
[10] Giulietti D, Galimberti M, Giulietti A, Gizzi L A, Numico R, Tomassini P, Borghesi M, Malka V, Fritzler S, Pittman M, Ta Phouc K, Pukhov A 2002 *Phys Plasmas* **9** 3655
[11] Giulietti D, 2007 *Il Nuovo Saggiatore* **23** 67; (http://www.lnf.infn.it/acceleratori/plasmonx/)
[12] Alesini D et al. 2003 *Nucl. Instrum. Meth. Phys. Res. A* **507** 345-349
[13] Giulietti A, Vaselli M, Cornolti F, Giulietti D, Lucchesi M 1987 *Rev. Roum. Phys.* **32** 23–27
[14] Gamucci A, Galimberti M, Giulietti D *et al.* 2006 *Appl. Phys. B* **85** 611–617
[15] Esarey E, Sprangle P, Krall J, Ting A 1997 *IEEE J. Quantum. Electr.* **33** 1879–1914
[16] Sprangle P, Esarey E, Ting A 1990 *Phys. Rev. A* **41** 4463–4469
[17] Gizzi LA, Anelli F, Benedetti C *et al.* 2009 *Il Nuovo Cimento C* **32** 433–436
[18] Grittani G, Master's Degree Thesis, Pisa University, 2013
[19] http://newscenter.lbl.gov/news-releases/2009/06/25/bella-accelerating-electrons/
[20] Clayton CE, Ralph 1. JE, Albert F *et al.* 2010 *Phys. Rev. Lett.* **105** 105003
[21] Velardi L, Delle Side D and Nassisi V, 2014 *Nucl. Instrum. Phys. Res. A* **735** 564-567
[22] Agosteo S, Anania M P, Caresana M, Cirrone G A P, De Martinis C, Delle Side D et al., 2014, *Nucl. Instrum. Meth. Phys. Res. B* in press (doi: 10.1016/j.nimb.2013.12.035)

TNSA ion acceleration at 10^{16} W/cm^2 sub-nanosecond laser intensity

L Torrisi[1,5], M Cutroneo[1], L Calcagno[2], M Rosinski[3] and J Ullschmied[4]
[1]Dep. of Physics-SdT, University of Messina, V. S. d'Alcontres 31, 98166 S. Agata (ME), Italy
[2]Dep. of Physics & Astronomy, University of Catania, V. S. Sofia 64, 95124 Catania, Italy
[3]Institute of Plasma Physics and Laser Microfusion, 23 Hery Street 01-497 Warsaw, Poland
[4] IPP ASCR - PALS, Za Slovankou 3, 182 00 Prague 8, Czech Republic

E-mail: Lorenzo.Torrisi@unime.it

Abstract. Micrometric thin targets have been irradiated in vacuum in TNSA (Target Normal Sheath Acceleration) configuration at PALS Laboratory in Prague by using 10^{16} W/cm^2 laser intensity, 1315 nm wavelength, 300 ps pulse duration and different laser beam energies and focal positions. The plasmas produced were characterized by using ion collectors, semiconductor SiC detectors, X-ray streak camera and Thomson parabola spectrometer. Time of flight techniques, time resolved imaging and ion deflection spectrometry were used to characterize the laser-generated non-equilibrium plasma and the electric field driving ion acceleration developed at the rear side of the target. The maximum ion acceleration can be obtained for optimal film thickness depending on the laser energy and on the kind of irradiated targets. Special targets containing nanostructures, showing high absorption and low reflective coefficients, induce resonant absorption effects enhancing the electric acceleration field. The maximum kinetic energy measured for proton ions was above 5.0 MeV and the ion distributions can be fitted with Coulomb-Boltzmann shifted functions.

1. Introduction

The properties of the non-equilibrium plasma generated by pulsed laser-matter interaction in vacuum, using high laser intensities and TNSA (Target Normal Sheath Acceleration) configuration, depend strongly on three key parameters: the laser properties, irradiation conditions and target structure. The main laser properties are the intensity, wavelength, pulse duration and polarization direction [1]. The irradiation conditions are characterized by the laser spot dimensions, focal position of the laser with respect to the target surface, pressure in the vacuum irradiation chamber, use of prepulse or presence of gas leak, etc. [2]. Not less important is the role of the target structure because its material, thickness, geometry, and microscopic structure influence significantly the plasma composition, its temperature and density, and the number of fast electrons. All that plays a substantial role at optimising the ion

[5] To whom correspondence should be addressed

acceleration, through reduction of reflection effects, increase of absorption, and induction of resonant absorption of electromagnetic waves [3].

Advanced targets enhancing the absorption coefficient permit to transfer more energy from the laser light to the target and to the plasma and, consequently, to develop a strong ion acceleration field at the rear side of a thin target. At the laser intensity of 10^{16} W/cm^2 the ponderomotive energy transferred to electrons from the electromagnetic wave is sufficiently high for the electrons to cross the thin target and produce TNSA effect, as it will be presented and discussed.

2. Experimental set-up

The experiments were carried out with the use of the PALS laser in Prague with laser pulse duration of 300 ps and fundamental wavelength of 1.3 μm, and of a very fast and sophisticated plasma diagnostics permitting detailed analysis of the accelerated ions [4]. The plasma was generated by the laser beam focused to a focal spot 70 μm in diameter on different planar solid targets mounted on a suitable holder. Thin targets with thickness from 0.1 μm up to 50 μm were employed, such as metals of different atomic numbers, polymers and ceramics, and advanced targets containing nanostructures. Generally, the laser beam was directed perpendicularly to the target surface with a laser pulse energy $E_L \approx 100$ J $\div$ 600 J using different focal positions (FP) with respect to the target surface, from +500 μm (focus inside the target) up to -500 μm (focus in front of the target). Vacuum in the experimental chamber was at a level of 5×10^{-6} Torr.

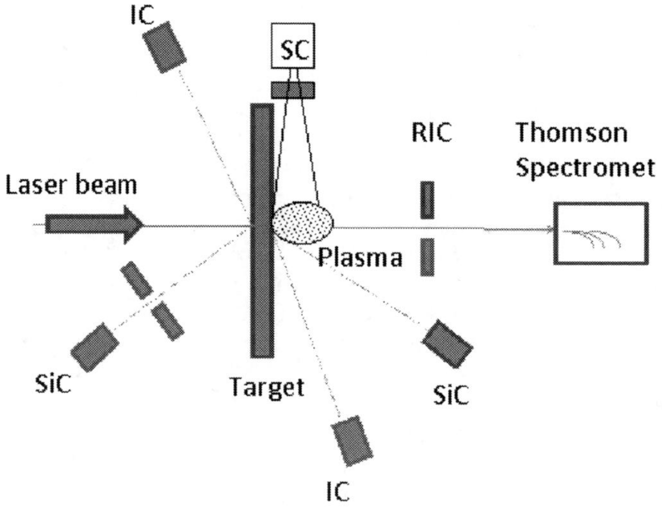

Figure 1. Schematic of the experimental set-up.

Ion collectors (IC) and ring ion collectors (RIC) with four different sectors using thin Al absorber films were employed to reduce the background coming from photons, electrons and ions and to separate protons from heavier ions. SiC detectors with 100 μm depletion layer Schottky barrier were employed in time-of-flight (TOF) configuration to detect energetic ions [5]. A Ni_2Si surface metallization is employed; it absorbs protons with an energy lower than 50 keV and Al ions with an energy lower than 350 keV. The SiC depletion layer depth permits to detect very well protons up to a maximum energy of about 10 MeV and Al ions with energy above 100 MeV. An X-ray streak camera (SC), with 2 ns exposition time and 30 ps time resolution, mounted in a side view, was employed for the FP distance evaluation. Details on the streak camera used are given in literature [6]. A Thomson parabola spectrometer (TPS) was employed as a fast measuring tool for analysis of the magnetically and electrically deflected ions in dependence on their mass-to-charge ratio. The spectrometer exploited a multi-channel plate coupled to a phosphorous screen and a CCD camera. The detected parabolas were identified by using a simulation program based on Opera 3D-Tosca taking in account the real dimension of the spectrometer and real values of the magnetic and electric field considering also their edge effects. Details on the IC, SiC, streak camera and Thomson parabola spectrometer employed in this experiment are given in previous papers [5-7].

The detectors were placed at different angles with respect to the target surface, both in backward and in forward directions, in order to study the ion emission in different irradiation conditions.
A typical scheme of the experimental set-up, for irradiation at 0° incidence angle and TNSA approach, is shown in Fig. 1.

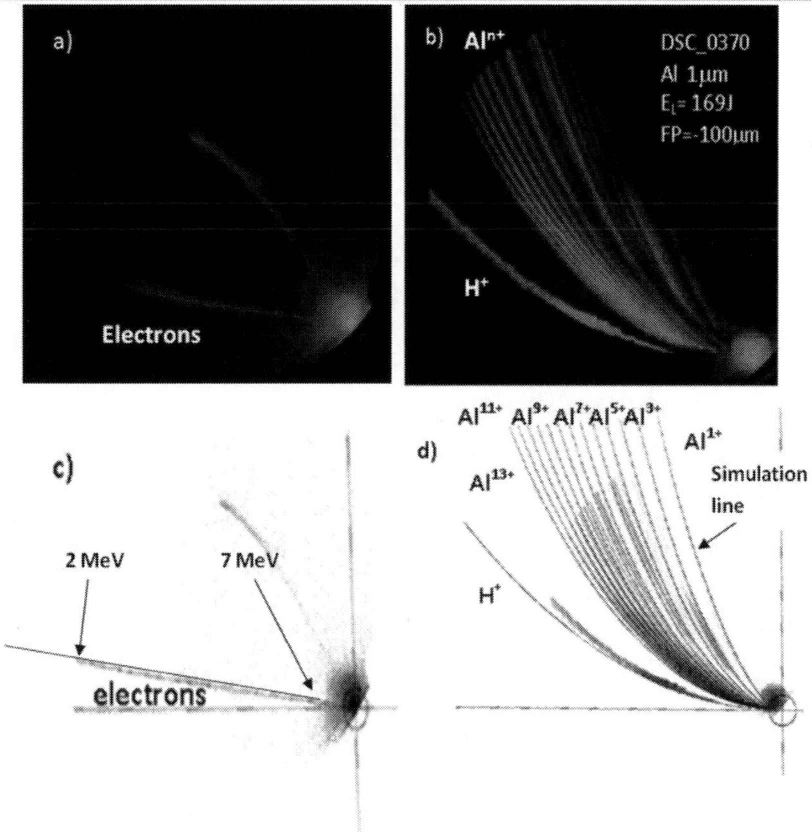

Figure 2. TPS spectra of fast electrons (a) and Al ions (b) and corresponding simulations (c, d).

3. Results

In order to confirm that the electrons emitted from a thin target in forward direction are responsible for the double layer creation and ion acceleration, TPS and SiC detectors were placed at 0° to monitor the electrons generated during the laser-matter interaction.

Fig. 2 shows two typical sets of parabolas obtained for 1 μm Al target irradiation in TNSA conditions at 170 J laser pulse energy and −100 μm FP distance, for electrons (a) and ions (b), and the corresponding simulated parabolas (line simulations overlapped to the experimental data) for electrons (c) and ions (d), respectively. The electrons were analysed by using electric deflectors charged to 2,8 kV, and a very weak magnetic field of only 50 G. The comparison of the experimental parabola (a) and the simulation one (c) shows that the electrons from 7 MeV up to 2 MeV are detected. Such high-energy electrons accelerated in the forward direction along the normal to the target surface are responsible for the electric field developed in a double layer at the rear side of a thin TNSA target.

The TPS ion diagnostics was performed using a 1.4 kV/cm electric and 0.06 T magnetic fields. The TPS spectrum for ions (b) for the case of the 1 μm Al film target shows protons and all the 13 charge states of aluminum. The recognition of the ion species, charge state and energy was based on the simulation data. In this case the maximum proton energy was about 3.3 MeV and the Al ion energy of about 3 MeV/charge state, in agreement with the values obtained from direct ion detection by SiC detectors.

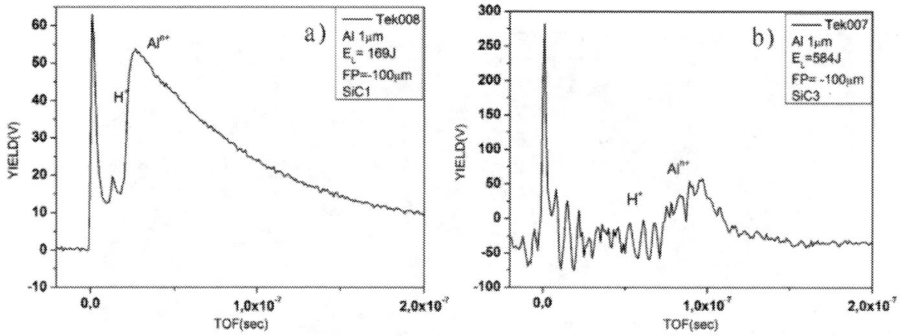

Figure 3. Typical SiC TOF signals for ion detection in forward (a) and in backward (b) directions.

Fig. 3 shows two typical examples of SiC signals for radiation detection in the forward (a) and backward (b) directions. The signals are acquired during the laser irradiation at 170 J pulse energy and -100 μm FP of 1 μm Al TNSA target. They show a photopeak, a ns-scale electron signals and slower ion peaks of which the faster ones are due to the protons. The measured spectra demonstrate that a high electron yield emission occurs both in the forward and backward directions. The kinetic energies of the detected electrons are in the range 1 keV-100 keV. The maximum proton energies in the forward and backward directions are 3.5 MeV and about 1.8 MeV, respectively.

The use of advanced targets with a structure that permits to decrease the surface reflectivity, to increase the absorption radiation and to induce resonant absorption effects, and the use of special target geometries may enhance the laser energy transferred to the plasma and increase the electric field of ion drive acceleration developed on the rear side of the TNSA target.
The surface roughness, porous structures, metal/polymer multilayer and polymer containing nanostructures were investigated at various laser energies, focal positions and target thicknesses in order to reach the maximum ion acceleration in forward direction.

Measurements have indicated that an increase in the proton energy of the order of 20-30% can be obtained by changing the reflectivity and absorption coefficient of thin films [3]. An increase of the order of 30% in energy and yield has been obtained also by using metallic nanometric structures (carbon nanotubes, Cu, Ag and Au nanoparticles) embedded in thin polymeric films (PE, acrylic resins). These nanostructures induce surface plasmon resonance absorption of the incident radiation, enhancing thus the final plasma temperature and density [8]. For the laser intensity of 10^{16} W/cm^2 the flat target thickness is crucial for obtaining the maximum proton energy and it must be optimised to reach the maximum ion acceleration, according to literature [9, 10].

Fig. 4 shows a comparison of the SiC TOF signal (a) and the corresponding spectrum of TPS parabolas (b) indicating that protons are accelerated in forward direction up to 4.5-MeV for an advanced 10-μm Al target irradiated at a pulse energy of 600 J. This high ion acceleration has been achieved by using special target irradiation conditions.

The Al target surface was treated with sandblasting using micrometric powder grain size accelerated at 100 m/s launched against the Al surface in order to generate high roughness, responsible of the reflectivity decreasing of the irradiation laser light. The target thickness was chosen as optimal for flat targets in order to maximize the proton acceleration in TNSA conditions using the experimental conditions above described.

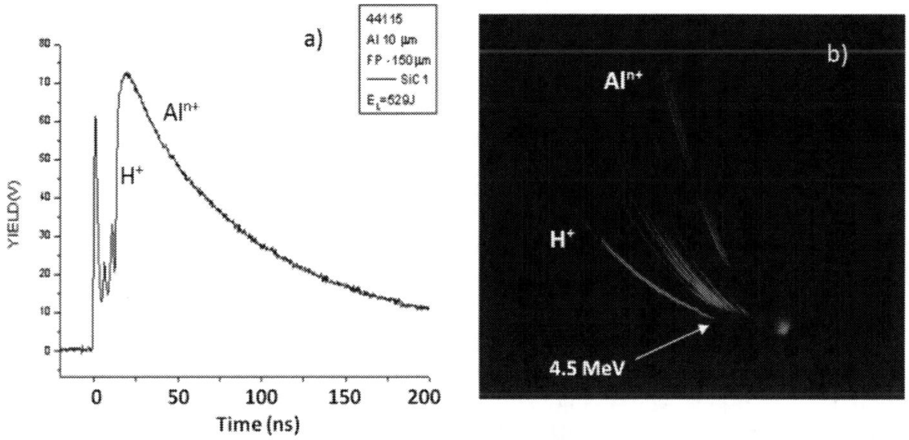

Figure 4. SiC signal (a) and TPS parabola spectrum (b) indicating a maximum proton energy of 4.5 MeV.

As seen in Fig. 5a, at 10 μm target thickness the proton energy reaches the maximum value. If the target is too thin or thick the electron density of plasma decreases. Moreover the protons loose energy in thick plasmas and their final energy decreases. The results illustrated by Fig. 4 were obtained for the focal position of – 150 μm in front of the target surface, at which the conditions for beam self-focusing are fulfilled. The self-focusing is due to the pre-plasma formation with high refraction index in front of the target surface, which results in reduction of the laser spot dimension on the target surface to values of the order of a single wavelength and in substantial enhancement of the irradiation laser intensity [2].

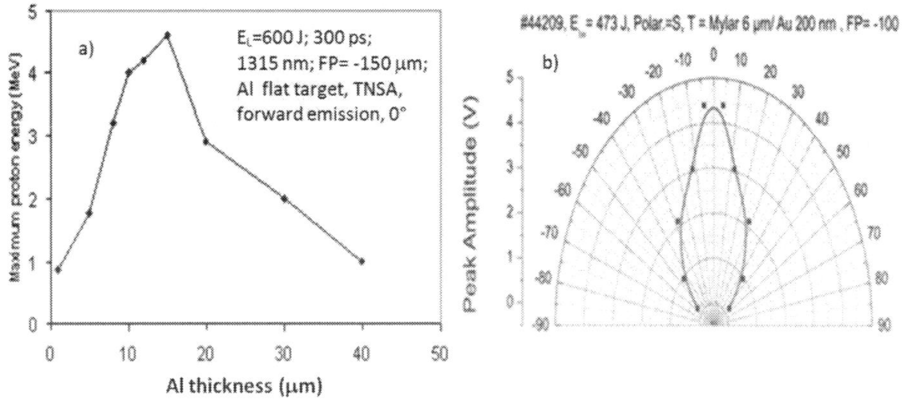

Figure 5. Maximum proton energy vs. Al thickness for flat targets irradiated a t 600 J and FP = -150m (a) and angular distribution of ions emitted in forward direction (b).

The angular distribution measurements of emitted ions performed with IC, SiC and X-ray streak camera demonstrated that under the condition of optimal thickness the forward emission occurs mainly along the normal to the target surface and that the ion stream in the forward direction is narrower than in the backward one. Fig. 5b shows a typical ion angular distribution of the ions emitted in forward direction when laser irradiates a mylar foil in TNSA conditions, 6 μm in thickness, covered by 200 nm of Au, by using 473 J pulse energy with -100 μm laser beam focal position. The proton and Au ion emission occurs along the normal to the target surface with a maximum angular aperture of about ± 30°.

4. Discussions and conclusions

The measurements reported in this article, performed by using the PALS facility, demonstrated that TNSA approach using 600 J laser pulse energy permits to accelerate ions in forward direction to high kinetic energies. Fast electrons of the order of 10 MeV were detected and their yield increased when using targets made of heavy metallic elements. These electrons are responsible for the high electric field driving ion acceleration. Less energetic electrons are also ejected from the target, together with the ion emission, as witnessed by SiC detectors.

About 4.5 MeV of energy per ion charge state can be obtained when irradiating thin Al targets prepared in such a way that reflectivity and absorption effects are enhanced. Thus, although the same laser and irradiation conditions were employed, significant improvement of the ion acceleration has been achieved in the last years by optimising the target composition, structure and geometry. With thick targets, for example, it was not possible to achieve more that 1 MeV per charge state still several years ago [11].

Target structure optimization plays an important role because it permits to enhance the laser energy transfer to the plasma and to increase the ion acceleration driving electric field developed in the double layers at the rear side of a thin target. Further improvements could be achieved by using polarized laser beams that permit to enhance absorption effects. When using Au nanospheres embedded in polymer, for example, the maximum proton energy was increased up to 5.0 MeV, as reported in forthcoming publication [12]. Also the advanced plasma diagnostics plays an important role at ion acceleration studies. SiC detectors and Thomson parabola spectrometry proved to be very useful in the measurements of the kinetic ion energy, charge state, angular distribution and at evaluation of the maximum electric field driving the ion acceleration.

Of course, in order to increase significantly the ion acceleration it is necessary to increase the laser intensity I, and the laser wavelength λ, because the acceleration field is proportional to the parameter $I\lambda^2$ [13]. Nowadays ion acceleration of the order of 50 MeV/charge state can be obtained in the large laser facilities using PW fs lasers, at intensities exceeding 10^{20} W/cm^2. However also in these cases the target optimization as for its geometry, composition and structure plays an important role at attempting to increase further the ion energy, yield and directivity.

Acknowledgements

The authors wish to thank the staff of the PALS Laboratory in Prague and all other participants of the experiment headed by Prof. L. Torrisi, and gratefully acknowledge the financial support given to the project by LASERLAB-EUROPE (GA 284464, EC's Seventh Framework Programme) and by the Czech Ministry of Education, Youth and Sports (project PALS RI, LM2010014).

References

[1] D. Giulietti and L.A. Gizzi, *La Rivista del Nuovo Cimento* 21, 1-103, 1998
[2] L. Torrisi, D. Margarone, L. Laska, J. Krasa, A. Velyhan, M. Pfeifer, J. Ullschmied, L. Ryc *Laser and Particle Beams* 26, 379-387, 2008
[3] L. Torrisi, M. Cutroneo, F. Caridi and C. Gentile *Laser and Particle Beams* 31(01), 37 – 44, 2013
[4] PALS web site, Prague 2013: http://www.pals.cas.cz/laser/
[5] M. Cutroneo, P. Musumeci, M. Zimbone, L. Torrisi, F. La Via, D. Margarone, A. Velyhan, J. Ullschmied and L.Calcagno, *J. Material Research* 28(01), 87 – 93, 2012
[6] L. Láska, J. Badziak, K. Jungwirth, M. Kálal, J. Krása, E. Krouský, P. Kubeš, D. Margarone,

P. Parys, M. Pfeifer, K. Rohlena, M. Rosinski, L. Ryc, J. Skála, L. Torrisi,J.Ullschmied, A. Velyhan, J.Wolowski *Rad. Eff. & Defects in Solids: Inc. Plasma Sci. & Plasma Tech.* 165(6–10), 463–471, 2010

[7] M. Cutroneo, L. Torrisi, L. Andò, S. Cavallaro, J. Ullschmied, J. Krasa, D. Margarone, A. Velyhan, E. Krousky and M. Pfeifer, *Acta Polytechnica* 53(2):138–141, 2013

[8] M. Cutroneo, L. Torrisi, D. Margarone, A. Picciotto, *Applied Surface Science* 272, 50-54, 2013

[9] L. Torrisi, M. Cutroneo, L. Andò, and J. Ullschmied, *Phys. Plasmas* **20**, 023106, 2013

[10] J. Fuchs, P. Antici, E. d'Humie`res, E. Lefebvre, M. Borghesi, E. Brambrink, C.A. Cecchetti, M. Kaluza, V. Malka, M. Manclossi, S. Meyroneinc, P. Mora, J. Schreiber, T. Toncian, H.P. Epinand,
P. Audebert, Nat. Phys.2, 48–54, 2006

[11] L. Laska, S. Cavallaro, K. Jungwirth, J. Krasa, E. Krousky, D. Margarone, A. Mezzasalma,M. Pfeifer, K. Rohlena, L. Ryc, J. Skala, L. Torrisi, J. Ullschmied, A. Velyhan, G. Verona-Rinati *Eur. Phys. J. D* **54**, 487–492, 2009

[12] L. Torrisi et Al., submittted to Journal of Applied Physics, 2014.

[13] E L. Clark, K. Krushelnick, M. Zepf, F. N. Beg, M. Tatarakis, A. Machacek, M.I.K. Santala, I. Watts, P.A. Norreys, A.E. Dangor, *Phys. Rev. Lett.* 85, 1654–1657, 2000

Quasi-monoenergetic proton beams by laser-plasma X-rays

D Delle Side[1,3], J Krása[2], V Nassisi[1], L Velardi[1]
[1]LEAS – Dipartimento di Matematica e Fisica "*Ennio De Giorgi*", Università del Salento and INFN Section of Lecce, Via per Arnesano s.n., 73100, Lecce, Italy
[2]Institute of Physics, ASCR, v.v.i., Na Slovance 2, 182 21 Praha 8, Czech Republic

E-mail: domenico.delleside@le.infn.it

Abstract. We report the details of a technique for the production of proton beams with very low energy spread exploiting the short soft X-rays obtained by laser ablation. These beams have been generated by the dissociation and ionization of an hydrogen buffer gas induced by the laser-plasma X-rays and then accelerated by means of an electrostatic accelerator. Their properties have been analyzed through the time-of-flight method applying different accelerating voltages. The resulting energetic spread ranges between 6 and 11%, as a function of the applied voltage. Such a system could be extremely useful for producing quasi-monoenergetic proton beams.

1. Introduction

Laser induced plasmas (LIPs) revealed over the years as a prolific field, both for scientific and applicative purposes. Since the early days of lasers, researchers all over the world investigated the effect of focusing high power radiation onto solid targets in high vacuum, obtaining as a result high density plasmas. In these experiments, plasmas expanding rapidly along the normal to the target surface were known to originate the emission of energetic ions since the early 1960's[1]. These findings evolved rapidly, suggesting that these beams could be used for the development of new ion sources[2,3]. Indeed, laser ion sources (LIS) proved to be an useful tool both in the scientific and industrial fields[4-7]. One of the main problems that researchers have to consider when planning the use of a LIS is the broad energy distributions of the ejecta, particularly when used as injectors for particle accelerators. In this case, ad hoc systems[8] should be used to reduce the energy spread of the ion beams.

In this work, we addressed the issue of the energetic spread of ion beams originating from LIPs, tackling the problem from a different point of view. In particular, as proton beams are a field of current scientific interest[9], we developed a technique through which a population of quasi-monoenergetic proton beams is effortlessly obtained. In order to reach our scope, we exploited the short soft X-rays emitted in LIPs, using them to dissociate and ionize a rarefied hydrogen buffer gas; then we were able to accelerate the resulting ions within an electrostatic accelerator. In this context, the LIP is used as a source of vacuum ultraviolet (VUV) and soft X-ray (SXR) radiation. In the past, in effect, it has been shown that LIPs offer the possibility to easily generate this kind of radiation[10].

[3] To whom correspondence should be addressed

The efficiency of the production of such radiation and its emission spectrum depend on several factors laser beam and target properties[11]. We show that our setup is well suited for generating proton beams with small energetic spread and easily tunable characteristics.

2. Materials and methods

The measures reported here were performed at the LEAS laboratory in Lecce (Italy). The setup used was based on the PLATONE device[12], which consists of a laser ion source (LIS) coupled to a double stage electrostatic accelerator. The LIS uses a KrF excimer laser (Lambda Physik, Mod. COMPEX 205) and a stainless steel vacuum chamber as accelerating chamber (AC), see figure 1. The laser generates pulsed beams of 248 nm wavelength and a duration (FWHM) of 23 ns. The laser beam, guided by a 15 cm focal length lens, enters the AC through a quartz window with an angle of 70° with respect to target normal. The accelerating chamber has inside an expansion one (EC) that enables the hydrodynamic expansion of the plasma before the ion extraction. The EC forms a hermetic contact with the target support T. At the opposite EC side there is a 1.5 cm diameter hole in order to extract ions. A base of the EC, together with T, is fixed to the AC by an insulating flange (IF) that allows the application of a positive high voltage to the EC for both extraction and electrostatic acceleration purposes. The distance T-EC hole is fixed at 18 cm ($L_{T\text{-}EC}$). At a distance of 3 cm from the EC hole ($L_{EC\text{-}GE}$), it is placed a grounded electrode (GE), having the central part drilled by a hole of the same diameter of the EC one. After the GE, at a distance of 98 cm ($L_{GE\text{-}FC}$), it is placed a further third electrode, connected to a power supply of negative polarity. This electrode, biased at -100 V, was used as Faraday cup (FC). The vacuum was operated through the use of two turbo-molecular pumps, reaching a pressure of 8.6×10^{-6} mbar. A fast photodiode sensible to the UV radiation coming from a beam splitter near the entrance window was used to trigger measurements on a digital oscilloscope.

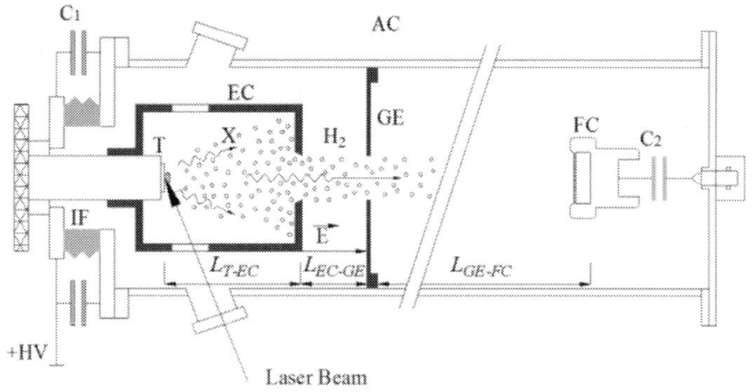

Figure 1. Cross section of the PLATONE accelerator. IF: insulation flange; AC: accelerating chamber; GE: ground electrode; EC: expansion chamber; FC: Faraday cup; C_1: stabilizing capacitors; C_2: insulation capacitor. The small circles represent H_2, while corrugated lines the X-rays.

The target used was a commercial tantalum disk, pure at 99.9%. Tantalum was chosen because, together with the VUV radiation, offers a high conversion efficiency of laser light into SXRs[13]. Moreover, due to its high atomic mass, it allows to easily separate time resolved currents of lighter ions from those of Ta ions. Since tantalum is a known hydrogen absorber material[14], before the main experiments, the target has been laser cleaned in order to remove surface contaminants via laser induced desorption. This would allow to sensibly reduce the contribution originating from these contaminants. In particular, we operated around 36000 laser shots at an irradiance of 3.5×10^7 W/cm^2, obtained defocusing the laser beam. The laser pulses heated the samples in vacuum and the temperature rise $\Delta T(r,z,t)$ at distance r from the spot center, depth z and time t was estimated through the equation[15]

$$\Delta T(r,z,t) = \frac{I_{\max}}{K}\left(\frac{D}{\pi}\right)^{1/2}\int_0^t \frac{p(t-\tau)}{\tau^{1/2}(4D\tau+d^2)}\exp\left(-\frac{z^2}{4D\tau}-\frac{r^2}{4D\tau+d^2}\right)\mathrm{d}\tau, \qquad (1)$$

where d is the average spot radius (0.3 cm in our case), I_{max} is the peak irradiance value, $p(t) \sim t\exp(-\beta t)$ is the temporal pulse shape (where β takes into account for the FWHM length of the pulse), while K and D are respectively tantalum thermal conducibility and diffusivity. Equation (1) is obtained in the hypothesis of a large absorption coefficient, so that it could be assumed that light absorption happens mainly at the surface. This represents a good approximation for the case under consideration. Solving numerically equation (1), we found that ΔT could be safely considered zero at 0.2 s (the time between two consecutive laser pulses). In other words, there is no residual temperature rise. Consequently, the surface temperature of the target as a function of time follows the shapes showed in Fig. 2, approaching zero as t reaches 0.2 s. The heating of the target induces thermal desorption of surface contaminants. Moreover, the use of an UV laser gives the opportunity to stimulate also photodesorption[16].

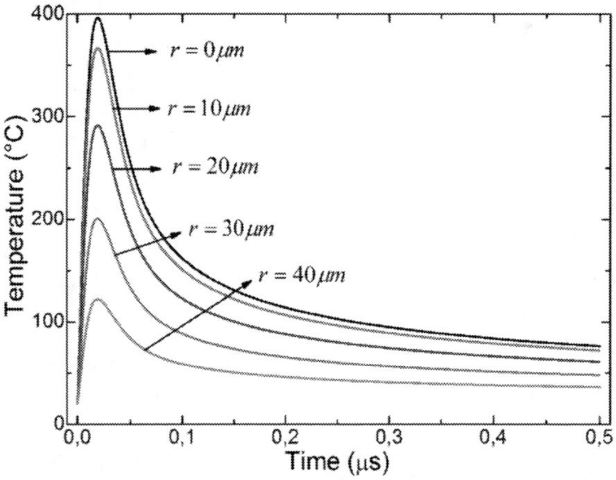

Figure 2. Temperature on the target surface ($z = 0$) at various distances from the spot centre as a function of time. The environment temperature considered was 20 °C.

After the cleaning, a series of measures were performed both in vacuum and with a constant flux of molecular hydrogen (Linde Gas Hydrogen 6.0) maintained within AC using a finely graduated valve. In presence of the hydrogen, the internal pressure was raised to 5.0×10^{-4} mbar. At such a pressure, plasma expansion is not significantly influenced by the presence of the background gas[17]. Furthermore, we refocused the laser beam on the target, obtaining an irradiance of 3.0×10^{9} W/cm^2 on a 0.001 cm^2 spot, while on EC it was applied a positive high voltage up to 40 kV.

Our aim is to dissociate and ionize the background H_2 available in the EC-GE gap through the SXRs generated from the LIP and then to accelerate the resulting ions. In these conditions, in effect, we found[13] that the LIP SXRs have an average energy around 180 eV, a value suitable to reach our scope.

The data was collected on a digital oscilloscope and then processed. In general, the time t_{TOT} spent by ions flying from T to FC could be divided in 3 different contributions[18],

$$t_{TOT} = t_{T-EC} + t_{EC-GE} + t_{GE-FC}\,. \qquad (2)$$

In effect, this originates from the three different motions an ion experiences in our setup: one with constant velocity from T to the EC hole; a uniformly accelerated one from the EC hole to GE; and a uniform motion again from GE to FC. Since we are interested in considering only the protons generated in the gap EC-GE by the LIP induced SXRs, we could safely consider t_{T-EC} to be zero (a rough estimation of the time spent by SXRs to reach the EC hole), while protons start to be accelerated in EC-GE with a non zero initial velocity v_0, transferred by the interaction with SXRs. Consequently, we could rewrite the total drift time as

$$t_{TOT} = \frac{mL_{EC-GE}}{\alpha eV}\left(\sqrt{v_0^2 + 2\alpha\frac{eV}{m}} - v_0\right) + \frac{L_{GE-FC}}{\sqrt{v_0^2 + 2\alpha\frac{eV}{m}}}, \tag{3}$$

where L_{EC-GE} and L_{GE-FC} are the distances the protons travel in the corresponding sections of our setup as in figure 1, m is the proton mass, e the elementary charge, V the voltage applied to EC and α is a parameter that takes into account the Debye shielding effect on the protons plasma. In effect, this is known to induce lower acceleration efficiencies with respect to the applied potential[18,19]. In equation (3) the only unknowns are v_0 and α. Although α is known to slightly reduce increasing V, we could in first approximation consider it fixed and use the recorded current peak times of our interest at each accelerating potential to estimate the unknowns. In effect, fitting the couples $(V, t_{TOT,peak})$ with equation (3), an estimation for v_0 and α could be obtained. With this knowledge, solving equation (2) for t_{GE-FC}, we could analyze the data through the procedure described in [20] for time-of-flight (TOF) analysis. This corresponds to consider protons as originating at GE. In particular, protons TOF signals has been fitted using the function

$$S(t; L_{GE-FC}) = \frac{L_{GE-FC}^2}{t^5}\sum_i S_{0,i}\exp\left[-\frac{m}{2k_B T_i}\left(\frac{L_{GE-FC}}{t} - u_i\right)^2\right], \tag{4}$$

where t_{GE-FC} was relabelled as t, k_B is the Boltzmann constant, while T_i, u_i and $S_{0,i}$ are respectively the Knudsen layer temperature, the center of mass velocity and the signal amplitude of the expanding proton population. Moreover, the energy spread σ_E at full width half maximum (FWHM) has been computed using the relation[20]

$$\sigma_E = \frac{\Delta E}{E_{peak}} \approx 2\frac{\Delta t}{t_{peak}}. \tag{5}$$

Since it is quite impossible to remove carbon impurities from the signals collected on FC (the carbon comes from the walls of the chamber), we proceeded with the experiments taking care of performing a comparative evaluation of the effect of a double laser pulse (i.e. a pulse followed by a second one after a few ms) over the single pulse.

3. Results and discussion

In the following, we'll be interested in reporting the results obtained for protons. For this reason, TOF spectra will be shown emphasizing their details using logarithmic scales in abscissas. Experimental results showed that without the application of an extraction potential, it wasn't possible to detect other protons than those coming from the target. In effect, the corresponding voltage had to be raised to 5 kV on T to observe the appearance of an additional fast peak on the currents recorded through FC (data not shown), when the H_2 was fluxed in AC. In particular, increasing the applied voltage resulted in the occurrence of fast and sharp protons signals when a background hydrogen is present, together with a slower population of protons coming from the target, as shown in figure 3. The estimation of the initial velocity of these protons through equation (3) resulted in $v_0 = 140 \pm 54$ km/s, corresponding to an energy of 101 ± 30 eV. The α parameter obtained was 0.28 ± 0.01.

Figure 3 clearly reports also the dramatic effect of the double pulse on the TOF signals. In single pulse waveforms, in effect, it is evident the occurrence of slower peaks (with respect to protons) due to carbon impurities. These peaks are severely reduced in the waveforms obtained with the double pulse technique. Despite of this, no other significant effect was observed on TOF signals in double pulse mode. For this reason, it seems that this option could be used mainly when it is advantageous to obtain beams with a low impurities content.

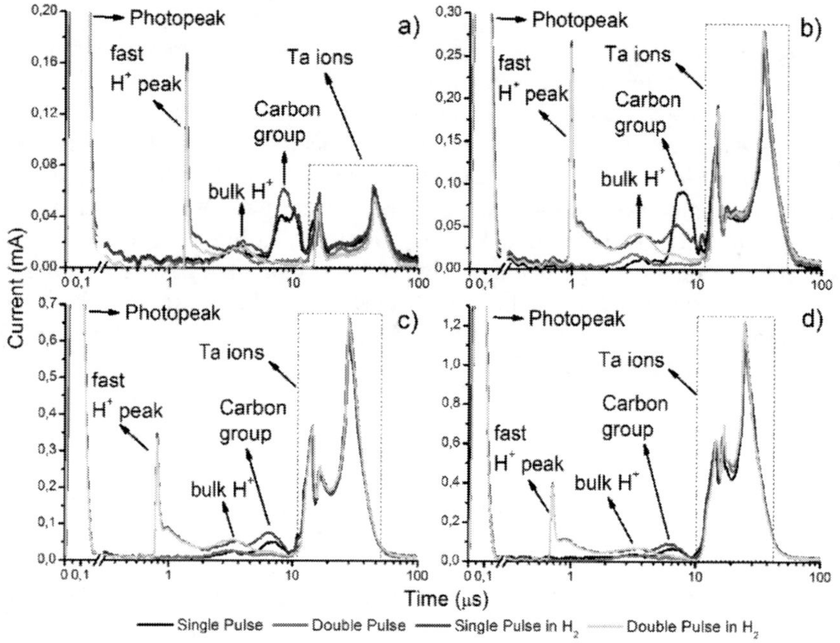

Figure 3. TOF signals for different extraction voltages: a) 10 kV, b) 20 kV, c) 30 kV and d) 40 kV. Abscissas are in logarithmic scale after the break.

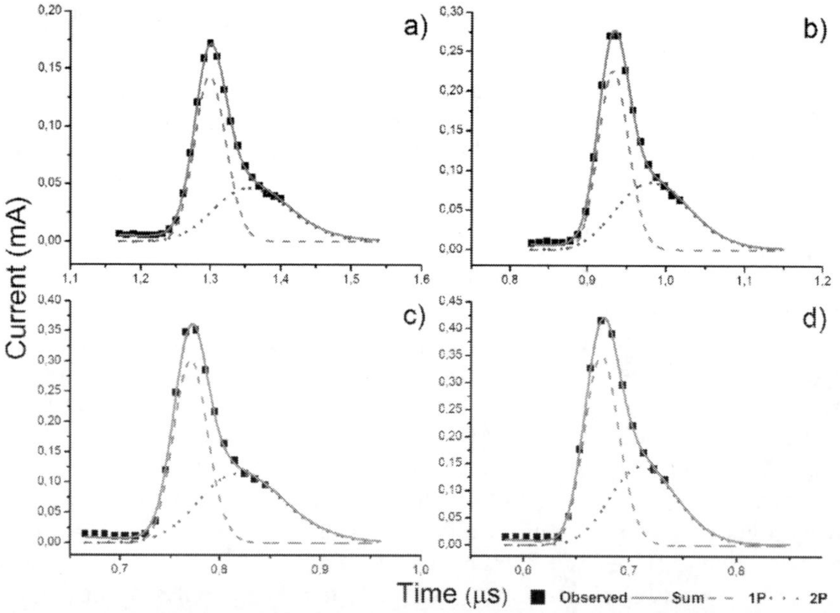

Figure 4. Details of the observed and computed fast proton signals for different extraction voltages: a) 10, b) 20, c) 30 and d) 40 kV. Waveforms in single pulse mode only, time has been rescaled as explained in section 2.

Careful analysis of the fast proton TOF signals by means of a fit using equation (4) revealed that they are composed by two different populations: one highly peaked around a maximum value, the other with a broader velocity distribution (respectively 1P and 2P in figure 4). The narrower peaks (1P) result from direct ionization by laser-plasma SXRs of the hydrogen available in the gap EC-GE near the EC hole, in effect they have a duration comparable to that of the photopeak. The broader peaks (2P), although resulting in the same way from SXRs interaction with ambient H_2, should come from the volume between the target and the extraction hole. This population is extracted thanks to the penetration inside EC of the electric field between EC-GE[12], that causes the observed broader velocity distribution.

Figure 5 reports, as a function of the extraction/accelerating voltage, some of the properties of the fast proton peaks obtained by the fits (in single pulse mode). It is interesting to note the small energetic spread of the proton beams (Fig. 5a), computed through relation (5), which is smaller than those obtained in [8] with a dedicated system. Its value starts from about 6% at 5 kV and increases up to about 11% at 40 kV with a trend similar to a power law ($\sim x^{1/3}$, in the range of voltages tested). The center of mass energy, as expected, depends linearly on the applied voltage (figure 5b). The most notable feature of this linear dependence is in the slope z of the line. In effect, one would expect that it is roughly equal to the charge state of the proton, instead it is about a third. Such a small acceleration efficiency is not surprising, since it is an expected characteristic of laser ion sources[18,19]. This depends on the fact that the protons plasma is very efficient in screening the applied voltage to preserve its neutrality. It is also interesting to note that the value of slope is in good accordance with the α parameter considered above. Furthermore, the intercept, taken with its error, is in accordance with the initial energy obtained from the fit with equation (3).

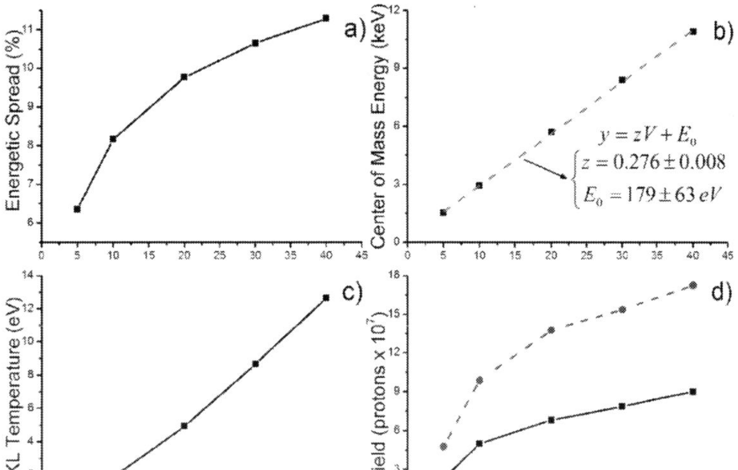

Figure 5. Properties of the fast protons as a function of the applied voltage: a) energetic spread, b) center of mass energy (together with a linear regression), c) Knudsen layer temperature and d) fast protons yield.

The Knudsen layer of the expanding protons plasma front has a temperature that increases faster than linearly with respect to the applied voltage (Fig. 5c). Indeed, in the range of voltages explored it follows a power law of type $x^{3/2}$. This behaviour is the result of the fact that we are considering the protons as being generated at GE. Consequently, the complex interactions of the protons plasma with the accelerating potential in the gap EC-GE result in the KL temperature dependence. It is worth noticing that double pulse waveforms are characterized by slightly smaller temperatures (data not shown), that translates also to slightly smaller energetic spreads. This should be related to the absence of the carbon impurities from the target surface that would otherwise induce small differences in the generation of the X-rays. The yield per pulse (Fig. 5d) is not very high, it is of the order of 10^7 protons per pulse, with an increase with respect to the applied voltage that is quite logarithmic. A similar trend is shown also by the total yield of the fast protons, obtained by computing the total area of the reconstructed proton signals. In this case, the yield reaches a maximum of 1.8×10^8 protons per pulse at 40 kV of applied voltage. It should be noted, however, that the low yields obtained have to be considered a result of the long observation distance, since it is known that the total charge Q extracted from a LIS depends inversely on it as $Q \propto L^{-2}$ [21]. Consequently, using shorter distances would result in sensibly higher yields.

4. Conclusions

We have shown an effective method to obtain a population of quasi-monoenergetic proton beams with a simple laser ablation setup. In the voltage range examined, the energetic spread varied from 6% up to 11%. This result is important since it offers a viable and simple solution to fulfil a need diffused in the scientific field. These beams, for example, could be used for calibration of a Thomson parabola spectrometer. Moreover, the properties of the beams could be easily determined as a function of the applied voltage. Future work will be devoted to a detailed study of the parameters (e.g. target geometry, material and interaction conditions) that could lead to a further reduction of the energy spread.

Acknowledgment

We are grateful to Dr. Domenico Doria for the insightful discussion with him.

References

[1] Linlor W I 1963 *Appl. Phys. Lett.* **3**, 210.
[2] Peacock N J and Pease R S 1969 *J. Phys. D: Appl. Phys.* **2**, 1705.
[3] Byckovsky Y A, Eliseev V F, Kozyrev Y P and Silnov S M, Sov. Patent 324 938, Oct. 1969.
[4] Okamura M, Pikin A, Zajic V, Kanesue T and Tamura J 2009 *Nucl. Instrum. Meth. Phys. Res. A* **606**, p. 94-96.
[5] Noli F, Lagoyannis A and Misaelides P 2008 *Nucl. Instrum. Meth. Phys. Res. B* **266**, p. 2437–2440.
[6] Rosinski M, Badziak B, Parys P, Wołowski J and Pisarek M 2009 *Appl. Surf. Sci.* **255**, p. 5418–5420.
[7] Chrisey D B and Hubler G K, *Pulsed Laser Deposition of Thin Films* (John Wiley and Sons, New York, 1994).
[8] Tambini J 1996 *SPIE Proc.* **2767**, p. 67.
[9] Velardi L, Delle Side D, Krása J, Nassisi V 2014 *Nucl. Instrum. Meth. Phys. Res. B* **735**, p. 564.
[10] Giulietti D and Gizzi L A 1998 *Riv. Nuovo Cimento* **21**, p. 1.
[11] J. M. Bridges, C. L. Cromer and Thomas J. McIlrath 1986 Appl. Optics **25**, 2208-2214.
[12] Nassisi V, Delle Side D and Velardi L 2013 *Appl. Surf. Sci.* **272**, p. 114.
[13] Siciliano M V, Lorusso A, Velardi L and Nassisi V 2009 *X-Ray Spectrom.* **38**, p. 544-547.
[14] Ferrin P, Kandoi S, Udaykumar Nilekar A, Mavrikakis M 2012 *Surf. Sci.* **606**, p. 679.
[15] Ready J F, *Effects of High-Power Laser Radiation* (Academic Press, Orlando, 1971).
[16] Chuang T J 1983 *Surf. Sci. Rep.* **3**, p. 1-105.
[17] Geohegan D and Puretzky A A 1995 *Appl. Phys. Lett.* **67**, p. 197.
[18] Yeates P, Costello J T and Kennedy E T, 2010 *Phys. Plasma* **17**, 123115.
[19] Trinczek M, Werdich A, Mironov V, Guo P, Gonzalez Martınez A J, Braun J, Crespo Lopez-Urrutia J R and Ullrich J 2006 *Nucl. Instrum. Meth. Phys. Res. B* **251**, p. 289.
[20] Krása J 2013 *Appl. Surf. Sci.* **272**, p. 46.
[21] Krása J, Parys P, Velardi L, Velyhan A, Ryć L, Delle Side D and Nassisi V 2014 *Laser Part. Beams* **32**, p. 15-20.

Distance-of-flight spectra of charge density of ions generated with a low laser intensity

J Krása[1], L Velardi[2,3], A Lorusso[2], D Delle Side[2,3], V Nassisi[2,3]

[1]Institute of Physics, ASCR, v.v.i., Na Slovance 2, 182 21 Praha 8, Czech Republic
[2]LEAS, Dipartimento di Matematica e Fisica, Università del Salento
[3]INFN Sezione di Lecce, Via Arnesano, sn – 73100 Lecce, Italy

E-mail: krasa@fzu.cz

Abstract. A charge density profile of ions emitted from a laser ion source is derived from a time-resolved current of ions collected with a time-of-flight detector. This relationship between the density profile over path of ions and an ion current observed far from the ion source is based on a similarity relationship for ion currents with "frozen" charges observed at different distances from the target. In this contribution, we present distance-of-flight spectra of the charge density of ions emitted from various plasmas driven by a Compex 205 KrF excimer laser (λ=248 nm, τ_{FWHM}=23ns) which works at laser irradiances between 10^8 and 10^{10} W/cm^2.

1. Introduction

In the time-of-flight spectroscopy, the ions emitted from a laser-produced plasma are detected at the end of their flight path by means of various types of detectors. In the simplest case, the ions strike a Faraday collector, which delivers a time-resolved signal to an oscilloscope. With the purpose of obtaining information on the space properties of the expanding plasma, a number of Faraday collectors are located at various angles with respect to the target surface normal to obtain angular distributions of plasma parameters, but losing the obtained time-resolution due to e.g. time integration of ion currents.

In this paper, we present a novel method for transforming the time-resolved signal, $S(t,L)$, of a Faraday collector positioned at a distance, L, from the irradiated target into the ion charge density profile $q(x,\tau)$ at a selected time, τ, after the laser-target interaction. These distance-of-flight spectra provide a new perspective on plasma expansion.

2. Theory

The signal S_x of a time-of-flight detector measuring the impacted ions in x-direction depends on its response to their number, velocity or energy expressed by the term α:

$$S_x(L,t,\alpha) \propto v_x^\alpha f(\vec{v})\, d\vec{v}\,, \tag{1}$$

where $f(\vec{v})$ is the 3-dimensional velocity distribution function and

$\alpha = 0$ if the response is proportional to the number (density) of incident particles,
$\alpha = 1$ if the response is proportional to the stream (current) of incident particles,
$\alpha = 2$ if the response is proportional to the deposited energy by particles.

Substituting L/t into $f(\vec{v})d\vec{v}$ we get the number of particles hitting the detector's area ds per dt, which is then proportional to $f(L/t)\,L\,t^{-4}$ and $S_x(L,t,\alpha) \sim L^{1+\alpha}\,t^{-(4+\alpha)}\,f(L/t)\,dtds$. Since the shifted Maxwell-Boltzmann distribution well describes the motion of ionized species, the detector signal can be expressed as:

$$S_x(L,t) = \kappa \frac{L^{\alpha+1}}{t^{\alpha+4}} \exp\left[-\frac{m}{2kT}(L/t-u)^2 \right], \tag{2}$$

where m is the mass of species constituting the current j, κ is the normalizing term, u is the velocity of their centre-of-mass motion directed to the ion collector, k is the Boltzmann constant and T is the ion temperature [1].

The relationship (2) enables us to determine a time t_{PEAK} at which the detector's signal reaches a maximum:

$$t_{PEAK} = \frac{L}{(\alpha+4)}\sqrt{\frac{m}{2kT}}\left(\sqrt{2(\alpha+4)+\frac{mu_{CM}^2}{2kT}} - \sqrt{\frac{mu_{CM}^2}{2kT}}\right). \tag{3}$$

The equation (3) reveals an α dependence of t_{PEAK} given by the detector type as mentioned above. It is evident that the term $(\alpha+4)^{-1}$ has substantial influence on the value of t_{PEAK}: the ion current reaches its peak value at time $t_{PEAK\text{-}J} \cong 0.8\times t_{PEAK\text{-}Q}$, where $t_{PEAK\text{-}Q}$ is the time at which the ion charge density reaches its maximum. The influence of the detector type on the time-of-flight spectra is shown in Fig. 1. The dashed line is the current density of ions $j(L,t)$ emitted from an Ag plasma and the solid one is the corresponding time-resolved charge density $q(L,t) = j(L,t)/v$, i.e. $\alpha = 0$. The arrow shows the delay of the second peak of the ion charge density with respect to the second peak of the ion current. The ratio of both values is nearly $t_{PEAK\text{-}J} / t_{PEAK\text{-}Q} \cong 0.8$. Moreover, Fig. 1 shows an example of the effect of ion velocity on the ion current amplitude. When a detector, which is sensitive to the ion energy, i.e. $(\alpha+4) = 6$, is employed, then not only the detector signal is enhanced proportionally to the energy of ions but also the peak velocity $v_{PEAK\text{-}E} = L / t_{PEAK\text{-}E}$ is enhanced by factor of ~1.5 with respect to the $v_{PEAK\text{-}Q} = L / t_{PEAK\text{-}Q}$. The ion-energy sensitive semiconductor detectors having $(\alpha+4) = 6$ have been advantageously used in measuring of multi-MeV ions emitted from a laser-produced plasma. [2].

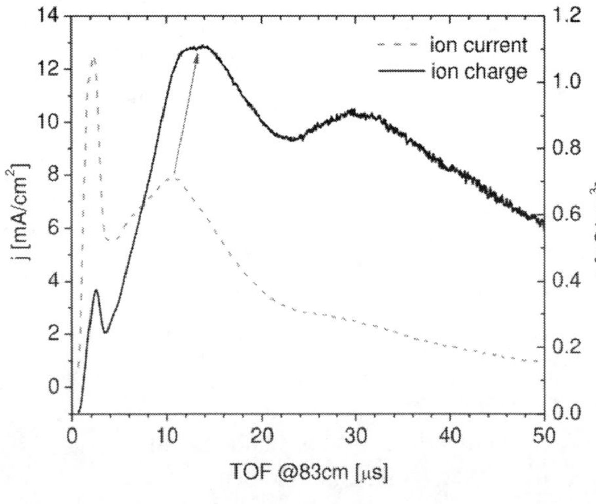

Figure 1. Time-resolved ion current (dashed line) and corresponding charge density (solid line) of ions emitted from Ag plasma produced with using a 350-ps iodine photo-dissociation laser PERUN operating at the second harmonic (λ=657 nm) and delivering onto an Ag target intensity of 5.5×10^{14} W/cm^2 [3].

Taking into consideration that the time-of-flight varies with the distance as $t = L/v$, the relationship (1) takes a form for ion current:

$$j(L,t)\,L^3 = \kappa\, v^5 f(v). \tag{4}$$

Then the similarity relationship for currents of ions with "frozen" charges that are detected at different distances x and L is

$$j(x,\tau)\,x^3 = j(L,t)\,L^3, \tag{5}$$

where $x/\tau = L/t$. The similarity relationship (5) allows to determine the space-resolved ion charge density $q_{IC} = j_{IC}/v$ by using the transformation of the time-of-flight spectrum $j_{IC}(L,t)$ to the distance-of-flight spectrum (DOF) $q_{IC}(x,\tau)$, where the flight times t, τ are related into the flight distances L, x as $L/t = x/\tau$, and L is the position of the IC on the x-axis. Thus, dividing both sides of the equation (3) by $v=x/\tau$, the DOF spectrum takes a form:

$$q(x,\tau) = j(L,\tau L/x)\,\tau L^3/x^4, \tag{6}$$

where time τ is a parameter. The equation (6) transforms $j(L,t)$ observed at a distance L from the target to $q(x,\tau)$ at arbitrary chosen instant τ after the laser-target interaction [4].

Supposing that ions have a Maxwell-Boltzmann velocity distribution and substituting it in (1) we obtain:

$$q(x,t) = \kappa\tau^{-4}\exp\left[-\beta^2\left(\frac{x}{\tau} - u_{CM}\right)^2\right], \tag{7}$$

where $\beta^2 = m/2kT$.

The distance-resolved ion charge density (7) reaches a maximum at a distance:

$$x_{PEAK} = \frac{u_{CM}\tau}{2}\left(1 + \sqrt{1 + 2/(u_{CM}\beta)^2}\right). \tag{8}$$

3. Experimental results and discussion

The reported measurements were performed with a Compex 205 KrF excimer laser ($\lambda = 248$ nm, $\tau_{FWHM} = 23$ ns) at LEAS, Lecce which works at laser irradiances from 10^8 to 10^{10} W/cm^2. The laser pulse struck the target at 70° with respect to the target surface normal. The ion current was detected with the use of Farady cups along the target surface normal.

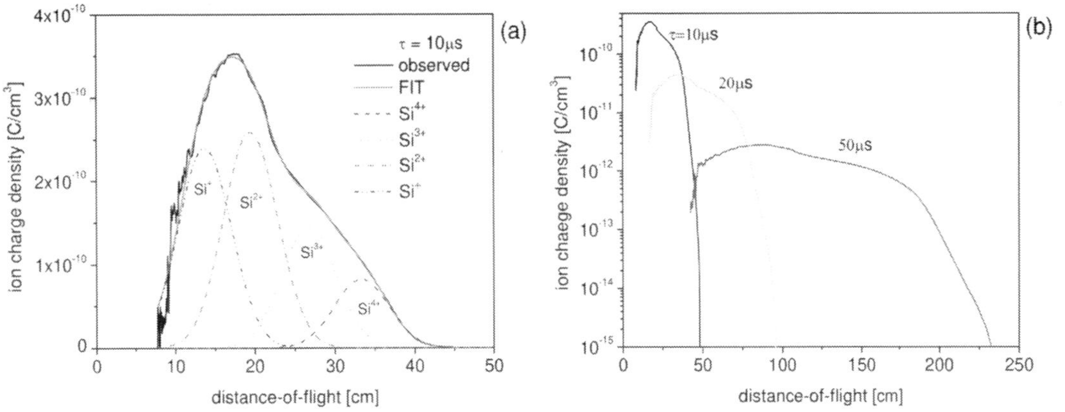

Figure 2. Charge density profile of Si^{q+} ($1 \le q \le 4$) ions at 10 μs after the laser-target interaction (a) and development of ion charge density profile in time (b). The laser fluence was 8 J/cm^2.

Fig. 2 shows a DOF dependence of charge density of Si ions. Partial peaks corresponding to Si^{q+} ($1 \leq q \leq 4$) ions were deconvolved from the observed signal using equation (2) [5]. Distances between the peaks are mainly given by their centre-of-mass velocity $u_{CM\text{-}q}$ as (8) shows. The influence of the term $u_{CM\text{-}q}\beta_q$ on $x_{PEAK\text{-}q}$ is less than ~20% to ~1%, for q=1 to q=4, respectively. The velocity $u_{CM\text{-}q}$ reflects the hydrodynamic and electrostatic forces accelerating ions in the plasma [5].

Fig. 3 shows a comparison of DOF spectra of ions emitted from Al, Au, C, Cu, CuBe, Mo, Nb, Ni, Pb, Ta, and W plasmas produced with fluence of 3 J/cm^2. Fig. 3 demonstrates two well-known groups of ions: fast ionized impurities chemisorbed onto the irradiated front surface, which occupy the space in a vacuum chamber from ~40-cm distance-of-flight at 10 μs after the laser-target interaction. Fig. 3 also clearly shows that for the most used target material except of C and Pb, the peak velocity of the ionized carbon is ~5–8×10^6 cm/s and the highest velocity of protons is ~2.1×10^7 cm/s, i.e. the maximum proton energy is 220±50 eV. The cut-off proton energy is lowest for the C and Pb targets.

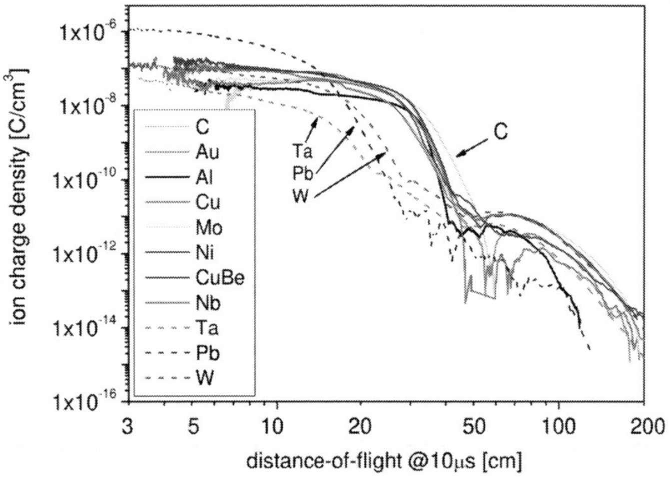

Figure 3. Comparison of DOF spectra of charge density of ions emitted from Al, Au, C, Cu, CuBe, Mo, Nb, Ni, Pb, Ta, and W plasmas which were produced with fluence of 3 J/cm^2.

The slower carbon ions as impurities occupy the space in a range of 40 to 80 cm from the target, as Fig. 4a shows in detail. Fig. 4b shows details of leading edges of charge density profiles of selected target materials as Nb, Mo, Cu. Au, and CuBe. There is no doubt, that except of the key parameter, which is the free electron density of the irradiated element or alloys affecting the laser pulse absorption [6,7], the melting point and other target properties affect the ion yield, as it can be emphasized by the charge density profile of the CuBe plasma. The charge density profiles of Nb and Mo are nearly identical. Similar result was obtained for Au and CuBe plasmas. The melting point of Cu, Au, and CuBe is 1083 °C, 1062 °C, and 860–1000 °C, respectively.

Another advantage of the presented novel method consists in the possibility to determine the charge density of the slowest ions having the lowest charge (in this case their q = 1), as Fig. 3 shows. Except of lead ions, the maximum ion charge density of the other species moving with the slowest velocity varies from ~6×10^{-8} to ~2×10^{-7} C/cm^{-3}. Only slow Pb^+ ions with velocity of 5×10^5 cm/s reached the highest charge density of ~1.2×10^{-6} C/cm^{-3} probably owing to a low value of the melting point that is 327°C.

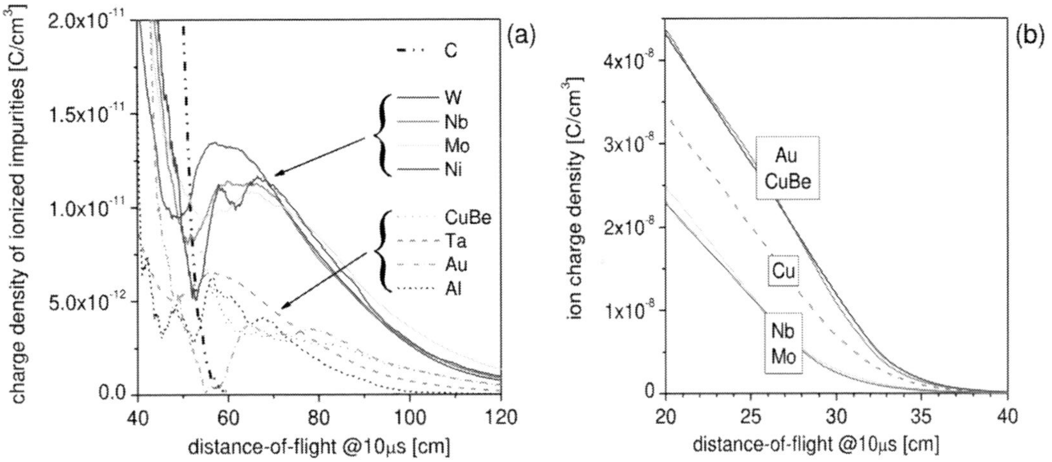

Figure 4. Details of Fig. 3: ionized impurities as H^+ and C ions emitted from selected targets (a); leading edges of DOF spectra of the charge density of ions emitted from Au, CuBe, Cu, Mo, and Nb plasmas (b)..

4. Summary

We presented a new method of the transformation of time-resolved ion currents into distance-of-flight charge density profiles. The separation of the ion charge density from the ion current and the obtained distance-of-flight spectra provide an easy way to compare the ion production efficiency in various laser-generated plasmas. Moreover, the time evolution of the distance-of-flight profiles of ion charge density was obtained.

Acknowledgements

The research leading to these results has received funding from the Czech Science Foundation (Grant Nos. P205/12/0454), and the Czech Republic's Ministry of Education, Youth and Sports (Project OPVK 3 - CZ.1.07/2.3.00/20.0279).

References

[1] J. Krása, Appl. Surf. Sci. **272** (2013) 46 – 49.

[2] D. Margarone, J. Krása, L. Giuffrida, A. Picciotto, L. Torrisi, T. Nowak, P. Musumeci, A. Velyhan, J. Prokůpek, L. Láska, T. Mocek, J. Ullschmied, B. Rus, J. Appl. Phys. **109** (2011) 103302.

[3] J. Krása, L. Láska, K. Rohlena, M. Pfeifer, J. Skála, B. Králiková, P. Straka, E. Woryna, J. Wolowski, Appl. Phys. Lett. **75**, 1999, 2539-2541.

[4] J. Krása, P. Parys, L. Velardi, A. Velyhan, L. Ryć, D. Delle Side, V. Nassisi, Laser Part. Beams (2014) **32**, 15-20.

[5] J. Krása, A. Lorusso, V. Nassisi, L. Velardi, A. Velyhan, Laser Part. Beams **29** (2011) 113–119.

[6] F. Caridi, L. Torrisi, D. Margarone, A. Borrielli, Laser Part. Beams **26** (2008) 265–271.

[7] L. Torrisi, F. Caridi, L. Giuffrida, Laser Part. Beams **29** (2011) 29-37.

On the Emittance dependence on anode morphology of laser induced ion beams

L Velardi[1], D Delle Side and V Nassisi
LEAS Laboratorio, Dipartimento di Matematica e Fisica, INFN sezione di Lecce, Università del Salento, Italia

E-mail: luciano.velardi@le.infn.it

Abstract. In this work, we studied the characteristics of ion beams generated by PLATONE accelerator in different anode configurations. The accelerator is a laser ion source with two gaps which accelerate the ions in cascade. The laser is an excimer KrF able to work at irradiances of 10^8-10^{10} W/cm^2. The target ablated was disk of Cu. The accelerating voltage applied in this work was 60 kV. The emittance evaluation was performed by the pepper pot method utilising radio-chromic films, EBT Gafchromic, as sensible targets. The study was performed by varying the geometric configuration of the anode (the extracting electrode), modifying the hole morphology. A plane and curved grids were mounted in order to change the extraction configuration. The results were compared with the ones obtained with the extraction hole without any grid. For the normalized emittance the lowest value found was 0.20 π mm mrad.

1. Introduction

In the last decades, new techniques by interaction between high power femtosecond laser pulses and thin foils[1] are employed to produce ion beams. In contrast with other techniques, these give the advantage to obtain highly collimated and energetic particle beams from the rear of the target surface. Two mechanisms are involved in the ion acceleration: target normal sheath acceleration (TNSA) [2] and radiation pressure acceleration (RPA) [3]. Despite of the high quality beams obtained through TNSA and RPA systems, older and well known techniques, such as pulsed laser ablation (PLA), still play a fundamental role for applications, since the former have extremely high total costs of ownership and require large space for its working.

The PLA technique works at intensities of the order of 10^8-10^{10} W/cm^2, producing hot plasmas[4] from which it is possible to extract ion beams whose energy can be increased up to some tens of keV by applying post acceleration[5,6]. Nowadays, ion beams of moderate energy have a wide range of applications, from scientific to industrial ones[7-9]. Consequently many laboratories, as well as the LEAS, are involved to develop accelerators of very contained dimensions, easy to be installed in little laboratories and hospitals.

In this work, we characterize ion beams provided by a laser ion source (LIS) accelerator composed by two independent accelerating sectors, using an excimer KrF laser to get PLA from a pure Cu target. Using a pepper pot system and a Faraday cup, we characterized the geometric quality of the

[1] To whom any correspondence should be addressed.

Journal of Physics: Conference Series **508** (2014) 012005 doi:10.1088/1742-6596/508/1/012005

beams, evaluating the emittance and the extracted charge for different morphologic configurations of the extracting electrode, in order to understand how these changes influence the resulting beams. The emittance is an important parameter that characterizes charged particle beams. It is the measure for the average spread of particle coordinates in position and momentum phase space. So it is necessary to have low emittance to increase the quality of the particle beam, when it is required a paraxial beam with low divergence.

2. Experimental setup

The *Platone* accelerator is a LIS source with an electrostatic system to extract and accelerate the ions from the plasma plume. It consists of a KrF excimer laser operating in the UV range ($\lambda = 248$ nm, $\tau = 25$ ns) to get PLA from solid targets and a vacuum chamber device for the expansion of the plasma plume. The maximum output energy of the laser is 600 mJ. The angle formed by the laser beam with respect to the normal to the target surface is 70°. Focalizing the laser beam by a thin lens of 15 cm length, the spot area onto the target surface was 0.005 cm^2, obtaining irradiances of the order of 10^8-10^9W/cm^2. The target was a pure disk (99.99%) of Cu.

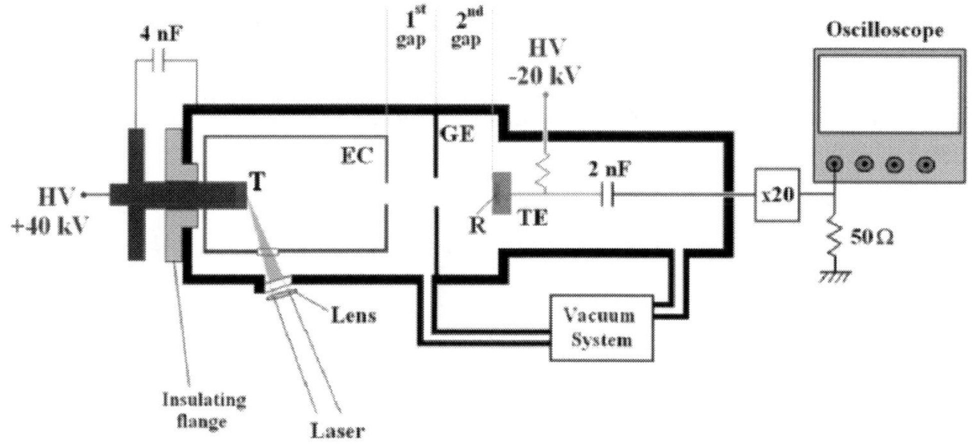

Figure 1. Schematic drawing of the LIS accelerator (T: Target support, EC: Expansion Chamber, GE: Ground Electrode, R: Radiochromic, TE: Third Electrode).

The accelerating system consists of three parts, Fig. 1: i) an expansion chamber (EC) placed around the target support (T) at a positive high voltage (HV) of +40 kV; ii) a pierce ground electrode (GE) placed at 3 cm distance from EC; iii) a third electrode (TE, 30 mm diameter) placed at 2 cm from GE connected to a power supply of negative bias voltage of 20 kV.

The EC and GE holes have both a 15 mm diameter. In this way, applying 40 kV at EC and 20 kV at TE, it is possible to generate an intense accelerating electric field in two gaps, between EC-GE and GE-TE, able to extract and accelerate the positive ion of the plasma plume. Four capacitors of 1 nF, between EC and ground, stabilize the accelerating voltage during the fast ion extraction.

TE is also utilized as Faraday cup collector. It is connected to the oscilloscope by a HV capacitor (2 nF) to separate the oscilloscope from the HV and a voltage attenuator, x20, to suit the electric signal to oscilloscope voltage. The value of the capacitors applied to stabilize the accelerating voltage (4 nF) and to separate the oscilloscope from the HV (2 nF) are calculated assuming a storage charge higher that the extracted one. Under this condition, the accelerating voltages during the charge extraction is constant as well as the oscilloscope is able to record the real signal.

TE is not able to support the suppressing electrode on the cup collector and therefore secondary electron emission, caused by high ion energy, is present leading to register more charge on FC. In order to compare the efficiency of the extraction and the geometric quality of the beams we varied the configuration of the EC (the extracting electrode) by modifying the hole. We mounted grids (with an attenuation factor of 20%) and so we obtained three configurations (see Fig. 2). Figure 2 shows the photos of the anode: a) extraction hole without grid; b) extraction hole with a plane grid; c) extraction hole with an inward curved grid, radius 0.8 cm.

Figure 2. Photos of the extraction hole without a grid (a), with a plane grid (b) and with an inward curved grid (c).

Using TE as Faraday cup and a pepper pot system[10], we studied the geometric quality of the beams and the extracted charges. Figure 3 shows a sketch of the system used to evaluate[10] the emittance by pepper pot technique. The emittance is related to the area of the ellipse in the trace plane. The mask we used was made of aluminum, in order to evacuate the charges impinging on its surface. It has 5 holes of 1mm in diameter and it was fixed on the GE. One hole is in the centre of the mask and 4 holes are at 3.5mm from the centre. We used as photo-sensible screen radiochromic films (R) Gafchromic EBT, placed on TE.

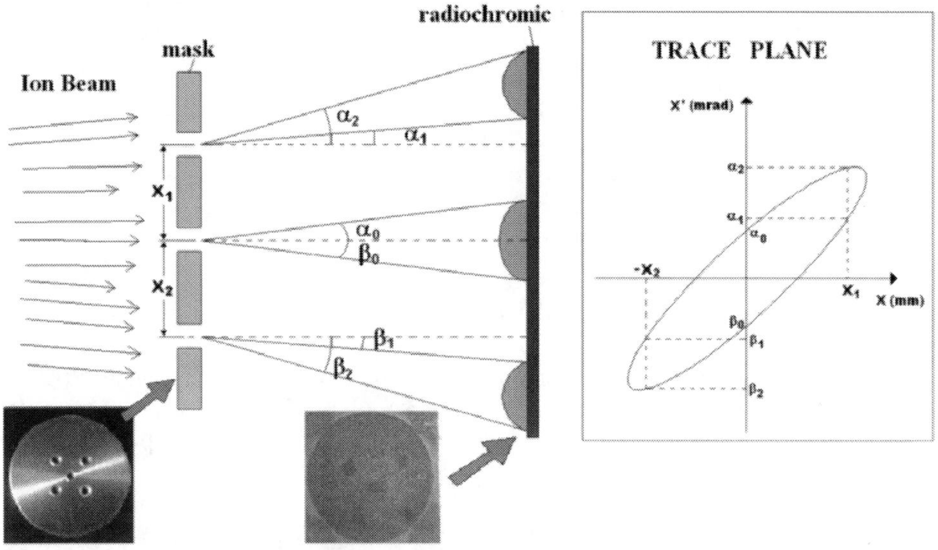

Figure 3. sketch of the system used to measure the emittance value by pepper pot technique.

Radiochromic detectors involve the direct impression of a material by the absorption of energetic radiation, without requiring latent chemical, optical, or thermal development or amplification. A radiochromic film changes its optical density as a function of the absorbed dose. This property and the relative ease of use, led to adopt these detectors as simple ion beam transverse properties diagnostic tools. So, the ion beam after the mask imprinted the radiochromic film and then it was possible to measure the divergence of all beamlets. The divergence values allowed to determine the beam area in the trace plane (TPx). For a z-axis beam propagation, the x-plane emittance ε_x is $1/\pi$ times the area Ax in the TPx occupied by the points representing the beam particles at a given value of z.

3. Results

We applied 250 laser shots to imprint the radio-chromic films. We measured the emittance for the three different anode configurations. By radiochromic images we drew the ellipses in the trace plane and we calculated the area in the TPx. Moreover, by Liouville's theorem it is known that the area occupied by the particle beam in phase plane (PPx) is an invariant quantity and the normalized emittance is:

$$\varepsilon_{nx} = \beta\gamma\varepsilon_x \tag{1}$$

where $\beta = v/c$ and γ is the Lorentz factor.

The results at the maximum accelerating voltage are shown in Fig. 4. In the figure it is inserted the table of the emittance values we measured. The results obtained show that the lowest emittance value, 435 mm mrad, is provided by the anode without any grid. Considering equation (1) we found the normalized emittance value, which resulted $\varepsilon_{nx} = 0.20\ \pi$ mm mrad.

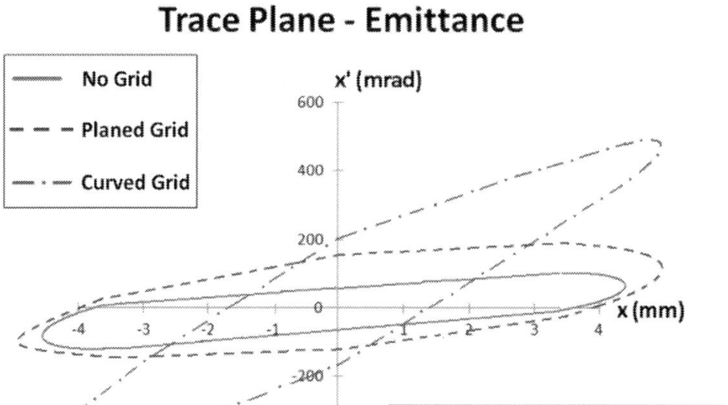

	No Grid	Plane Grid	Curved Grid
Voltages (kV)	40+20	40+20	40+20
ε_x (mm mrad)	435	795	1316
ε_{nx} (π mm mrad)	0.20	0.36	0.60

Figure 4. Emittance diagram in the trace plane for different anode configurations.

The experimental results were confirmed by a simulation study, performed with Comsol Multiphysics v3.5 software [11]. In Fig. 5 it is shown the simulation of Cu^{1+} beams (black lines) under 40+20 kV of accelerating voltage. The gray lines represent the electric streamline distribution.

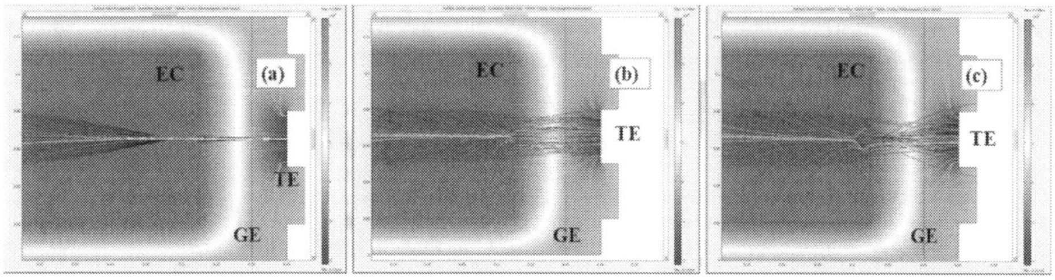

Figure 5. Comsol simulations of Cu^{1+} beams (black lines) under 60 kV of accelerating voltage for different anode configurations. The gray lines represent the electric streamline.

The simulations showed, in accordance with the results obtained experimentally, that: i) the hole without grid (a) focus the ion beam near the extraction hole and reduces its divergence and emittance; ii) the hole with the plane grid (b) enlarges the beam and forms a paraxial beam with a good emittance value, but larger than that obtained with configuration (a); iii) the behaviour of the hole with the curved grid (c) is similar a convergent lens which diverges the beam after the focus position, increasing its emittance.

The results presented in this work confirm and extend a finding presented in a previously published work[12], showing that the PLATONE setup is a very promising LIS that could be considered to feed large accelerators.

4. Conclusion

In this work we investigated how the morphology of the extraction anode in a LIS affects the beam qualities. We have shown that the presence of a grid on the extraction hole of EC modifies both the emittance and the total extracted charge. Increasing the voltage of the first accelerating gap, we increased substantially the efficiency of the extracted current due to the rise of the electric field and extracting volume inside the EC. We measured the geometric characteristics of the beam by means of the pepper pot technique and we found that the optimal configuration both for emittance and charge extraction is that without any grid. In this configuration, we obtained the lowest value for the normalized emittance, $\varepsilon_{nx} = 0.20\ \pi$ mm mrad and the maximum charge extracted, 1.1 μC. Finally, the findings obtained in this work suggest that improvements on the quality of the beams could be obtained only by changes in the geometry of the apparatus.

References

[1] Hegelich M, Karsch S, Pretzler G, Habs D, Witte K, Guenther W, Allen M, Blazevic A, Fuchs J, Gauthier J C, Geissel M, Audebert P, Cowan T, and Roth M 2002 *Phys. Rev. Lett.* **89** 085002
[2] Wilks S C, Langdon A B, Cowan T E, Roth M, Singh M, Hatchett S P, Key M H, Pennington D, MacKinnon A J, and Snavely R A 2001 *Phys. Plasmas* **8** 542
[3] Liseykina T V, Borghesi M, Macchi A, Tuveri S 2008 *Plasma Phys. Control. Fusion* **50** 124033
[4] Belloni F, Doria D, Lorusso A and Nassisi V 2004 *Rev. Sci. Instrum.* **75** 4763
[5] Torrisi L, Cavallaro S, Giuffrida L, Gammino S, Andò L 2010 *Rad. Eff. and Def. in Solids* **165** 509
[6] Nassisi V and Pedone A 2003 *Rev. Sci. Instrum.* **74** 68
[7] Lorusso A, Velardi L, Nassisi V, Paladini F, Visco A M, Campo N, Torrisi L, Margarone D, Giuffrida L, Rainò A 2008 *Nucl. Inst. and Meth. in Phys. Res. B* **266** 2490
[8] Noli F, Lagoyannis A and Misaelides P 2008 *Nucl. Inst. and Meth. in Phys. Res. B* **266** 2437
[9] Rosinski M, Badziak B, Parys P, Wołowski J and Pisarek M 2009 *Appl. Surf. Sci.* **255** 5418
[10] Wang J G,. Wang D X and Reiser M 1991 *Nucl. Inst. and Meth. in Phys. Res. A* **307** 190
[11] http://www.comsol.com/
[12] Nassisi V, Velardi L and Delle Side D 2013 *Appl. Surf. Sci.* **272** 114

Journal of Physics: Conference Series **508** (2014) 012006 doi:10.1088/1742-6596/508/1/012006

The LICPA accelerator of dense plasma and ion beams

J Badziak[1], S Jabłoński[1], T Pisarczyk[1], T Chodukowski[1], P Parys[1], P Rączka[1], M Rosiński[1], E Krousky[2], J Ullschmied[3], R Liska[4], M Kucharik[4], L Torrisi[5]

[1]Institute of Plasma Physics and Laser Microfusion, Euratom Association, Warsaw, Poland
[2]Institute of Physics AS CR Prague, Czech Republic
[3]Institute of Plasma Physics AS CR Prague, Czech Republic
[4]Czech Technical University, FNSPE, Prague, Czech Republic
[5] Dipartimento di Fisica, Universita di Messina, Messina, Italy

E-mail: badziak@ifpilm.waw.pl

Abstract. Laser-induced cavity pressure acceleration (LICPA) is a novel scheme of acceleration of dense matter having a potential to accelerate plasma projectiles with the energetic efficiency much higher than the achieved so far with other methods. In this scheme, a projectile placed in a cavity is irradiated by a laser beam introduced into the cavity through a hole and accelerated along a guiding channel by the thermal pressure created in the cavity by the laser-produced plasma or by the photon pressure of the ultraintense laser radiation trapped in the cavity. This paper summarizes briefly the main results of our recent LICPA studies, in particular, experimental investigations of ion beam generation and heavy macroparticle acceleration in the hydrodynamic LICPA regime (at moderate laser intensities ~ 10^{15}W/cm^2) and numerical, particle-in-cell (PIC) studies of production of ultraintense ion beams and fast macroparticles using the photon pressure LICPA regime (at high laser intensities > 10^{20} W/cm^2). It is shown that in both LICPA regimes the macroparticles and ion beams can be accelerated much more efficiently than in other laser-based acceleration scheme commonly used and the accelerated plasma/ion bunches can have a wide variety of parameters. It creates a prospect for a broad range of applications of the LICPA accelerator, in particular in such domains as high energy density physics, ICF research (ion fast ignition, impact ignition) or nuclear physics.

1. Introduction

Laser-induced cavity pressure acceleration (LICPA) is a recently proposed [1,2] scheme of acceleration of dense matter having a potential to accelerate plasma macroparticles or ion beams with the energetic efficiency significantly higher than that achieved with other laser-based methods known so far. In the LICPA scheme, a projectile (e.g. foil target or small disc) placed in a cavity is irradiated by a laser beam introduced into the cavity through a hole; the projectile is then accelerated in a guiding channel (the cylindrical or conical one – Fig. 1) by the pressure created in the cavity by the laser-produced hot plasma expanding from the irradiated side of the projectile and from the cavity walls, or by the photon (radiation) pressure of the ultra-intense laser pulse trapped in the cavity. An important part of the scheme is the guiding channel, which plays a role similar to that of a barrel in a

Journal of Physics: Conference Series **508** (2014) 012006 doi:10.1088/1742-6596/508/1/012006

conventional gun, allowing the accelerating forces to act for an extended period of time as well as collimating and compressing the accelerated matter.

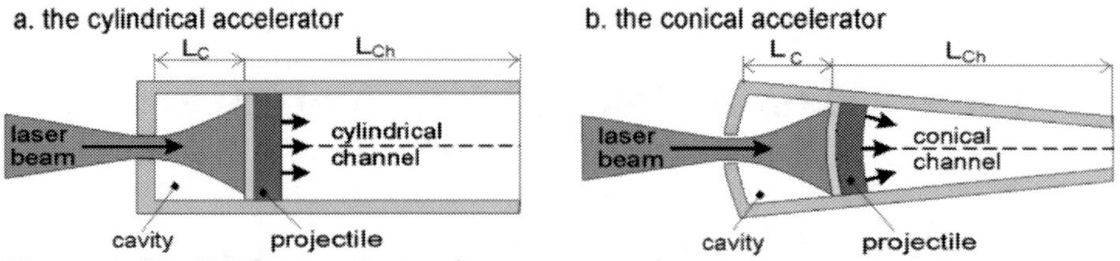

Figure 1. The LICPA accelerator in two geometries.

In dependence on the laser intensity and the pulse length, different regimes of the LICPA accelerator operation can be distinguished [2]. At relatively low laser intensities and moderately long laser pulses (say, at $I_L < 10^{17}$ W/cm^2 and $\tau_L > 10^{-11}$ s) we can speak about a pure hydrodynamic LICPA regime, as the dominating force driving the projectile is due to the hydrodynamic pressure of hot plasma produced and confined in the accelerator cavity. On the other hand, at very high laser intensities and short laser pulses of circular polarization (say, at $I_L \geq 10^{20}$ W/cm^2, $\tau_L \leq 10^{-11}$ s), the photon pressure LICPA regime occurs, as the acceleration of the projectile is predominantly due to the photon (radiation) pressure of the laser pulse trapped in the cavity. For intermediate laser intensities (say, 10^{17}W/cm$^2 \leq I_L < 10^{20}$ W/cm^2) we may speak about a mixed LICPA regime, as in this case there is a complicated interplay between hydrodynamic and photon pressure and a large amount of non-thermal hot electrons is produced.

This paper summarizes briefly the main results of our recent LICPA studies, in particular, experimental investigations of ion beam generation and heavy macroparticle acceleration in the hydrodynamic LICPA accelerator and numerical, particle-in-cell (PIC) studies of production of ultraintense ion beams and fast macroparticles using the photon pressure-driven LICPA accelerator.

2. The hydrodynamic LICPA accelerator

Main properties of the hydrodynamic LICPA accelerator were investigated experimentally and numerically in [2]. In particular, it was shown that the basic, unique feature of the accelerator is extremely high energetic efficiency of acceleration $\eta_{acc} = E_p/E_L$ (E_p is kinetic energy of the projectile, E_L is energy of the laser beam) which for a light (CH) plasma projectile can reach the values even above 50% [2]. Here, we will present exemplary results demonstrating that the hydrodynamic LICPA accelerator can be an efficient tool for generation of high-current-density moderate-energy ion beams as well as acceleration of heavy (~ 10 μg) macroparticles.

2.1. Generation of high-current-density ion beams

The experimental investigation of ion beam generation was performed on the kilojoule PALS laser facility in Prague [3]. The ion beams produced from a 25-μm CD_2 foil target in the cylindrical (Fig. 1a) or conical (Fig. 1b) LICPA accelerator were compared with the ones generated from the CD_2 foil in the conventional laser-planar target interaction scheme (the L-T scheme) and the scheme where the foil was placed at the entrance of the cylindrical or conical channel without a cavity (the L-Ch scheme). Parameters of the forward emitted ion fluxes were measured by the time-of-fight (TOF) method using two ion collectors placed 30cm from the CD_2 target at angles 0^0 and 5^0, respectively, in respect to the target normal and the laser beam axis. The CD_2 target was irradiated by the 3ω (λ = 0.438 μm) PALS laser beam of 0.3 ns duration and energy up to ~ 300J. A comparison of parameters of ion beams produced in the different schemes was accomplished for the same laser intensities on the CD_2 target being in the range $10^{14} - 10^{15}$ W/cm^2.

Journal of Physics: Conference Series **508** (2014) 012006 doi:10.1088/1742-6596/508/1/012006

Fig. 2a presents the TOF ion signals, recorded by the ion collector situated on the laser beam axis, for ions generated from the cylindrical LICPA accelerator and from the planar target irradiated in the L-T scheme. In the presented signals we can see only the thermal (slow) component of the ion beam of keV ion energies as the number of fast ions is too small here to produce a fast component visible in the scale presented in the figure. A difference between parameters of ion beams produced in the LICPA accelerator and generated in the L-T scheme is very pronounced: for LICPA the peak current density of thermal ions j_i is about order of magnitude higher and the mean ion energy $\overline{E}_i$ a factor 4 – 5 higher than for the L-T scheme, so the peak ion beam intensities $I_i \approx (1/\overline{z}e)j_i\overline{E}_i$ differ by a factor 20 or more for these schemes ($\overline{z}$ is the average ion charge state, e is the electron charge). The peak ion current density as a function of laser energy for three considered schemes is presented in Fig. 2b. It can be seen that for all laser energies for which the acceleration schemes were compared, the peak current densities for ion beams produced in the LICPA accelerator are always much higher than those for ion beams generated in the L-T scheme or the L-Ch scheme.

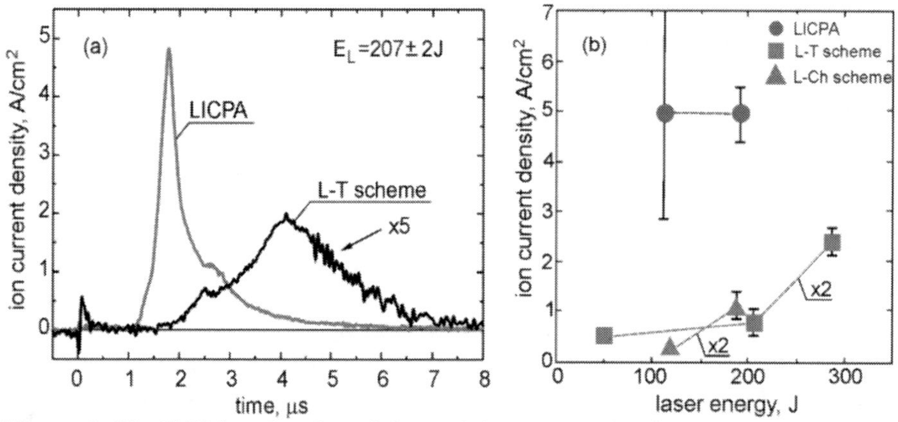

Figure 2. The TOF ion signals and the peak ion current density at the distance of 30cm from the target as a function of laser energy for the cylindrical LICPA scheme, the cylindrical L-Ch scheme and the L-T scheme.

Similar, as presented in Fig. 2b, differences in parameters of ion beams driven in the LICPA scheme and the L-Ch or L-T schemes we observed for the conical configurations. An interesting new feature was the appearance of a strong fast ion component in the TOF signal, which was especially clearly visible for high laser energies (~ 300 J). The mean energy of the fast ions, mostly deuterons (D^+), was more than a factor 5 higher for LICPA than for the L-T scheme, while their maximum energies were by a factor 10 higher for LICPA and reached a few hundred keV for C ions and ~ 150 keV for deuterons. The estimated current density of deuterons of energy 10 – 150 keV at the exit of the conical channel attained very high value $j_{Ds} \approx 1.2 \times 10^5$A/cm^2 and the current density per steradian approached $(j_{Ds}/\Theta) \approx 3.4 \times 10^9$A/cm^{2}sr [4]. The deuteron beam of such parameters heating a deuterated target can, potentially, produce neutrons from DD fusion. Really, when at the exit of the conical LICPA accelerator we placed the massive Al target covered by 10-μm CD_2 layer, we recorded a neutron yield ~ 10^4 neutrons/shot [4]. When such a target was used in the L-Ch or L-T schemes no neutrons were recorded.

To conclude, our experiment has demonstrated that parameters of ion beams produced in the hydrodynamic LICPA accelerator driven by a sub-ns laser pulse of intensity ~ $10^{14} - 10^{15}$ W/cm^2 are significantly higher than the ones of ion beams generated in the conventional laser-planar target interaction scheme. In particular, the current density and intensity of the ion beam are by an order of magnitude higher and the mean and maximum ion energies for both thermal and fast ions are by a factor ~ 4 – 10 higher than for the conventional scheme. The current density of 10 – 150 keV deuteron beam leaving the conical LICPA accelerator can reach an extremely high value ~ 1GA/cm^{2}sr which makes it possible to produce fusion neutrons from the interaction of the beam with a CD_2 target. Parameters of ion beams produced in the accelerator can be further increased by a careful selection of the target thickness and optimizing the accelerator geometry and at laser intensities on the target ~ 10^{16} W/cm^2 efficient generation of intense ion beams with MeV ion energies seems to be feasible.

2.2 Acceleration of heavy macroparticles

Acceleration of heavy (~ 1 – 100 μg or more) macroparticles to hyper-velocities ~ 100 – 1000 km/s is of high importance for high energy density physics [5,6] and inertial confinement fusion (ICF) [7,8] and has been a challenge for a long time. At present, only laser-based methods using laser-induced ablative acceleration (AA) [7] are capable of accelerating a heavy macroparticle to such high velocities [7,8,9]. Unfortunately, the energetic acceleration efficiency η_{acc} for AA is low, in practice below a few percent [6,9,10]. As a result, to accelerate a multi-μg macroparticle to velocities above 100 km/s a big multi-kJ laser driver is needed. The LICPA accelerator has a potential to decrease the required laser driver energy by an order of magnitude. To check this hypothesis we performed an experiment at the PALS laser facility.

In the experiment, an aluminum disc covered by 5 μm thick CH ablator was irradiated by the 0.3-ns, 3ω PALS beam of energy 200J in free space (the AA scheme) or in the cavity of the cylindrical LICPA accelerator (Fig. 1a) of L_c= 0.4mm, L_{Ch} = 2mm and the channel diameter D_{Ch} = 0.3mm. We used Al discs of various thicknesses equal 10μm, 20μm, 50μm and 75μm which correspond to the mass m_p of the accelerated macroparticle equal 2μg, 4μg, 10μg and 15μg, respectively. The macroparticle velocity was measured using the time-of-flight (TOF) method employing the ion collector placed on the laser beam axis at the distance of 25 cm from the irradiated target. Additionally, we recorded the density distribution of Al plasma leaving the accelerator channel using three-frame interferometry [2]. The macroparticle velocity measured by the TOF method was found to be in a rough agreement with that estimated from the interferometric measurements [2]. Fig. 3 presents the macroparticle velocity v_p and the macroparticle acceleration efficiency $\eta_{acc} = E_p/E_L$ as a function of the macroparticle mass (E_p is the macroparticle kinetic energy and E_L is the laser beam energy). For a comparison, the values of v_p and η_{acc} for the macroparticle of $m_p \approx$ 10μg driven by the ablative acceleration are also shown in the figure. It can be seen that the velocity of macroparticle driven by LICPA decreases from ~ 130 km/s to ~ 70 km/s and the acceleration efficiency increases from ~ 9% to ~ 15% with an increase in m_p. So far, multi-μg macroparticles could be accelerated to such velocities only with the use multi-beam, multi-kJ laser facilities [6].

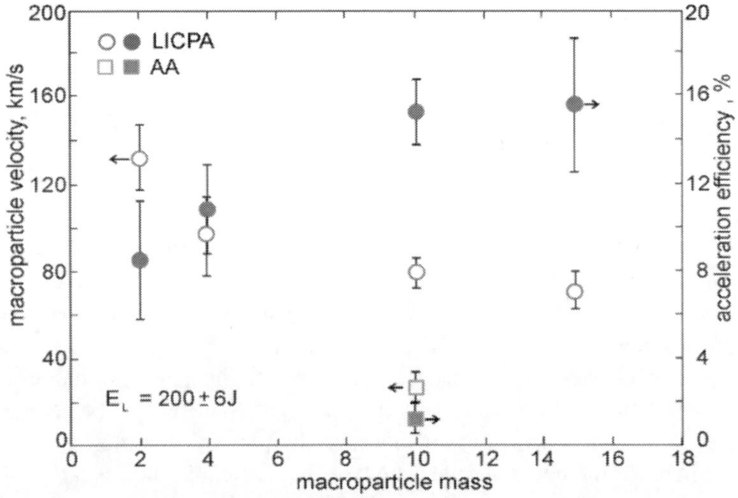

Figure 3. The velocity and the energetic efficiency of acceleration of macroparticle as a function of the macroparticle mass.

In conclusion, we have found that: (a) a heavy macroparticle of the mass ~ 10μg has been accelerated to the velocity ~ 100 km/s with the laser driver of moderate (sub-kJ) energy, (b) the energetic acceleration efficiency reaches ~15% and is the highest achieved so far for heavy (>1μg) macroparticles, (c) the acceleration efficiency for LICPA is by more than order of magnitude higher than that for ablative acceleration, (d) by collision of the LICPA-driven macroparticle with a solid target, an ultra-strong shock wave can be generated and production of sub-Gbar pressures with the use of sub-kJ laser drivers seems to be feasible.

3. The photon pressure-driven LICPA accelerator

The photon pressure-driven LICPA accelerator required very high laser intensities ($>10^{20}$W/cm^2) which are hardly attainable with currently operating laser facilities, so this LICPA regime has been studied only numerically so far [2]. Here, we will present exemplary results of our 1D PIC numerical simulations demonstrating possibilities of using LICPA for ICF-related applications, in particular, for generation of ultraintense ion beams for the ion fast ignition of ICF target [11-13] and for acceleration of a heavy macroparticle for impact ignition fusion [8,10,14].

3.1. Generation of ultraintense ion beams for ion fast ignition

In the ion fast ignition scenario [11-13], a DT fuel compressed to the density ~ 300 – 500 g/cm^3 is ignited by an intense short-pulse (5 – 20 ps) beam of light ions (usually protons or carbon ions) of the mean energy $\overline{E}_i \sim$ 10 – 50 MeV/amu [13]. These energies are relatively low, however other parameters of the beam should be extremely high: intensity $I_i \geq 10^{20}$ W/cm^2, energy fluence $F_i \geq 2$ GJ/cm^2, total beam energy $E_{ib} \geq 20$ kJ [11-13]. In addition, the driver-ions energy conversion efficiency η should be sufficiently high ($\eta \geq 15\%$) to limit the driver energy and cost to a reasonable value. The only way of ion acceleration known so far which promises attaining such ion beam parameters is laser-driven acceleration, in particular, the one using the radiation pressure acceleration (RPA) mechanism [15,16].

For a quantitative estimation of the possibility of achieving the ion beam parameters required for the fast ignition in the LICPA and the conventional RPA schemes we performed 1D simulations of acceleration of carbon ions using the relativistic 1D PIC code [17]. In the simulations, a laser beam interacted with a fully ionized carbon plasma target consisting of a homogenous layer of the thickness $L_T = 8$ μm and the ion number density $n_i = 10^{23}$ cm^{-3}, and an exponential pre-plasma layer on the front side of the target with the density gradient scale length equal to 0.25 μm. We assumed that a circularly polarized laser pulse has wavelength of 1.05 μm, a super-Gaussian profile $I(t) = I_L \cdot \exp(-t^6/\tau^6)$, with $I_L = 2.5\times10^{21}$ W/cm^2, 5×10^{21} W/cm^2 or 8×10^{21} W/cm^2, and FWHM equal to 2 ps. In the LICPA scheme, the laser light reflected from the carbon foil back towards the laser was assumed to be partially reflected from the accelerator cavity wall placed at a distance L_c from the foil in the upstream direction, and redirected back towards the carbon foil. Simulations were done for the reflection coefficient $R_c = 0.64$ and the cavity length $L_c = 120$ μm.

Fig. 4 presents the dependence of the carbon ion beam intensity I_i and fluence F_i as well as the laser-ions energy conversion efficiency on laser intensity I_L, predicted by the PIC simulations for the acceleration length $l_{acc}= 200$ μm. It can be seen that the beam intensity is a factor 10 higher and the beam fluence and the conversion a factor 2 – 4 higher for LICPA than those for RPA. This enhancement of the ion beam parameters in the LICPA scheme is due to the circulation of the laser pulse in the cavity which results in a higher laser-ions energy conversion efficiency. Moreover, a longer time of pushing the ion bunch by the radiation pressure in the cavity results in additional compression of the bunch and, as a consequence, in much higher intensity of the bunch than for the RPA case. Since the bunch time duration lies in the sub-picosecond range, the bunch intensity is even higher than the laser pulse intensity. In the considered range of laser intensities (2.5 – 8) × 10^{21}W/cm^2, both LICPA and RPA produces quasi-monoenergetic ion beams of similar energy dispersions $\Delta E_i/\overline{E}_i \sim 0.3-0.4$, but of significantly different mean ion energies $\overline{E}_i$: from 7.2 MeV/amu to 41 MeV/amu for LICPA and from 2.2 MeV/amu to 18 MeV/amu for RPA. In turn, the laser-ions energy conversion efficiency η changes from 23% to 41% for LICPA and from 7% to 18% for RPA in this intensity range. As a result, for LICPA all the ion beam parameters are well above the threshold for the ignition at the laser intensity of 5×10^{21} W/cm^2, which corresponds to 100kJ of laser energy at the laser focal spot on the target of 35 μm. To reach the ignition threshold in the case of using RPA, the laser intensity and energy should be a factor 2 – 3 higher.

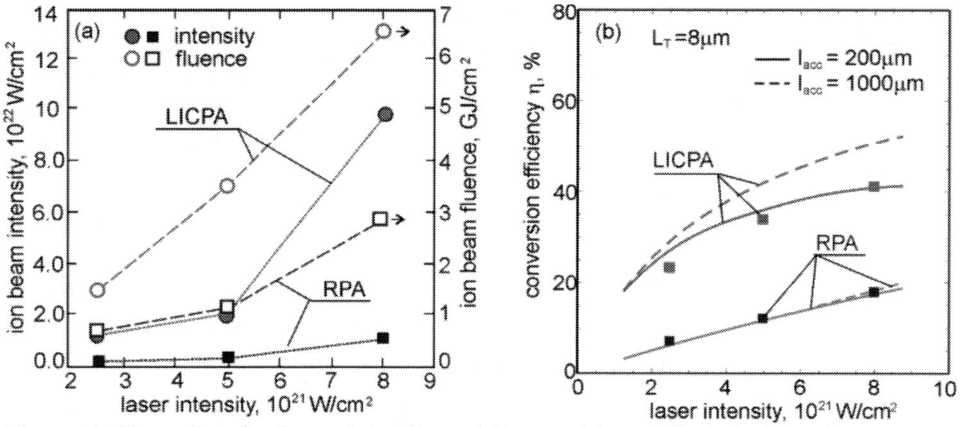

Figure 4. The carbon ion beam intensity and fluence (a) as well as the laser-ions energy conversion efficiency (b) predicted by the PIC simulations (dots). In (b), the is also shown the conversion efficiency calculated for l_{acc}= 200μm (solid lines) and l_{acc}= 1000μm (dashed lines) from the generalized light-sail model [2].

In conclusion, the photon pressure-driven LICPA accelerator employing a picosecond 100 kJ laser driver can produce – with a high energetic efficiency – quasi-monoenergetic carbon ion beams of mean ion energies ~ 20 – 30 MeV/amu, energy fluencies > 2G J/cm2 and intensities >10^{21} W/cm^2. The beam parameters are several times higher than those achieved in the conventional RPA scheme and they meet fairly well the ion fast ignition requirements.

3.2. Acceleration of a heavy macroparticle for impact ignition fusion

In the impact ignition scenario of inertial fusion, a rapid heating and igniting of the compressed DT fuel is the result of the impact of a macroparticle of the mass m_p~ 10^{-6} – 10^{-4} g accelerated to the velocity above 10^8 cm/s [8,14]. Here, we consider the impact ignition scheme proposed by Caruso and Pais [14] in which a ~ 1μg macroparticle made of Au and accelerated to ~ 5 × 10^8 cm/s collides with the DT fuel compressed to the density ≥ 200 g/cm^3. Due to the collision, the macroparticle is rapidly collapsing to high densities (>1000 g/cm^3) and a large fraction of its energy is transferred to the fuel in very short time (~ 10^{-11} s). As a result, a hot spot is created which ignites the fuel. In the considered case, the minimum kinetic energy of the macroparticle is ~ 10 – 20 kJ and the minimum energy fluence of the macroparticle is ~ 500 MJ/cm^2 [14].

We investigated the possibility of acceleration of a gold macroparticle to the required velocities and energies in the LICPA and RPA schemes using the same PIC code and the calculation scheme as those for acceleration of carbon ions. The gold (Au^{10+}) plasma target of the homogenous layer thickness L_T = 5 μm and the preplasma thickness of 0.25 μm was irradiated by a circularly polarized super-Gaussion laser pulse of τ_L = 10 ps, 5 ps or 2.5 ps (FWHM), λ = 1.06 μm and the focal spot diameter d_L = 80 μm (the target mass corresponding to this diameter ≈ 0.5 μg). The laser pulse energy E_L was assumed to be of 100 kJ, 200 kJ or 300 kJ which correspond to the 10-ps pulse intensity I_L of 2×10^{20} W/cm^2, 4×10^{20} W/cm^2 and 6×10^{20} W/cm^2, respectively, and 2 or 4 times higher intensities for the shorter pulses. For LICPA we assumed $L_c = 300$ μm and $R_c = 0.75$.

The macroparticle energy fluence as a function of laser energy or the pulse duration for LICPA and RPA is presented in Fig. 5. There is also marked the ignition threshold fluence calculated with the use of the hydrodynamic numerical code in [14]. For LICPA, the macroparticle energy fluence reaches the threshold value at the laser energy ~ 150 kJ, while for RPA the energy fluence is well below the threshold even at E_L = 300kJ. In the considered laser energy range of 100 – 300 kJ and at τ_L= 10ps, the macroparticle velocity v_p and the acceleration efficiency η_{acc} reach for LICPA the values from 7.6×10^8 cm/s to 1.6×10^9 cm/s and from 12% to 21%, respectively, while for RPA v_p and η_{acc} are in the range of (2.4 – 6.9) x 10^8 cm/s and 1.2 – 3.3 %, respectively. Thus, at the laser energy ~ 150 kJ, the velocity and the energy fluence as well as the total kinetic energy of the macroparticle driven by LICPA achieve the values required for the ignition. In the case of using RPA, the ignition seems to be possible only at laser energies at least 3 – 4 times higher.

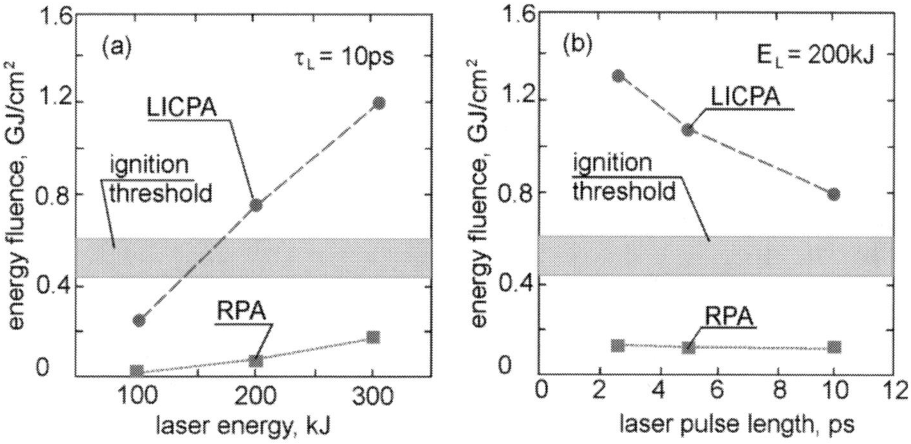

Figure 5. The energy fluence of Au macroparticle driven by LICPA or RPA as a function of laser energy (a) and the laser pulse length (b).

In conclusion, the LICPA accelerator can efficiently ($\eta \geq 10\%$) accelerate heavy (~ 1μg) macroparticles to high velocities (> 5×10^8 cm/s) and the acceleration efficiency for LICPA is almost an order of magnitude higher than that for RPA. In particular, the LICPA accelerator using a 10-ps laser driver of energy ~ 150 kJ can accelerate a 0.5 μg Au plasma macroparticles to velocities ~ 10^9 cm/s required for DT ignition in the impact ignition fusion scenario of Caruso and Pais.

The results obtained in this section should be treated as the first step to more comprehensive studies of the photon pressure-driven LICPA accelerator employing, in particular, 2D codes and a more detailed model of laser-plasma interaction in the accelerator cavity.

4. Summary

Our measurements performed at the kilojoule PALS laser facility have demonstrated that the hydrodynamic LICPA accelerator can be a highly efficient tool both for generation of high-current-density, moderate-energy ion beams and acceleration of heavy macroparticles and parameters of ion beams or macroparticles driven by LICPA can be much higher than the ones achieved with other laser-based method commonly used.

Our PIC simulations have shown that the photon pressure-driven LICPA accelerator can be highly efficient like the hydrodynamic one and, in particular, can produce ultraintense, quasi-monoenergetic ion beams of multi-MeV to GeV ion energies with the efficiency of tens percent. It can also efficiently

accelerate heavy macroparticles to velocities required for impact ignition fusion. The energetic efficiency of acceleration (both ion beams and macroparticles) in the LICPA accelerator is significantly higher than that in the conventional RPA scheme.
The LICPA accelerator can, potentially, be driven by lasers covering a very broad range of laser energies (from 1J to 1MJ) and intensities (from 10^{10} W/cm^2 to 10^{23}W/cm^2 or higher) as well as pulse lengths (from ns to fs) and laser wavelengths (from UV to IR). As a result, the accelerator can produce dense, fast or ultrafast projectiles of a wide variety of parameters. The unique efficiency of the LICPA accelerator coupled with its flexibility in using various laser drivers and in producing projectiles of various parameters create a potential for a broad range of the accelerator applications, in particular in high energy density physics, ICF research, nuclear physics or material science.

5. References

[1] Badziak J, Borodziuk S, Pisarczyk T, Chodukowski T, Krousky E et al. 2010 *Appl. Phys. Lett.* **96** 251502.
[2] Badziak J , Jabłoński S, Pisarczyk T, Rączka P, Krousky E et al. 2012 *Phys. Plasmas* **19** 53105.
[3] Jungwirth K, Cejnarova A, Juha L, Kralikova B, Krasa J, et al. 2001 *Phys. Plasmas* **8**, 2495.
[4] Badziak J, Parys P, Rosiński M, Krousky E, Ullschmied J and Torrisi L 2013, *Appl. Phys. Lett.* **103**, 124104.
[5] Drake R P 2006 *High-Energy-Density Physics* (Springer, Berlin,).
[6] Cauble R, Phillion D W, Hoover T J, Holmes N C, Kilkenny J D and Lee R W 1993 *Phys. Rev. Lett.* **70**, 2102.
[7] Atzeni A and Meyer-ter-Vehn J 2004 *The Physics of Inertial Fusion* (Clarendon, Oxford,).
[8] Murakami M and Nagatomo H 2005 *Nucl. Instrum. Meth. Phys. Res.* A **544** 67.
[9] Karasik M, Weaver J L, Aglitskiy Y, Watari T, Arikawa Y et al. 2010 *Phys. Plasmas* **17**, 056317.
[10] Azechi H, Sakaiya T, Watari T, Karasik M, Saito H et al. 2009 *Phys. Rev. Lett.* **102**, 235002.
[11] Temporal M, Honrubia J H, and Atzeni S, 2002 *Phys. Plasmas*, **9**, 3098.
[12] Fernandez J C , Honrubia J H, Albright B J, Flippo K A, Cort D 2009 *Nucl. Fusion* **49** 065004.
[13] Davis J, Petrov G M and Mehlhorn T A 2011 *Plasma Phys. Control. Fusion* **53** 045013.
[14] Caruso A and Pais V A 1996 *Nucl. Fusion* **36** 745.
[15] Esirkepov T, Borghesi M, Bulanov S V, Mourou G, and Tajima T 2004 *Phys. Rev. Lett.* **92** 175003.
[16] Macchi A, Cattani F, Liseykina T V and Cornalti F 2005 *Phys. Rev. Lett.* **94**, 165003.
[17] Badziak J and Jabłoński S 2010 *Phys. Plasmas* **17** 073106.

Basic features of electromagnetic pulse generated in a laser-target chamber at 3-TW laser facility PALS

M De Marco[1], M Pfeifer[1], E Krousky[1], J Krasa[1], J Cikhardt[2], D Klir[2] and V Nassisi[3,4]

[1]Institute of Physics, ASCR, v.v.i., Na Slovance 2, 182 21 Praha 8, Czech Republic
[2] Czech Technical University, Technická 2, 166 27 Prague 6, Praha 6 Czech Republic
[3]LEAS, Dipartimento di Matematica e Fisica, Università del Salento
[4]INFN Sezione di Lecce, Via Arnesano, sn – 73100 Lecce, Italy

E-mail: demarco@fzu.cz

Abstract. We describe the radiofrequency emission taking place when 300 ps laser pulses irradiate various solid targets with an intensity of 10^{16} W/cm^2. The emission of intense electromagnetic pulses was observed outside the laser target chamber by two loop antennas up to 1 GHz. Electromagnetic pulses can be 800 MHz transients, which decay from a peak electromagnetic field of $E_0 \cong 7$ kV/m and $H_0 \cong 15$ A/m. The occurrence of these electromagnetic pulses is associated with generation of hard x-rays with photon energies extending beyond 1 MeV. This contribution reports the first observation of this effect at the PALS facility.

1. Introduction

If a high intensity laser pulse irradiates a target, the production of plasma can be accompanied by the emission of an intense electromagnetic pulse (EMP) within and outside the laser target chamber [1-10]. The EMP causes electromagnetic interference and can disrupt or damage nearby electronic diagnostics and control units. It is supposed that the EMP arises in the laser target chamber where the short-pulse laser produces energetic electrons that are able to escape the target until the built electrostatic field is low enough [3]. The short pulse of electrons can produce very high transient currents and intense EMP with broad spectrum ranging from several tens of MHz to several GHz. The occurrence of small-scale structures in a laser-target chamber contributes to the fraction of the EMP that is at higher frequencies [4]. Just for effective mitigation of diagnostic-generated electromagnetic interference it is essential to know the characteristics of the EMP. The interference between the EMP and ion collector (IC) signal can also be identified in Fig. 2 of [11] which were produced by a 500 ps laser pulse irradiating a target with intensity of ~10^{14} W/cm^2. The aim of the work reported here is to characterize the EMP, which has been produced by the laser facility PALS delivering intensity up to 3×10^{16} W/cm^2 onto a target in 300-ps pulses.

2. Experimental arrangement

The 3-TW PALS laser system irradiated 10 μm silicon and 200 μm plastics targets at 0° and 30° with respect to the target normal. The emission of ions was measured with the use of TOF ion collectors and EMP was detected with the use of 12 cm diameter Moebius loop antennas [14]. Other devices used are TEKTRONIX DPO4104 2.5 GS/s 1 GHz oscilloscope and Rg-58 coaxial cables. Fig. 1 shows the scheme of the target chamber and the loop antenna. An ion ring collector IC with internal diameter of 3.8 cm and outer one of 5.0 cm was used to measure the ion component of the plasma expanding into the vacuum; the schematic drawing of the used collector is in [12].

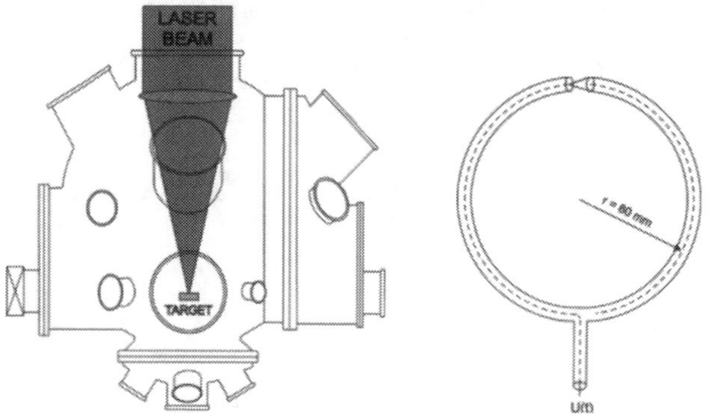

Figure 1. Schemes of the target chamber and the loop antenna.

3. Experimental results

Fig. 2a shows a typical example of a strongly degraded ion collector signal caused by the EMP interference. The observed EMP wavefront at ~12 ns prior the XUV-signal clearly reveals a propagation delay of the IC signal. This delay is proportional to the length of the used coaxial cable, as Fig. 2b shows. This experimental finding can be useful in some other experiments because this fact allows us to eliminate the EMP interference, e.g. by inserting a delay unit (e.g. a long Rg-58 coaxial cable) between the oscilloscope and IC, as it was proposed by one of our team (M.P.). The propagation delay of the IC signal passing through a 40-m coaxial cable on the interference is shown in Fig. 2b. It reveals that the EMP is not a part of the IC signal because it is not coupled to the active part of the TOF detector and, thus, it is not propagating with the signal induced by XUV radiation and ions. Nevertheless, the acceptable very low capturing of these fast transient events in the harsh electromagnetic environment occurring in the vicinity of the beam-target interaction chamber has been performed when the oscilloscope was protected inside a Faraday cage.

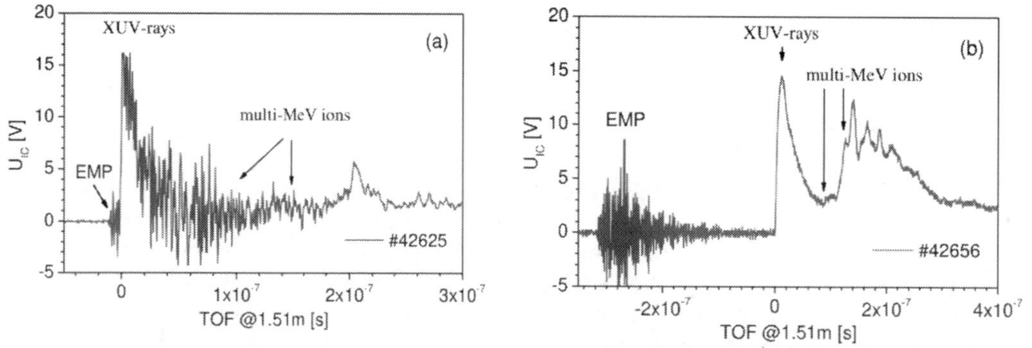

Figure 2. Interference between EMP and IC signal (a). Separation of the EMP from the IC signals inserting a delay unit between oscilloscope and IC (b). The $(CD_2)_n$ target was exposed to $I_L = 2\times10^{16}$ W/cm^2.

The transient signals in the vicinity of the target chamber were recorded using the Moebius loop antennas. All the EMP waveforms observed in the PALS experiment have a similar structure, as Fig. 3

Journal of Physics: Conference Series **508** (2014) 012007 doi:10.1088/1742-6596/508/1/012007

shows; the antenna was placed vertically to the ground above the target chamber at a distance of 0.5 m. The analysis of the waveform based on a band pass FFT filter method shown that its first part lasting about 50 ns (labelled by ΔT_1 in Fig. 3) has a frequency spectrum ranging from ~370 MHz to ~ 610 MHz. The central part that is stronger (labelled by ΔT_2 in Fig. 3) has a frequency spectrum from ~690 MHz to ~ 850 MHz. Although the magnitude and frequency of the EMP generated within an ideal spherical, cylindrical or rectangular target chamber can be calculated, the behaviour of a real target chamber is different due to its shape (see Fig. 1) and the effect of equipment inside the chamber [1]. Such a target chamber has multiple, distinct, resonant frequencies which can only be determined experimentally.

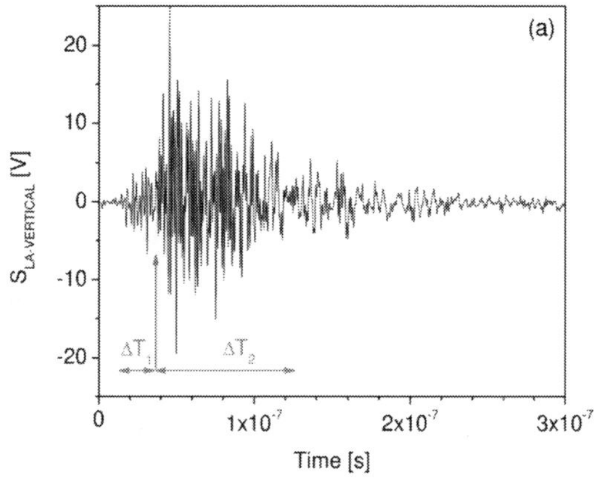

Figure 3. Example of vertical loop antenna waveform; the antenna was positioned above the laser target chamber at a distance of 0.5 m. A 10-μm thick Si single crystal foil was exposed to the laser intensity of 2×10^{16} W/cm^2.

The distribution of the electromagnetic field in the laboratory room has a complex structure mainly due to the different EMP spread through the wall and the windows of the laser target chamber, as Fig. 4 shows. The comparison of the FFT spectra clearly demonstrates that the glass window is quite more transparent for the EMP. Thus, if the detector is positioned near this window, than the induced noise signal can reach a high value and the detector signal is degraded. The measurements show that the signals passing through the glass windows are about three times higher than the signals passing through the steel wall.

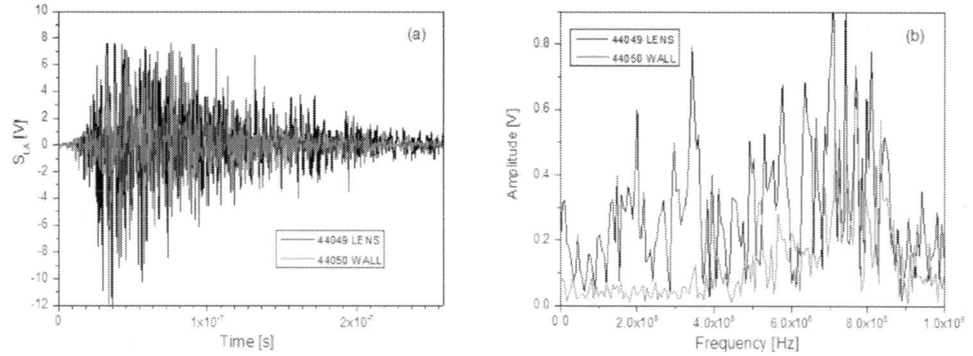

Figure 4. Loop antenna signals detected outside the target chamber beyond the laser-beam input glass window (near the focussing lens) and beyond the steel wall (a) and corresponding FFT spectra (b). A 10-μm thick Si single crystal foil was exposed to the laser intensity of 4×10^{16} W/cm^2.

From the experiment described in [15] we know that using the loop antenna positioned into the target chamber, the values of electric field inside the target chamber were determined to be in average

$E_0 \cong 7$ kV/m for $(CD_2)_n$ targets exposed to intensity of 10^{16} W/cm^2. The comparison of EMP signals detected at the same shot by the loop antenna placed inside the chamber (less or more in the middle of the chamber) [15] and by our loop antenna outside at 0.6 m behind the steel wall, shows that there is a wall attenuation of about 20 dB. Our measurements confirm the experiences with shielding and EMP mitigation strategy presented in [5,6] which is based on optimized grounding and detector configurations.

4. Conclusions

Our experiment has shown that the laser system PALS delivering onto a target the intensity of about 4×10^{16} W/cm^2 can generate not only protons up to 4-MeV [13] but also damaging intense EMP. The frequency spectrum of the detected EMP ranges up to 1GHz. As expected, also in the case of the PALS target chamber, the EMP behaviour is different from one in ideal spherical chamber. In fact due to the effects of small-scale structures in the target chamber as well as of equipment inside the chamber such as the target positioner, cameras, Faraday cups, and other detectors, there is a shift in the ideal spherical chamber resonant frequency [1]. Bunche structure of the generated EMP, where bunches have different mean frequencies and appear at different times, could be due to bursts of plasma escaping the target, and propagating through the target chamber and then interact with the target chamber wall. The most intense EMP outside the target chamber was observed near the large area glass window (corresponding attenuation is ~10dB).

A topic for future investigation will be using various solid target materials and novel plasma diagnostics allowing us e.g. measuring the target holder current and plasma expansion with the application of femtosecond interferometry.

Acknowledgements
The authors gratefully acknowledge the support of the staff of the PALS laser facility without whose assistance this work would not has been possible. The research leading to these results has received funding from the Czech Science Foundation (Grant Nos. P205/12/0454), the Czech Republic's Ministry of Education, Youth and Sports (grant for the PALS RI - Project IC: LM2010014, and for the OPVK 3: CZ.1.07/2.3.00/20.0279), and from the LASERLAB-EUROPE (grant agreement n° 284464, EC's Seventh Framework Programme).

References

[1] M. J. Mead, et al., *Rev. Sci. Instrum.* **75** (2004) 4225 - 4227.
[2] J. Raimbourg, *Rev. Sci. Instrum.* **75** (2004) 4234 - 4236.
[3] F. S. Felber, *Appl. Phys. Lett.* **86** (2005) 231501.
[4] C. Stoeckl, et al., *Rev. Sci. Instrum.* **77** (2006) 10F506.
[5] C. G. Brown Jr., et al., *J. Physics: Conf. Ser.* **112** (2008) 032025.
[6] D. C. Eder, et al., LDRD Final Report LLNL-TR-411183, LLNL, 2009.
[7] C. G. Brown, et al., *Rev. Sci. Instrum.* **83** (2012) 10D729.
[8] N. L. Kugland, et al., *Appl. Phys. Lett.* **101** (2012) 024102.
[9] A. L. Galkin, et al., *Contrib. Plasma Phys.* **53** (2013) 109 – 115.
[10] F. Consoli, et al., *Nucl. Instr. Meth. Phys. Res.* A **720** (2013) 149–152.
[11] E. Woryna, et al, *Appl. Phys. Lett.* **69** (1996) 1547-1549.
[12] E. Woryna et al. *Laser Part. Beams* **14** (1996) 293-321.
[13] J. Krasa, D. Margarone, *AIP Conf. Proc.* **1462** (2012) 139-142.
[14] C. E. Baum. Pulse Sensor Characteristics of the Moebius Strip Loop. Electromagnetic and Simulator Notes, Note 7, 1964. http://www.ece.unm.edu/summa/notes/Sensor.html.
[15] D. Kocon, et al. Proceedings of IBIC2013, Oxford, UK. MOPF05, pp 208-211.

Optical emission from Al target irradiated by FLASH

M Stránský and K Rohlena

Institute of Physics, Na Slovance 2, Prague 8-Libeň, 182 21 Czech Republic

E-mail: rohlena@fzu.cz

Abstract. The following text touches on some peculiarities in optical emission spectroscopy results from experiments on the free-electron laser FLASH [1, 2]. Aluminum targets were irradiated with 13.5 nm ~ 25 fs pulses at intensities of 10^{13} and 10^{16} W/cm^2 (20 and 1 μm foci). Surprisingly, only neutral atom lines for the case with wider focus and traces of ion lines in the tighter focus case were observed with the optical emission spectroscopy (200-600 nm range), [2]. The motivating idea behind this work is the suggestion in [1] by Zastrau that the optical spectrometer sees only emissions from a cold expanding lower-density ($< 10^{22}$ cm^{-3}) plasma plume. In this contribution the notion of UV range screening is analyzed in detail.

1. Introduction

In the high energy density plasma experiment on the free-electron laser facility in Hamburg (FLASH), see [3], in which aluminum targets were irradiated with 13.5 nm ~25 ns pulses at two different intensities 10^{13} and 10^{16} W/cm^2 (20 and 1 μm foci), the FEL radiation is absorbed by a volumetric photo-ionization mechanism. During this absorption process one L-shell electron is knocked into the conduction band, leaving the ionized atom transparent to the incoming radiation, as the L-shell absorption edge is shifted to energy (93.6 eV) which is higher than that of the incoming photons (92 eV) – induced transparency phenomenon first described in [4]. The L-hole has a life time of about 40-60 fs before it is filled by an Auger decay process. The intermediate stage of energized crystalline latice following the absorption process is often referred to as the warm dense matter (WDM, see [5] for a more accurate definition).

The non-thermal electrons in the conduction band then start to thermalize and their energy is redistributed over the lattice on the femto- to picosecond time scale. The lattice gradually disintegrates and finally, after a full thermalization, a non-ideal plasma is formed, which keeps expanding on a hydrodynamic pico- to nanosecond scale reducing further the density and temperature. However, at the very start of the expansion, as the density is reduced, and the band structure – the conduction band starts to disappear, and higher orbitals emerge. At the same time, the temperature also decreases, and some lower energy radiative transitions might occur between the newly appeared outer orbitals. For more details see Zastrau et al., [1], for the properties of the resulting non-ideal Al plasma governed by a complex equation of state, see [6]. In this experiment, the optical range (near UV and optical, 200-600 nm) was observed with a spectrometer. Cihelka at al. [2] note that the spectral lines referring to the transitions of ionized aluminum are either missing altogether or are weak (in the higher intensity case) while the neutral transitions dominate in this wave-length range. This intermediate time range during the expansion is studied in the following sections in order to explain the absence or weakness of the ionic lines in the optical wave-length range, see Fig. 1 for a typical spectrum as presented in

Journal of Physics: Conference Series **508** (2014) 012008 doi:10.1088/1742-6596/508/1/012008

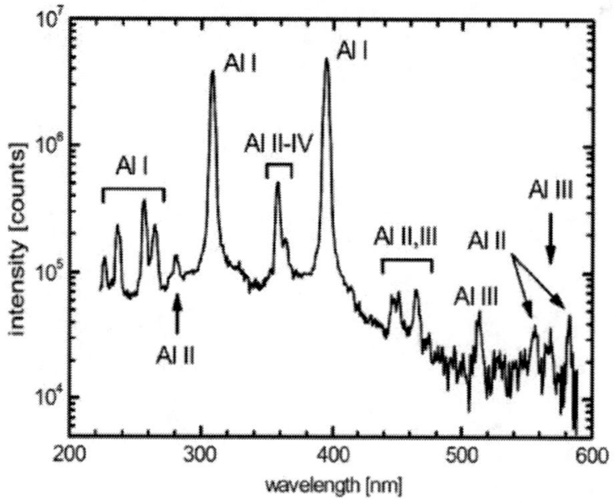

Figure 1. Time integrated optical spectrum of Al illuminated with the tighter (1 μm) focus of the FEL beam. Some ion lines are visible, but the short wavelengths segment of the spectrum is almost void of the ionized Al. Taken from [2].

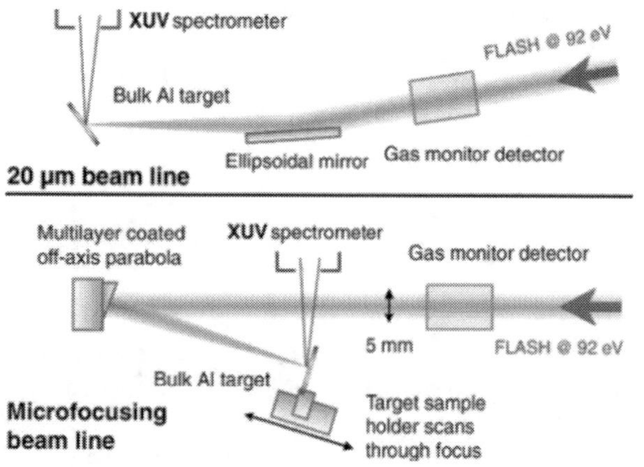

Figure 2. Experimental layout of optical spectra registration in the FEL vacuum chamber with the tighter ($\sim$ 1 μm) and the wider focusing ($\sim$ 20 μm), taken from [1]. The arrangement with the tighter focus had a slightly different geometry, see Fig. 3 and was installed in a separate chamber.

[2]. The experimental arrangement for measurement of the optical spectra emanating from the FLASH generated plasma on a solid Al target is described in [1], see Figs. 2 and 3.

2. Physical model

The plasma outflow was modelled using a 1D planar adiabatic expansion based on the self-similar model described in [7]. A slight complication is due to the presence of the recombination heat, which is released while the composition of the plasma follows the Saha equilibrium used to describe the ionization state during the adiabatic expansion. However, the additional heating can be incorporated in a modified adiabatic curve as it is expressed by the following energy balance

$$\frac{3}{2}N_T k_B dT = -\frac{N_T k_B T}{V}dV - \sum_{\alpha} \varepsilon_{i(\alpha)} \left(\frac{\partial N_\alpha}{\partial T}dT + \frac{\partial N_\alpha}{\partial V}dV\right) \tag{1}$$

where N_T is the total number of particles including electrons and all the ion species, $\varepsilon_{i(\alpha)}$ is the sum of ionization energies to reach the ionization stage α starting from the neutral atom, and the temperature T is assumed to be equal for all the particle species. The partial derivatives of N_α are obtained from the Saha equation. Equation (1) was integrated, and the resulting dependence on both the electron and the overall heavy particle concentrations was plotted in Fig. 4. This modified adiabatic curve was then used in the adiabatic expansion model.

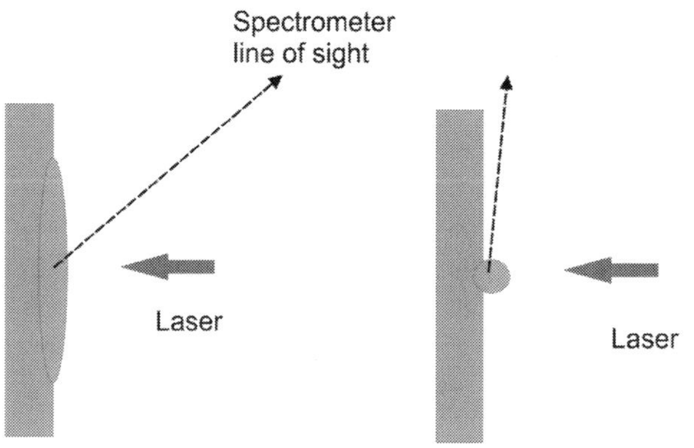

Figure 3. The measurement scheme showing the differences between the geometry of the wider and tighter focusing (out of scale).

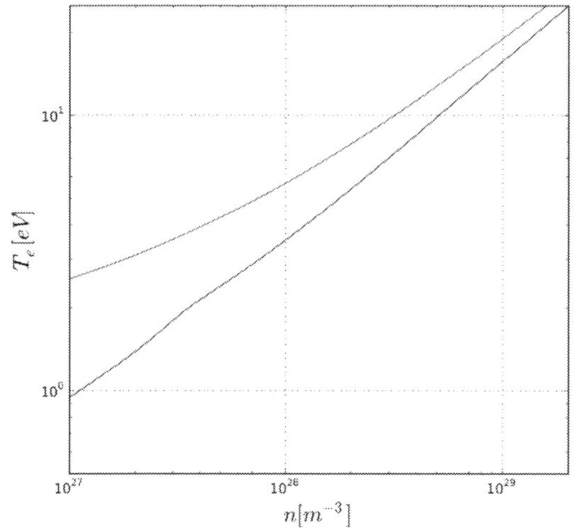

Figure 4. The modified adiabatic curve for Al plasma including the recombination heating starting at the electron concentration $1.5 \times 10^{23}\,\mathrm{cm}^{-3}$ and a temperature of 25 eV. The upper curve gives the temperature as a function of the electron concentration and the lower one as a function of the total heavy particle concentration with the electrons excluded. The slight kink visible on the lower curve marks the onset of the heating.

Of course, the Saha equilibrium as well as the adiabatic term in (1) are derived under the assumption of equation of state of an ideal gas. Neither other recently observed phenomenon occurring in a non-ideal plasma, see [8], namely the ionization potential lowering, has been included. This means that (1) is correct only after the expansion causes the diluted plasma to lose an essential part of its non-ideality. However, since the spectral measurement is integrated in time and the expansion time scale extends over a nanosecond scale, equation (1) is valid for most of the data collection interval.

3. Results

The expansion of the density profile as a function of time using the Schmalz [7] model is shown in Fig. 4. The ions have just a very restricted time window (few 10's of ps) to contribute to the emissions in a range of 200-600 nm. The ion line intensities at the critical density surface are less than 1/10 of the neutral lines for the wider focus case. Since the optical spectrometer was time-averaging the emissions over 2 ms interval, the neutral atoms have a much wider time window for the emissions, and therefore are the most prominent ones. In addition, an opacity calculation reveals that the expanding Al plasma is initially optically very dense (optical thickness >10000 for the typical 280 nm Al II line - see the trajectory of the critical surface in Fig. 5), and only a very thin plasma cooled by the expansion would be sufficiently transparent for the ion lines to be visible for an outside observer. That is why the spectrometer records only neutral lines corresponding to a temperature of 0.6-0.8 eV as it was indeed reported in [2].

In the case of the tighter focus (1 μm), in which traces of ion lines appear, Fig. 2, the expansion is governed by a different geometry, see Fig. 3. Whereas in the wide focus case the plasma expands in a 1D fashion forming a thin plasma layer before it cools down, the tighter focus case is more complex. The process is aided by the induced transparency mechanism (which happens at the WDM stage still before the plasma is formed, [4]), which might help to drill a slightly deeper heated hole into the target as against the linear absorption. The resulting hole is up to 10-20 μm deep, and it thus takes a longer time for the heated plasma to pour out. Assuming it would do so at the sound speed, the expansion would take long enough time ($\sim$1 ns) to cause some of the ion (Al II) lines to be registered in the 2 ms averaged spectrum. A rough estimate renders an intensity ratio of 1/170 for the Al II lines over the neutral atom lines. The measured spectra gave an intensity ratio around 1/100.

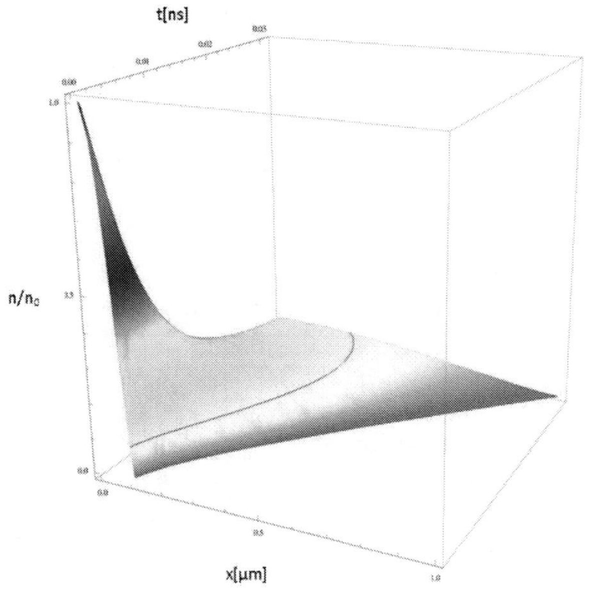

Figure 5. Density profile temporal evolution. n_0 means the initial number density of heavy paticles, the gray line marks the trajectory of the critical density surface belonging to $\lambda = 280$ nm (one of the strongest Al II lines in the examined spectral range).

4. Conclusions

Even though this topic concerns experiments with non-ideal plasmas, the peculiarity of the measured optical spectra can be attributed to the effects that are no longer non-ideal – an explanation involving the non-ideal plasma topics is not necessary in this case. The main reason for the absence of the ionic spectral lines in the wider focus case is a very narrow time window during which they can be recorded. In contrast, the time available for the collection of the part of the spectra belonging to the neutral lines is long enough to integrate a sufficient number of photons to render a clear signal, which is thus far outweighing the ionic lines. This is mainly due to the fact that the plasma is optically thick with respect to the ionic line wavelengths down to the very low temperatures at which only the atomic lines are emitted. In the higher intensity case of the tighter focus, when the screening of the hot outpouring plasma is less efficient and also the expansion of hotter plasma takes longer, the time window for the ionic lines is sufficiently long for the traces of the ionic lines to show up in the time-averaged spectrum.

Acknowledgments

A support by the grant No. P205/11/0571 of the Czech Science Foundation is gratefully acknowledged as well as the Czech Ministry of Education project OP VK2 No.

CZ.107/2.3.00/20.0087.

References

[1] Zaustrau U, Burian T, Chalupský J, et al. 2012 XUV spectroscopic characterization of warm dense aluminum plasmas generated by the free-electron-laser FLASH *L&P Beams* **30** 45–56

[2] Cihelka J, Juha L, Chalupský J, et al. 2009 Optical emission spectroscopy of various materials irradiated by soft x-ray free electron laser *Proc. SPIE* vol **7361** 73610P

[3] Tiedtke K, Azima A, et al. 2009 The soft X-ray free-electron laser FLASH at DESY: Beamlines, diagnostics and end-stations New J. Phys. **11** 023029

[4] Nagler B, Zastrau U, Faustlin R R, et al. 2009 Turning solid aluminium transparent by intense soft X-ray photoionization *Nature Phys* **5** 693–696

[5] Lee R W, et al. 2003 Finite temperature dense matter studies on next-generation light sources. J. Opt. Soc. Am. **B** **20**, 770

[6] Lomonosov I 2007 Multi-phase equation of state for aluminum L&P Beams **25** 567–584

[7] Schmalz R F 1986 Free unsteady expansion of a polytropic gas – self-similar solutions *Phys. Fluids* **29** 1389–1397

[8] 2013 Preston T R, Sam M. Vinko S M, et al. The effects of ionization potential depression on the spectra emitted by hot dense aluminium plasmas High Energy Density Physics **9** 258e263

MeV ion beams generated by intense pulsed laser monitored by Silicon Carbide detectors

L Calcagno[1], P Musumeci[1], M Cutroneo[2], L Torrisi[2], F La Via[3], J Ullschmied[4]
[1] *Dipartimento di Fisica, Università di Catania, V. S. Sofia 64, 95123 Catania, Italy*
[2] *Dip.to di Fisica & SdT., Università di Messina, V.le F. Stagno d'Alcontres 31, 98166 S. Agata (ME), Italy*
[3]*Istituto Microelettronica Microsistemi-CNR, Stradale Primosole, Catania, Italy*
[4] *Institute of Physics of the ASCR-PALS, Na Slovance 2, 18221, Prague, Czech Republic*

E-mail: lucia.calcagno@ct.infn.it

Abstract. The high energy ions produced with intense pulsed laser were analyzed with Silicon Carbide detectors. In order to realize high performances and radiation resistant detectors, high quality and thick epitaxial layer were grown on a substrate and a Schottky diodes were then realized. These detectors were employed to probe the plasma generated with a 300 ps laser at intensity of 10^{16} W/cm^2 operating at Prague Asterix Laser System Laboratory. They show a fast response and a high sensitivity to high energy ions. Metallic and polymeric thin films were irradiated and the produced plasmas were monitored in forward and backward directions. The analysis of the time-of-flight spectra evidences the emission of protons and ions at different energies. The spectra were deconvolved with a shifted Maxwell Boltzmann distribution. In our experimental conditions we detected protons in the energy range 1.2 - 3.0 MeV and heavy ions between 1.0 MeV up to 40 MeV depending on the target and the laser energy. The results were compared with the ones obtained by Thompson Parabola Spectrometer.

1. Introduction

In the last decade a growing interest was addressed to devices realized on Silicon Carbide (SiC) semiconductor [1,2]. The superior physical properties of this material allow to realize devices operating in harsh environments, at elevated temperatures, at high power and under high radiation fields. In that context, SiC radiation detectors have attracted a large attention due to the improvement of processes allowing to grow thick and high quality epitaxial layers [3,4]. The detectors realized on SiC show very interesting features when are used to diagnose the laser plasma [5,6].

The wide band gap allows to blind the visible, near UV and IR light and to operate under high intensity visible light. The high carrier mobility gives the possibility of fabricating fast (ns) devices, the high resistance to radiation allows to operate under high radiation fields. Once more, a further interesting feature of these detectors is the proportionality of response to the ion energy. In fact, the detected ions generate, into the active detector volume, e-h pairs loosing 7.8 eV for pair production. The current signal collected by the electrodes will be proportional to the number of ions, i.e. to their energy loss.

In the last years a large activity was devoted to the production of high energy protons and light ions by a plasma laser because of the many applications in different scientific fields, as new generation

ion source, nuclear and astrophysical field, biomedicine, microelectronics, metallurgy and nuclear fusion technique [7]. In order to demonstrate the capability of laser plasma accelerators, the characterization of the produced ion beams must be performed. In that context we have already shown that SiC semiconductor detectors can be used in the monitoring of ionizing radiation emitted from laser-generated plasmas [8].

In the present paper we report some results obtained in the characterization of laser plasma by using SiC detectors realized on epitaxial SiC layer. In order to test the detector sensitivity, plasmas obtained in different conditions and from different targets were monitored. To provide the possibility of a better evaluation of the data, the results obtained with SiC detectors were compared with the ones coming from a Thompson Parabola Spectrometer (TPS).

2. Detectors and experimental setup

The SiC detectors were built on a 4H-SiC epitaxial layer grown with chemical vapour deposition technique in HCl atmosphere, which allows to obtain very thick layer (about 100 μm) with low dopant concentration (about 10^{14} cm^{-3}) and with very low defect concentration [9]. Schottky diodes were fabricated on these epilayers. The backside ohmic contact was formed by a rapid thermal annealing of a 100 nm thick Ni film at 950 °C. A Ni layer was then deposited in a Ultra High Vacuum chamber on the front of the wafer and various diodes were realized using standard photolithography. These diodes were then further annealed at 600 °C to optimize the Schottky barrier properties by the formation of a nickel silicide layer (Ni_2Si), with a resulting thickness of 200 nm. In Fig.1a a schematic section of the employed Schottky diode is shown, in which the track and the range (R_p) of a generic incident ion beam is also reported.

The application of a reverse bias creates a depletion layer (w) that lies entirely in the lightly doped SiC epitaxial layer. In our experiments we use a reverse bias of 600 V which produces a depletion layer of about 75 μm, thus allowing the measurement of the energy loss of ions whose penetration range (R_p) is less than 75 μm. This range value corresponds to proton energy of 3.2 MeV [10].

In such configuration, at 600 V inverse bias, the detector response has high linearity with the radiation energy, high resolution (FWHM=34 eV), and 100% charge collection efficiency, as demonstrated detecting alpha particles emitted from radioactive sources [11]. The active area of the SiC detector was 2x2 mm^2 and it was glued onto high temperature resistance holder.

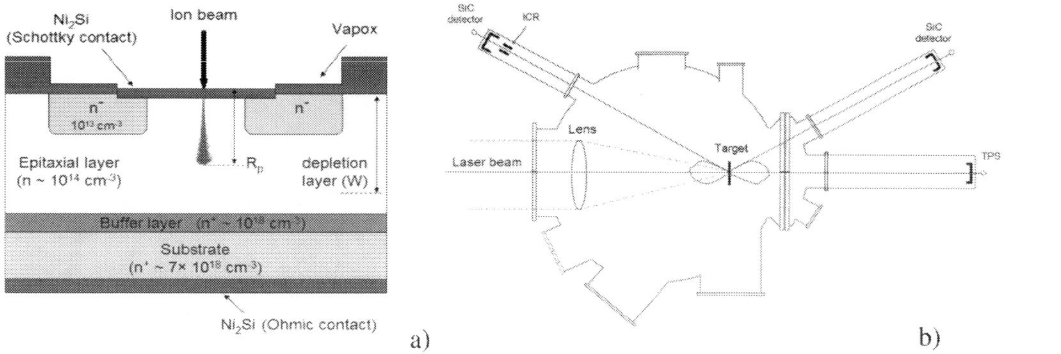

Figure 1. (a) Schematic cross-section of the SiC detector and of the ion trajectories (R_p is the ion range) (b) Schematic of the experimental setup for the target irradiation in the vacuum chamber at PALS laboratory.

The experimental measurements were carried out with the Iodine laser of Prague Asterix Laser System (PALS) Laboratory operating in single pulse at 1315 nm wavelength, 300 ps pulse duration, 70 μm laser spot diameter, 10^{16} W/cm^2 pulse intensity using laser energies ranging from 500 J to 600 J.

A schematic of experimental apparatus is shown in Fig.1b. Measurements were performed irradiating in normal incidence targets of aluminum (Al) and polyethylene (PE) and monitoring the plasma radiation emission both in forward direction (rear face of the irradiated target) and in backward direction (laser irradiated front face). SiC detectors were located at 30° angle with respect to the target normal in forward and backward direction. A Ring Ion collector (ICR) was also used and it was placed at 30° in backward. The laser focal position (FP) on the target surface was controlled though a X-ray streak camera permitting to measure its distance from the target surface. Micrometric step-motors permitted change of the focal position in order to check the self-focusing conditions for distances of the order of 100 μm in front of the target surface.

Time of Flight (TOF) technique was employed in our measurements and the spectra were recorded with a fast oscilloscope operating to 20 GS/s recording the laser shot as a start signal and the ion peaks as stop signals.

The ions emitted from plasma were also analyzed through a Thomson Parabola Spectrometer (TPS) located in forward direction along the normal to the target surface. The TPS is preceded by two pinholes, the first 1 mm in diameter and the second 100 μm in diameter at a distance of 16 cm, in order to finely collimate the detected ions from the plasma and to exactly determine the incidence direction. The spectrometer uses a 0.06 T horizontal magnetic deflection and + 2.8 kV (from the top) voltage for the vertical electric deflection. The distance between the electric deflectors and the Micro-Channel Plate (MCP) plane detector of the parabola ions is D=16.5 cm, as in similar apparatuses reported in literature [12].

3. Results and Discussion

The TOF spectra are usually composed by a relatively fast photopeak, followed, sometimes, by a signal due to the detection of plasma electrons, and a relatively long and structured peak due to the detection of plasma ions, from light to heavy ones depending on their velocity and charge state distributions. However if the detector is not fast enough, it becomes difficult to distinguish all the contributions in the TOF spectra.

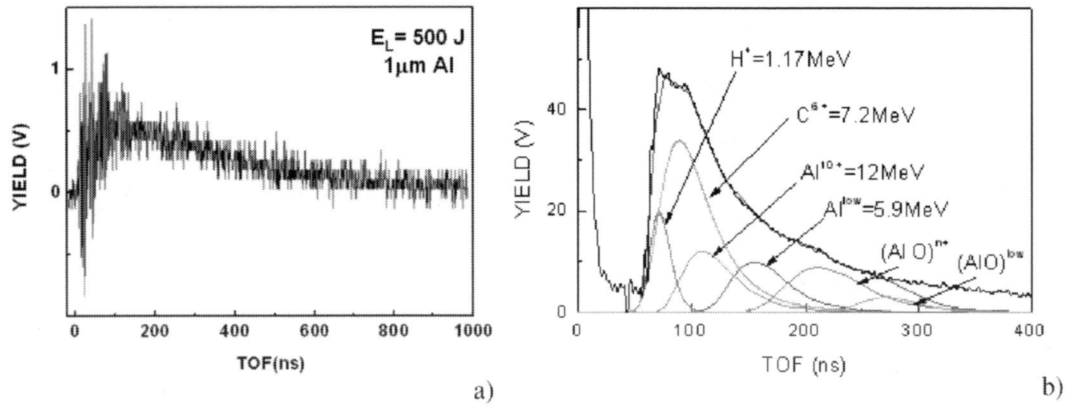

Figure 2. TOF spectra detected by a ion collector (ICR) (a) and by a SiC detector (b). The spectra are obtained by irradiating a thin Al target with a laser energy of 500 J. In the spectrum of SiC detector it is also shown the deconvolution obtained with the use of Coulomb-Boltzmann Shifted function (blue line).

The last statement is demonstrated by the spectrum shown in Fig.2a, which was obtained by an ICR detector when an Al target was irradiated at laser energy of 500 J. The spectrum shows a broad peak between 0-250 ns related to X-rays, UV radiation and fast ions, followed by a broad structure related to slow ions. Indeed, it is very hard to distinguish each component.

Journal of Physics: Conference Series **508** (2014) 012009 doi:10.1088/1742-6596/508/1/012009

The use of SiC semiconductor detector allows to separate the plasma fast protons/ions with respect to the X-UV photopeak in the TOF spectrum. The wide bandgap (3.2 eV) allows a high attenuation of the plasma-emitted UV and soft X-ray radiation (XUV) then the SiC main advantage consists in the possibility to detect simultaneously the hard X-rays and fast ions emitted from the laser-generated plasma. This statement is evidenced by the spectrum obtained by SiC detector reported in the Fig.2b. This spectrum is detected in forward direction and it is relative to 500 J pulse energy irradiating a thin Al foil (1.0 μm) at -100 μm Focal Position, with a target-SiC detector distance of 102 cm. The spectrum shows an intense photo peak at fast time of flight (≈ 20 ns), followed by a large structured peak extending from 60 ns to about 300 ns. This structure is due to the fast protons and to the different charge states of the Al ions. Carbon ion contaminations might also be present. The spectrum reported in Fig.2b was fitted with the Coulomb-Boltzmann-shifted (CBS) function, which allows to deconvolve the experimental spectrum in the different ion species and charge states contributions [13]. The CBS function takes into account thermal and Coulomb interactions between plasma particles. Inside the plasma, in fact, a high equivalent acceleration voltage due to charge space separation effects is developed. It modifies the ion kinetic energy proportionally to the ion charge state. The fit of the experimental spectrum permits to know the most probable equivalent temperature and acceleration voltage of the plasma core. In the same Fig.2b the deconvolution obtained with this CBS function is reported. The proton kinetic energy is evaluated to be close to 1.2 MeV and the maximum energy of the Al ions is 12 MeV. As the acceleration energy is 1.2 MeV/charge state, Al ions with charge state $n^+=10$ were produced in the plasma. Carbon ions at a maximum energy of 7.2 MeV ($n^+=6$) are also present.

We have tested the SiC detector with different type of targets in the same experimental conditions and in Fig.3 the spectra obtained at 600 J of laser energy with Polyethylene (PE) and Al targets, both 10 μm thin, are reported.

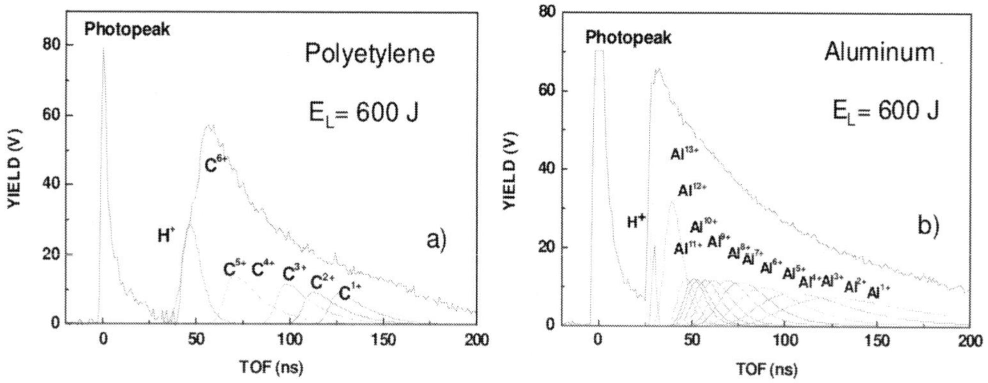

Figure 3. TOF spectrum achieved by a SiC detector with a polyethene (a) and an Al (b) target with a laser energy of 600 J. In both spectra the deconvolution with the use of Coulomb-Boltzmann Shifted function is reported.

The shape of the spectra is the same of the previous one, showing the feature of photopeak followed by the structure related to the ion components. In the same figures are reported the fits obtained with CBS function: we obtain an acceleration energy of 1.5 MeV/charge state and a temperature of 4 keV for polyethylene, while for the Al target we obtain 3.0 MeV/charge state, and the temperatures is 8 keV. This result reveals that in the same experimental condition the irradiation of Al target produces more energetic protons with respect to PE target. This result can be explained on the basis of the highly conductive metallic Al target that acts as an electron injector into the plasma, increasing the electron density. Such electrons increase the electric field of the plasma responsible of the high ion acceleration.

In addition the comparison between the results reported in Fig.2b and Fig.3b reveals that the energy of the produced ions increases with increasing the laser energy, as expected. In fact, the ion energy should be proportional to $I\lambda^2$, being I the laser intensity and λ the wavelength.
The plasma was also characterized by the TPS, which allows to determine the charge-to-mass ratios and the charge state distributions of the Al on the forward emitted plasma. Fig.4a shows the TPS spectrum obtained by the irradiation of an Al thin target with 600 J of laser energy, during the same SiC detection experiment. The spectrum shows a circular zone due to plasma photons and undeflected neutral particles arriving onto the MCP and different parabolas outgoing from this circular zone.

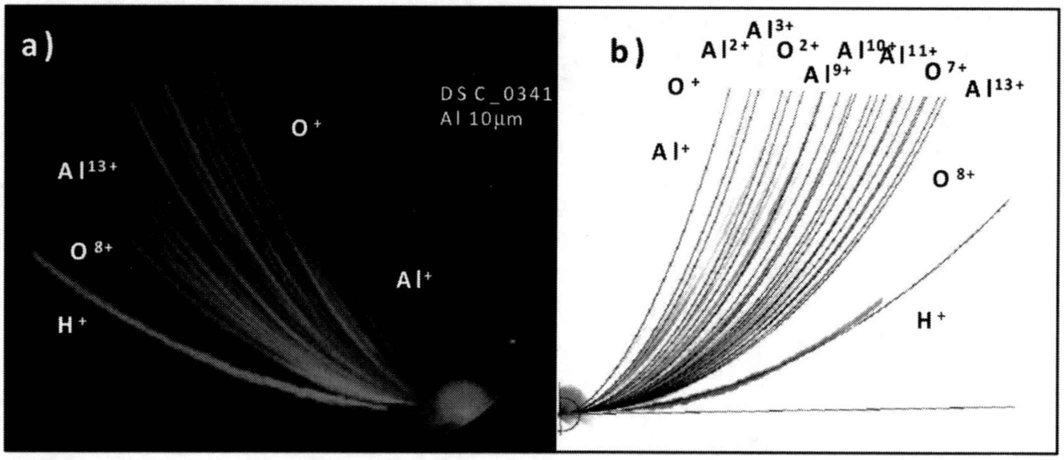

Figure 4. (a) Original image provided from TPS by irradiating the same Al target of Fig.3b and simulated spectra for the same shot obtained by executing Opera 3D-TOSCA code (b).

The experimental TPS parabolas are compared with the simulation data obtained by Opera-3D/TOSCA code [14]. The fixed parameters are the geometrical TPS dimensions, the magnetic field value, and the electric voltage applied to the electric deflector plates. In Fig.4b the simulation data are shown. The lower parabola is related to proton deflections and the others to Al ions from high charge state to the single charge state. The more vertically deflected parabola reveals the lower Al charge-to-mass ratio (on the top) and the less vertically deflected indicates the higher Al charge-to-mass ratio (on the bottom). The points of parabola lying nearest to the circular zone is due to the detection of highest energy ions. The distance between the protons parabola and the center circle is compatible with a proton maximum energy of 2.8 MeV, close to the value obtained with SiC detector (3.0 MeV). TPS measurements show thirteen parables related to all the charge states of Al and some ones associated to oxygen ions. The spectrum reported in Fig.3b was fitted without considering the oxygen ions as their intensity is too low.

The comparison between SiC detector and TPS evidences the high resolution of TPS to low energy ions, while at high energy the parabolas are close to each other because of the low applied magnetic field (0.06 T). Then, different detectors should be employed in order to have a complete energetic characterization of the laser accelerated ions and of the plasma proprieties in terms of maximum charge state, temperature, ion current and kinetic energy.

4.Conclusions

In the present paper we have shown that single crystal Silicon Carbide detector can be used to monitor the ions emitted from a laser generated plasma. The large energy gap of SiC allows well separate the photopeak and the ionic component of the plasma. These detectors were employed to characterize the plasma in different experimental conditions. The plasma emitted by polyethylene target at high laser energy allows to generate proton beams of 1.5 MeV. This energy increases to about 3.0 MeV by the use of metallic treated target. The results reveals that the SiC detectors are very sensitive to high energy ions and they are suitable for a characterization of ion beams generated by laser systems.

Moreover complementary techniques as Thomson Parabola Spectrometer permits a high resolution to low energy ions, evidencing that the use of different detectors gives a more complete characterization of the laser-generated plasma.

Acknowledgments
This work was performed at the PALS Research Infrastucture, benefiting from the Laserlab Europe project, N. pals001823 "High energy proton acceleration from thin advanced targets at PALS" coordinated by prof. L.Torrisi. The authors thanks the Mr. E.Krousky, M.Pfeifer, A.Velyan and J.Skala for the technical and scientific support provided during the experiments. Thanks also to G. Saccà and M. D'Andrea of the Istituto Nazionale di Fisica Nucleare (Sezione di Catania) for the technical assistance for the SiC detector electronics equipment.

References

[1] Park Y S SiC Materials and Devices, Semiconductors and Semimetal Vol. 52, edited by R. K. Willardson and E. R. Weber, Academic, San Diego, (1998)
[2] Matsunami H 2004 Jpn. J. Appl.Phys. 43 6835
[3] Calcagno L, Izzo G, Litrico G, Foti G, La Via F, Galvagno G, Mauceri M and Leone S 2007 J.Appl.Phys. 102 43523
[4] La Via F, Izzo G, Mauceri M, Pistone G, Condorelli G, Perdicaro L, Abbondanza G, Calcagno L, Foti G and Crippa D 2008 Journal of Crystal Growth 311 107
[5] Torrisi L, Foti G, Giuffrida L et al. 2009 J. Appl. Phys. 105 123304
[6] Margarone D, Krása J, Giuffrida L et al. 2011 J. Appl. Phys. **109** 103302
[7] Borghesi M, Fuchs J, Bulanov S V, Macckinnon A J, Patel P K and Roth M 2006 Fusion Sci. Technol. 49 412 .
[8] Cutroneo M, Musumeci P, Zimbone M, Torrisi L, La Via F, Margarone D, Velyhan A, Ullschmied J and Calcagno L 2013 Journal Material Research 28 87
[9] La Via F, Galvagno G, Foti G et al. 2006 Chem. Vapor Deposition 12 509
[10] Ziegler F, Biersack J P and Ziegler M D, 2008 SRIM-The Stopping and Range of Ions in Matter (SRIM, Annapolis, MD)
[11] Nava F, Bertuccio G, Cavallini A and Vittone E 2008 Meas.Sci.Technol. 19 102011
[12] Harres K, Schollmeier M, Brambrik E 2008 Rev. Sci. Instr. 79 93306
[13] Torrisi L, Gammino S, Andò L and Laska L 2002 J. Appl. Phys. 91 4685
[14] Chobam, OPERA-3D/TOSCA, 2012: http://www.cobham.com/about-cobham/aerospace-and-security/about-us/antenna-systems/kidlington/products/opera-3d/statics.aspx

ELIMED, MEDical and multidisciplinary applications at ELI-Beamlines

F Schillaci[1,2], A Anzalone[1], G A P Cirrone[1,2], M Carpinelli[3], G Cuttone[1], M Cutroneo[9], C De Martinis[10], D Giove[10] G Korn[2], M Maggiore[2,4], L Manti[5], D Margarone[2], A Musumarra[1], F M Perozziello[5,] I Petrovic[6], P Pisciotta[1], M Renis[7], A Ristic-Fira[6], F Romano[1], F P Romano[1,11], G Schettino[8], V Scuderi[1,2], L Torrisi[1,9], A Tramontana[1,7], S Tudisco[1]

[1] INFN-LNS, Catania, Italy
[2] Institute of Physics of the ASCR, ELI-Beamlines Project, Prague, Czech Republic
[3] Università di Sassari, Dipartimento di Chimica e Farmacia e INFN, Sassari, Italy
[4] INFN-LNL, Legnaro, Italy
[5] Università Federico II di Napoli, Dip.to di Fisica and INFN Sezione di Napoli, Napoli, Italy
[6] University of Belgrade, Vinca Institute of Nuclear Sciences, Mihajla Petrovica 12-14, Belgrade, Serbia
[7] Università di Catania, Catania, Italy
[8] National Physical Laboratory, Hampton Road — Teddington, TW11 0LW — UK
[9] Università di Messina, Messina, Italy
[10] Università di Milano and INFN Sezione di Milano, Milano, Italy
[11] IBAM – CNR, Catania, Italy

e-mail: francesco.schillaci@ct.infn.it

Abstract. ELI-Beamlines is one of the pillars of the pan-European project ELI (Extreme Light Infrastructure). It will be an ultra high-intensity, high repetition-rate, femtosecond laser facility whose main goal is generation and applications of high-brightness X-ray sources and accelerated charged particles in different fields. Particular care will be devoted to the potential applicability of laser-driven ion beams for medical treatments of tumors. Indeed, such kind of beams show very interesting peculiarities and, moreover, laser-driven based accelerators can really represent a competitive alternative to conventional machines since they are expected to be more compact in size and less expensive. The ELIMED project was launched thanks to a collaboration established between FZU-ASCR (ELI-Beamlines) and INFN-LNS researchers. Several European institutes have already shown a great interest in the project aiming to explore the possibility to use laser-driven ion (mostly proton) beams for several applications with a particular regard for medical ones. To reach the project goal several tasks need to be fulfilled, starting from the optimization of laser-target interaction to dosimetric studies at the irradiation point at the end of a proper designed transport beam-line. Researchers from LNS have already developed and successfully tested a high-dispersive power Thomson Parabola Spectrometer, which is the first prototype of a more performing device to be used within the ELIMED project. Also a Magnetic Selection System able to produce a small pencil beam out of a wide energy distribution of ions produced in laser-target interaction has been realized and some preliminary work for its testing and characterization is in progress. In this contribution the status of the project will be reported together with a short description of the of the features of device recently developed.

1. Introduction

Laser ion acceleration has gained interest over the last years and a wide variety of applications has been proposed to exploit the power of high-current multi-MeV ion (especially proton) beams produced in high-power (>10^{18} W/cm^2) short-pulse (30 fs – 10 ps) laser interaction with thin solid foils [1,2]. Several acceleration mechanisms have been already discussed in literature such as Target Normal Sheath Acceleration (TNSA) regime [3], Coulomb Explosion (CE) [4, 5], Radiation Pressure Acceleration (RPA) [6]. The highest proton energies with best characteristics have been experimentally reached using the TNSA regime as the laser energies required for the other regimes are not available yet. In TNSA an intense laser is focused on a several microns thick target where many processes involving energy transfer between the laser main pulse and the target electrons can occur and, as a consequence, a population of hot electrons with a mean free path larger than the target thickness is produced. Hence, they can cross the target itself setting up an intense electrostatic field due to the charge imbalance between positive ions at rest in the target and the electron sheath expanding at its rear surface, namely in the laser incidence direction [7]. In this field of several TV/m, predominantly protons are accelerated showing a large divergence angle [8] and an exponentially decaying energy spectrum with a spread up to 100 %, these characteristics depending on the laser as well as on target parameters [9]. This would require the development of an angular and energy selection system to make such beams suitable to different applications or the utilization of different mechanisms of laser proton acceleration. There are other acceleration regimes than TNSA which have been identified and investigated with simulations, as the high power laser required for this regimes are not available yet. A part from the TNSA, very promising results have been obtained for the Radiation Pressure Acceleration regime (RPA) which is different for thick (hole boring) and thin (light sail) targets [10 - 13]. Radiation pressure effects start to be progressively more important at very high intensities and they dominate over TNSA at laser intensities $I > 10^{23}$ W=cm^2. The RPA regime could be achieved within the ELI (Extreme Light Infrastructure) project whose main goal is to develop the first petawatt class lasers with very short pulses (from the femto down to the zepto-second regime) with excellent shot-to-shot reproducibility in order to accelerate particles in the ultra-relativistic energy range [14].

2. ELIMED Network

ELI is an European large scale infrastructure developed within the ESFRI process and based on three laser facilities located in different countries and dedicated to different research fields from nuclear physics to the attosecond physics. One of this facility is being implemented in Prague (Czech Republic) and it is named ELI-Beamlines and will be mainly dedicated to applications of the laser produced photon and particles beams in several fields such as high-resolution X-ray imaging, exotic physics etc. One of the secondary sources available at ELI-Beamlines will be dedicated to Multidisciplinary Application of laser-Ion Acceleration, thus the acronym used as name for the beam line is ELIMAIA. An experimental hall of the ELI-Beamline building has been already assigned for the development and the realization of a transport beam-line and will be available for users interested in studying the applicability of laser-driven ions in different fields, including future clinical applications, as suggested in [15]. Around the ELIMAIA beam line a network between groups of researchers from different European and non-European countries has been launched under the name of ELIMED (MEDical application at ELI-Beamlines). One of the motivation of the project is based on the fact that the actual hadrontherapy centres are based on conventional accelerators, such as cyclotrons or synchrotrons which are huge, complex and, hence, expensive machines, both in terms of financial and human resources. These are the reasons why only few centres are available around the world and the request of proton therapy treatments cannot be fully satisfied. On the other hand, laser-based accelerators could be a competitive alternative as they can be smaller in size and less expensive. Design of miniaturized systems have been already proposed in literature [16] in which the laser-driven ion part is combined to a conventional beam transport system, or in which the acceleration occurs inside a moving gantry [17].

In this context and with the aim to develop a new approach in hadrontherapy, the ELIMED network has been established. The project officially started with a Memorandum of Understanding (MoU) signed between ASCR-FZU and INFN-LNS within the ELI-Beamlines project [18]. The purpose of the MoU is to start a research program whose main aim is to study, design and realize an irradiation facility for dosimetric and radiobiological studies with the high energy proton/ion beams which will be produced at ELI-Beamlines. Several institutes have shown interest in the project and have joined the network. The INFN researchers will be involved in the development of new dosimetric techniques suitable for very high dose-rates, the characterization of laser-driven beams [19], the development of a proton/ion transport line and the carrying out of in-vitro radiobiological studies. Moreover, INFN researchers will be involved in the beam diagnostics with a new Thomson-like spectrometer, whose first prototype has been already accomplished and some results are presented in the following. The energy selection of the beam is an INFN concern as well, and it is crucial in order to produce a beam with spectral and spatial characteristics similar to the ones required for clinical applications. A magnetic system has been realized for this purpose and will be also presented.

3. The first prototype of high dispersive power Thomson Parabola Spectrometer

The Thomson Parabola Spectrometer is a widely used beam diagnostics device. The working principle, deeply described in literature [20-23], is based on parallel electric and magnetic fields acting on a well collimated ion beam and the Lorentz force splits the different ion species according to their q-over-m ratio. The result on the detection plane is a set of parabolic traces each of them corresponding to a well determined ion species. Moreover, electric and magnetic deflections are related to the energy and momentum of the particles. Hence a complete set of information on the beam is obtained in a single measurement.

A new Thomson Parabola have been developed at LNS with high dispersive power (up to 20 MeV protons) and high energy resolution. Its deflection sector is made of partially overlapping electric and magnetic fields in order to have a more compact system. The magnetic field is produced by two resistive coils set inside and H-shaped iron matrix ensuring a very good field uniformity. The resistive coils allow to increase the dynamics and the resolution of the device as the field intensity can be changed according to the current used, which is an advantage with respect to the use of permanent magnet based dipoles. The length of the electromagnet is 15 cm and the maximum achievable field is 2500 gauss, corresponding to a 86.2 A current. The possibility to tune the fields increase the energy range detectable from the spectrometer, as it can be used for very low energy particles reducing also the errors in energy evaluation. The electric field is produced by two 7 cm long copper electrodes starting at 6 cm from the magnetic field centre. The electrodes have been designed to stand an electric potential of 30KVolt, ensuring a very strong electric field which which means a very high dispersion power and hence good separation between different ion traces and an very high energy limit (about 40 MeV/amu). A scheme of the deflection sector is available in [24]. The beam is collimated using two pinholes: the first one is an 1 mm hole on a 2 cm thick double matrix of brass and lead which screens from neutral radiation coming from the plasma, the second one is a 100 μm hole on a 1 mm thick aluminium matrix. The distance between the two pinhole is 10 cm. The collimator allows to have the beam entering out-off the central device axis, in order to exploit the whole surface of the MCP detector set at the end of the spectrometer. In this way it is possible to increase the spatial resolution as the low energy part of ions traces, where different parabolas are more separated, is detected by the imaging system. The MCP is the main part of the imaging system which is also made of a phosphor screen and a conventional reflex camera.

The system is completed by a MATLAB based simulation and analysis tool. The simulation part of the tool solves the second order differential equations to describe the particle motion in the spectrometer from the collimator to the detector plane. Computational errors are reduced using the effective magnetic and electric lengths and a symplectic integrator to solve the particle motion. The semi-automatic analysis part of the tool allows to fit traces, detected with the imaging system using a second order polynomial function expressing a parabolic shape of the experimental curves. The tool gives a reference frame with electric (y) and magnetic (x) deflections, the q-over-m ratio for each trace and the maximum ion energy, corresponding to the experimental point closest to the spectrogram center, due mainly to neutral radiation produced in the laser-plasma interaction.

The spectrometer has been successfully tested at PALS laboratory (Prague, Cz) in the 2012, using the ASTERIX IV laser system [25] and, after some improvement it has been recently used for another experimental run at the same laboratory. An example of spectrogram is shown below.

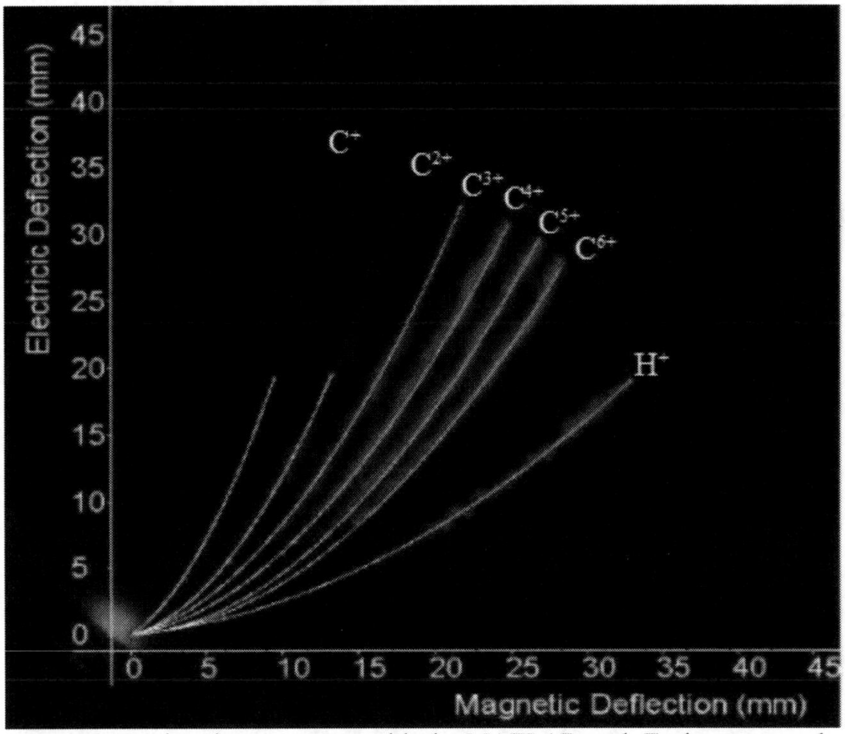

Figure 1. Analysed spectrogram with the MATLAB tool. Each traces on the image is fitted with a second order polynomial, the output of the code are the q/m ration and the electric and magnetic deflections.

The spectrogram has been acquired using a 50 μm thick target of PET (Polyethylene terephthalate) and the Thomson Parabola was set on the target normal in backward direction. Protons parabola is the first from the right side and their maximum energy is about 0.7 MeV. The other set of parabolas is due to carbon ions (C^{+6} to C^{+1}) with maximum energy of q*0.7 MeV, being q the charge state. In the spectrogram it is also evident that the central spot, due to neutral radiation passing the collimator, falls close to the edge of the detector.

During the last experimental run some slotted CR-39 have been used in series with the MCP in order to perform a calibration of the imaging system [26] and to have information on the particle energy spectra. Result will be reported elsewhere.

Moreover an energy calibration of the system will be performed within the end of this year at the LNS Tandem beam in the energy range 6-12 MeV.

4. The Energy Selection System

As stated above, protons from TNSA regime have a huge energy spread and, in view of the potential use of these beams for medical applications, a magnetic energy selection system [27] (ESS) has been realized in order to select particles in a narrow and controlled energy range. An energy calibration have been already performed at LNS using the Tandem beam in the energy range [4-12] MeV in order to characterize the device and validate the GEANT4 simulations. Data analysis is in progress and related results will be published elsewhere. The calibration is crucial for the preparation of the first experimental run already scheduled at the Center for Plasma Physics at Queen's University in Belfast (UK) where the TARANIS laser system is used to accelerate protons up to 12 MeV [28]. The ESS is a magnetic chicane and consists of four dipoles based on permanent magnets producing a field of ~0,85 T each. The second and the third magnetic fields are parallel to the first and fourth but oriented in the opposite direction. The fields configuration allows to radially defocus the beam in the first two dipoles. In this way it is possible to have a good separation between the particle trajectories at different energy in correspondence of the central pair of magnets where, by means of a moving slit collimator, the particles with the desired energy are selected. The particles passing through the slit are then refocused by the opposite gradient of the third and fourth dipoles. More details can be found in [29-31]. The field profile have been mapped along three selector axes and its integral is not exactly zero, as expected. Hence the beam coming out from the device won't be parallel to the selector axis. The issue, due to manufacturing errors, can be fixed displacing the forth dipole back and forth. During the calibration the right position of the last magnetic field have been also evaluated. The energy spread and the amount of particles passing through the slit depends on the size of the aperture. The lower the energy spread, the lower the number of particles will be transported through the energy selector as it needs to use a smaller slit size. On the other hand bigger slit size means lower selection efficiency but higher transmission efficiency. The energy of the proton beam can be tuned by moving transversely the slit position between 30 mm and 8 mm from the target normal axis. A roller guide system, where the central twin magnets are placed, allows to radially displace the two dipoles in order to increase dynamics and select the lowest energy particles. The lower energy limit is hence related to the reflection of particles on the first dipole and is less than 1 MeV. The energies at which the system can work vary within the wide range of 1 MeV and 60 MeV. The energy spread reachable by using 1 mm slit aperture ranges from 3 % for low energy (1 MeV) up to 20 % for the highest (60 MeV). The whole system is almost 600 mm long and will be placed into a dedicated vacuum chamber. Two additional collimators are placed both upstream and downstream the selector system; the first collimator is used to select the beam in angle in order to avoid spatial mixing of particle with different energies inside the device, the second one would refine the beam collimation and selection as well.

Selection and transmission efficiencies of the magnetic system can be improved if a beam transport element is designed in order to inject the beam in the chicane with proper characteristics. A magnetic lenses system is under studying for this purpose.

5. Conclusions

The ELIMED preparatory phase has already started and several researchers are already involved in the development and study of the detectors and transport devices. As described above, the Thomson Parabola for beam diagnostics have been used in two different experimental run at PALS Laboratory showing good response, the Energy Selector System characterization has been completed and it will be tested with optically accelerated ion beams within the end of the year. Hence, the actual status of the project makes us confident that the planned schedule will be followed without significantly delay for, at least, the first phases. The whole project is of outstanding importance as, if the goal will be reached, it would be possible to perform proof-of-principle experiments devoted to demonstrate new concepts for non-conventional hadrontherapy centres which would be less expensive in terms of realization and operating costs, as they would be less complex than conventional centres, and smaller in size. Hence, it would be possible to develop more facilities around the world which would allow a widespread availability of hadrontherapy centres as required by current demand for this type of treatment.

Acknowledgements
This work benefited from the support of the Czech Science Foundation (Project No. P205/11/1165), the CzechRepublic's Ministry of Education, Youth and Sports to the ELI-Beamlines (CZ.1.05/1.1.00/02.0061) and ECOP2 (CZ.1.07/2.3.00/20.0087) projects.

We also want to thank the PALS (Prague, Cz), FLAME (Frascati, It). technical staffs for the support during the experiments. A particular thank to the LNS (Catania, It) technicians for their support in the accomplishment and testing of the devices described in this contribution.

References

[1] Mourou G, Tajima T, Bulanov, S V 2006 *Rev. Mod. Phys* **78** 309
[2] Daido H, Nishiuchi M, Pirozhkov A S 2012 *Rep. Prog. Phys.* **75** 056401
[3] Wilks S C, Langdon A B, Cowan T E, Roth M, Singh M, Hatchett S, Key M H, Pennington D, MacKinnon A, Snavely R A 2001 E *Phys. Plasmas* **8** 542
[4] Bulanov S V, Esirkepov T Z, Khoroshkov V S, Kuznetsov A V, Pegoraro F 2002 *Phys. Lett. A* **299** 240
[5] Fourkal E, Velchev I, Ma C-M 2005 *Phys. Rev. E* **71**, 036412
[6] Esirkepov T, Borghesi M, Bulanov S V, Mourou G, Tajima T *Phys. Rev. Lett.* **92** 175003
[7] Passoni M, Bertagna L, Zani A 2010 *New J. Phys.* **12** 045012
[8] Zhou C T, He X T 2007 *Optics Letters* **32** 2444-2446
[9] Malka V, Fritzler S, Lefebvre E, d'Humières E, Ferrand R, Grillon G, Albaret C, Meyroneinc S, Chambaret J P, Antonetti A, Hulin D 2004 Med. Phys. **31** 1587
[10] Macchi A, Benedetti A 2010 *NIMA* **620**
[11] Henig A, Steinke S, Schnurer , Sokollik T, Horlein R, Kiefer D, Jung D, Schreiber J, Hegelich B M, Yan X Q, ter Vehn J M, Tajima T, Nickles P V, Sandner W, Habs D 2009 *Phys. Rev. Lett.* **103**, 245003
[12] Palmer C A J, Dover N P, Pogorelsky I, Babzien M, Dudnikova G I, Ispiriyan M, Polyanskiy M N, Schreiber J, Shkolnikov P, Yakimenko V, Najmudi Z 2011 *Phys. Rev. Lett.* **106** 014801
[13] Steinke S, Hilz P, Schnure M, Priebe G, Branzel J, Abicht F, Kiefer D, Kreuzer C, Ostermayr T, Schreiber J, Andreev A A, T P Yu A P, Sandner W 2013 *Phys. Rev. ST Accel. Beams* **16** 011303
[14] Authors, V. 2011 *ELI - Extreme Light Infrastructure. White Book. Science and Technology with Ultra-Intense Lasers*", Andreas Thoss Berlin
[15] Bulanov S V, Khoroshkov V. S. 2002 *Plasma Physics Reports* **28** 453
[16] Sakaki H, Hori T, Nishiuchi M, Bolton P, Daido H, Kawanishi S, Sutherland K, Souda H, Noda A, Iseki Y, Yoshiyuki T 2009 *Proceedings of PAC09*, Vancouver, BC, Canada TU6PFP009
[17] Murakami M, Hishikawa Y, Miyajima S, Okazaki Y, Sutherland K L, Abe M, Bulanov S V, Daido H, Esirkepov T Z, Koga J, Yamagiwa M, Tajima T 2007 *AIP Conference Proceedings* **1024**
[18] INFN, FZU 2012 *ELIMED: Memorandum of Understaning*
[19] Doria D, Kakolee K F, Kar S, Litt S K, Fiorini F, Ahmed H, Green S, Jeynes J C G, Kavanagh J, Kirby D, Kirkby K J, Lewis C L, Merchant M. J, Nersisyan G, Prasad R, Prise K M, Schettino G, Zepf M, Borghesi M 2012 *AIP Advances, Special Topic: Physics of Cancer* **2** 011209
[20] Rhee M J 1984 *Rev Sci Instrum* **55** 1229
[21] Schneider R F, Luo C M, Rhee M J 1985 *J. Appl. Phys.* **57** 1
[22] Harres K, Schollmeier M, Brambrink E, Audebert P, Blazević A, Flippo K, Gautier D C, Geissel M, Hegelich B M, Nürnberg F, Schreiber J, Wahl H, Roth M 2008 *Rev Sci Instrum* **79** 093306

[23] Jung D, Hörlein R, Kiefer D, Letzring S, Gautier D C, Schramm U, Hübsch C, Öhm R, Albright B J, Fernandez J C, Habs D, Hegelich B M 2001 *Rev. Sci. Instrum.* **82** 013306

[24] Cirrone G A P, Cuttone G, Maggiore M, Torrisi L, Tudisco S 2010 *Rad. Eff. Def. in Solids* **165** 6

[25] Jungwirth K, Cejnarova A, Juha L, Kralikova B, Krasa J, Krousky E, Krupickova P, Laska L, Masek K, Mocek T, Pfeifer M, Präg A, Renner O, Rohlena K, Rus B, Skala J, Straka P, Ullschmied J 2001 *Physics of Plasmas* **8** 5

[26] Prasad R, Doria D, Ter-Avetisyan S, Foster P S, Quinn K E, Romagnani L, Brenner C M, Green J S, Gallegos P, Streeter M J V, Carroll D C, Tresca O, Dover N, Palmer C A J, Schreiber J, Neely D, Najmudin Z, McKenna P, Zepf M, Borghesi M 2010 *NIMA* **623** 712

[27] Yogo A, Maeda T, Hori T, Sakaki H, Ogura K, Nishiuchi M, Sagisaka A, Kiriyama H, Okada H, Kanazawa S, Shimomura T, Nakai Y, Tanoue M, Sasao F, Bolton P R, Murakami M, Nomura T, Kawanishi S, Kondo K 2011 *App. Phys. Lett.* **98** 053701

[28] Dzelzainis T, Nersisyan G, Riley D, Romagnani L, Ahmed H, Bigongiari A, Borghesi M, Doria D, Dromey B, Makita M, White S, Kar S, Marlow D, Ramakrishna B, Sarri G, Zaka-Ul-Islam M, Zepf M, Lewis C L S 2010 *Laser and Particle Beams* **28** 451

[29] Cirrone G A P, Margarone D, Maggiore M, Korn G, Cuttone G, Schillaci F, Scuderi V, Tramontana A, Prokupek J, Velyhan A, Jia S B, Torrisi L, Gammino S, Renis M, Manti L, Romano F P, Limpouch J, Psikal J, Borghesi M, Bulanov S S, Bulanov S S 2013 *Proc. SPIE* 8779 87791I

[30] Maggiore M, Scuderi V, F. Schillaci, A. Tramontana, *AIP Conf. Proc.- II ELIMED Workshop* **1546** 34

[31] Cirrone G A P, Korn G, Maggiore M, Margarone D, Prokupek J, Velyhan A, Jia S. B, Cuttone G, Gammino S, Manti L, Renis M, Romano F, Tomasello B, Torrisi L, Schillaci F, Tramontana A 2013 *NIMA* **735** in press

Journal of Physics: Conference Series **508** (2014) 012011 doi:10.1088/1742-6596/508/1/012011

High energy X-ray emission driven by high voltage circuit system

M. Di Paolo Emilio[1,2] and L. Palladino[1,2]

[1]University of L'Aquila, Dep. MeSVA - Via Vetoio 67100 L'Aquila - Italy
[2]INFN Laboratori Nazionali del Gran Sasso, S.S. 17 bis km 18+910 67010 Assergi (AQ) - Italy

E-mail: mdipaolo@aquila.infn.it

Abstract. The x-rays are produced by high voltage discharge applied inside plasma source interaction chamber; the control circuit system of high voltage is composed of a power supply and a LC-inverter. The goal of the project is based on the production of x-rays by a plasma to increase the efficiency of a classic x-ray tube with hot filament. Preliminary results of high energy x-rays emission, the layout and simulation with p-spice of the high voltage system are presented.

1. Introduction

Different applications of soft X-rays (~70 eV - 2 keV) and hard x-rays (~3 keV - 25 keV) in biological, radiobiological and in medical imaging field, have encouraged the development of X ray sources based on the production of plasma. The x-rays emission is due to plasma obtained focalizing an high power Nd-YAG laser beam on different targets [1].

Plasmas produced by focusing pulsed laser beams on solid targets are currently recognized as one of the most promising alternative to conventional (X-ray tubes) and to synchrotron X-ray sources.

Optimization of pulsed x-rays sources (~ 3 keV - 25 keV), due to the plasma electrons accelerated by an electric field, it is necessary to make x-rays source comparable with the most advanced current x-ray tubes (i.e. microfocus) for applications in the field of mammography and non-destructive inspection. There are many variables that allow us to get this optimization but the electrical parameters are especially the basis of the choice for the high voltage system configuration.

The reason to obtain a high voltage through a LC inverter is because of the system is working with high voltage in the discharge phase, but during the charge of the capacitors the system is isolated which is essential for the safety and average life of the electrical components.

The aim of the electrical simulation is to obtain phenomenological mathematical model using P-spice to understand the behaviour of the system and to check the condition of functionality optimizing the emission of x-rays. Orcad Pspice is a full-featured, native analog and mixed-signal circuit simulator [2].

2. Experimental Apparatus

The x-rays of the higher energy region (about 25 keV) is produced for bremmsthralung effect. The relative configuration of the interaction chamber is shown in figure 1. The main feature is to apply the High voltage (30 50 kV) from the target, on which the laser is focalized, (cathode) to the anode (such as metallic materials Mo or W) where x-rays are produced [1].

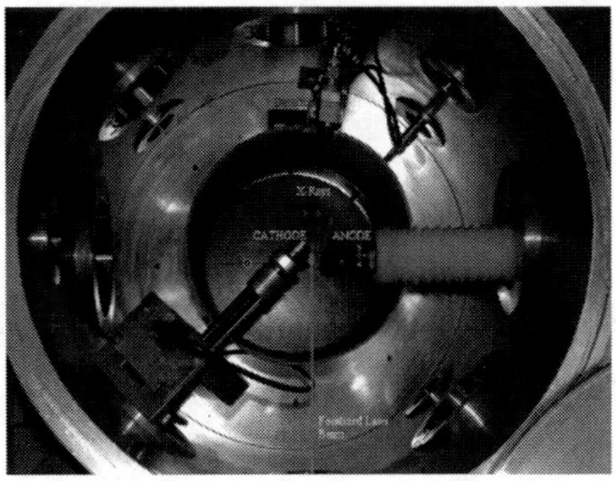

Figure 1. Interaction chamber (inside)

Figure 2. Spark Gap

The mechanism of x-rays production is based on the generation of a plasma obtained by focusing a laser beam NdYag (5J per pulse, duration 6 ns, 0.3 mrad divergence, 536 nm wavelength) on a target metal.

The apparatus for the High Energy X-rays generation is composed of interaction chamber in the vacuum ($10^{-6} mbar$) and two main circuits systems: a power supply that provides a maximum voltage of 40 kV and an LC-inverter circuit. From figure 1, Rogowsky loop is used to measure the intensity of the electronic current.

3. High Voltage System

High Voltage System is composed of following sub-circuits:

- High Voltage Power Supply
- Trigger circuit for Spark-Gap
- Trigger circuit for Thyratron
- LC Inverter

A customize high power supply is applied to spark gap for generating a discharge to produces x-rays. The Maximum voltage is around 40 kV. Spark Gap is a high power three-terminal switch which is used to be activated the discharge, shown in figure 2. It is composed of 3 poles: two poles are connected to the power supply and the central pole is connected to the output of the trigger circuit for spark gap.

A trigger circuit for Spark Gap (Figure 3) consists in a LC inverter and a Thyratron, which is used as high electrical power switch. The trigger circuit for Thyratron is shown in figure 4 and is used to generate the peak voltage sent to the Thyratron (Figure 5). The control signal of the Silicon Controlled Rectifier (SCR) (see figure 4) is obtained with a customize circuit as shown in figure 7. Moreover, a LC inverter (Figure 6) is used to activate the discharge by charging a couple of capacitors of 1.5 μF.

The P-spice models consist in the design of the circuits with the following considerations:

- Spark Gap and Thyratron are simulated as a switch.
- Periodic Pulse Voltage is used to simulate the command of activation of the discharge.
- Standard P-spice models of passive and active components are used.

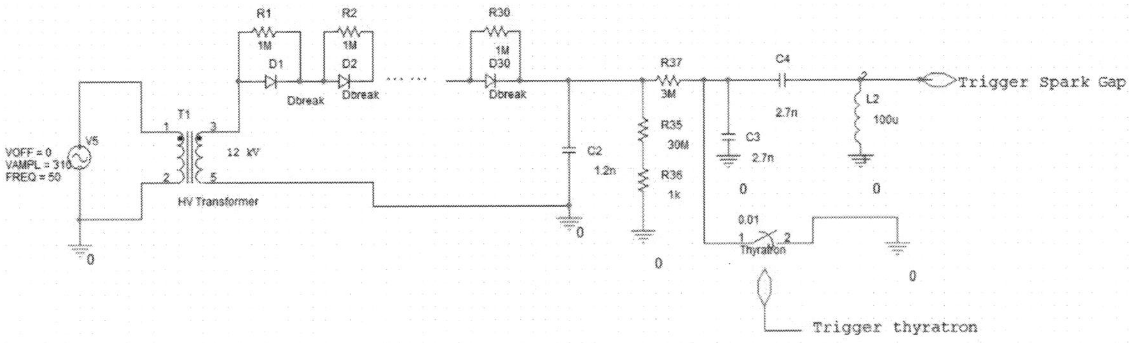

Figure 3. Trigger circuit for Spark Gap (Model)

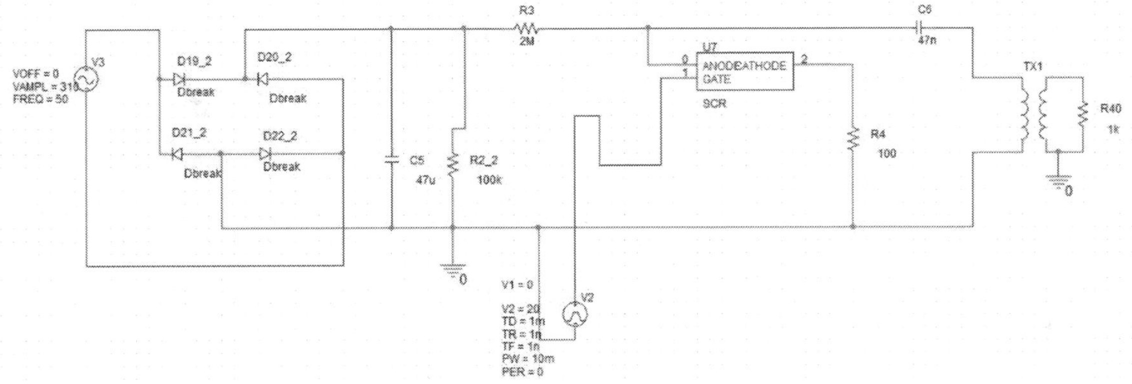

Figure 4. Trigger Circuit for Thyratron (Model)

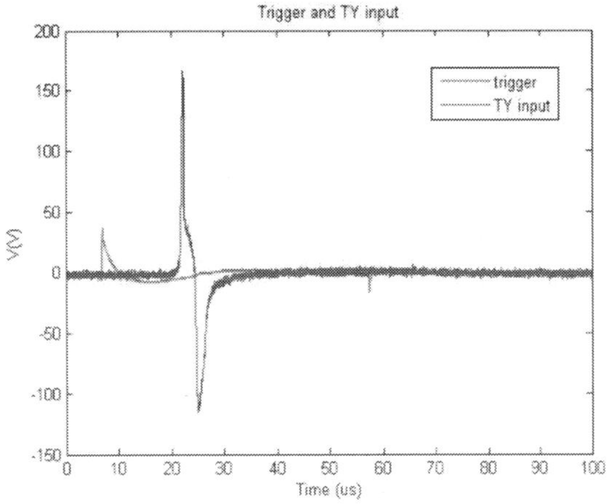

Figure 5. Output of trigger circuit for thyratron (blue line). In red line the control signal of SCR (Silicon Controlled Rectifier)

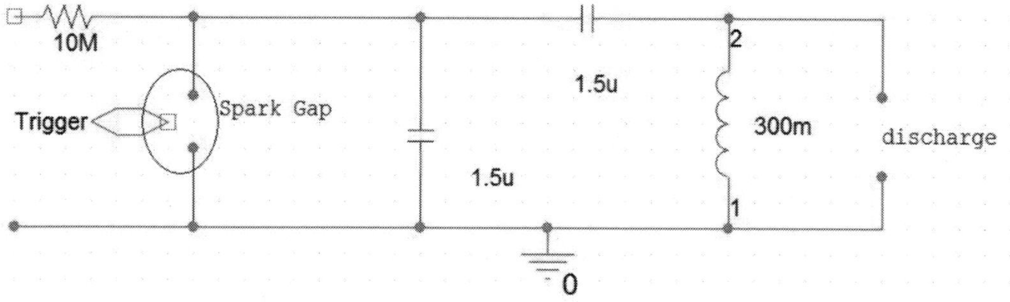

Figure 6. LC-inverter

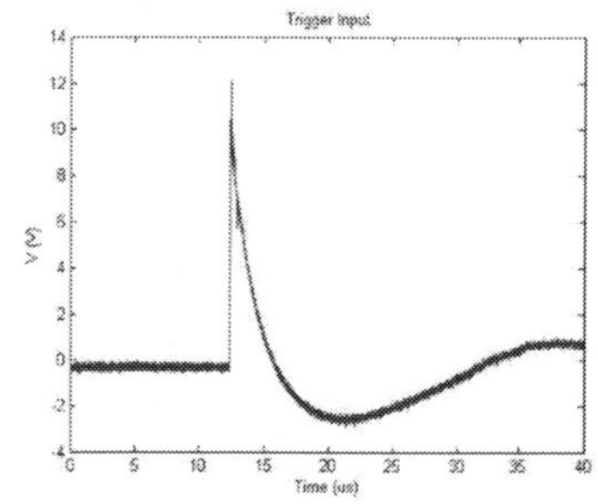

Figure 7. Control signal of SCR (Silicon Controlled Rectifier)

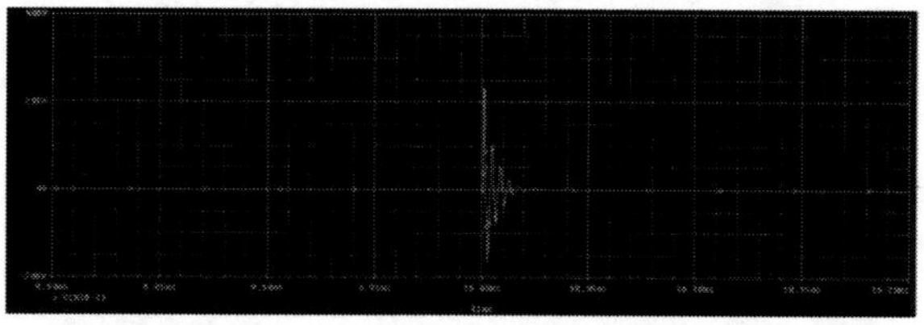

Figure 8. Output of trigger circuit for spark gap (Model)

From the p-spice analysis we can observe a signal with the maximum value around 20kV in the output of the trigger circuit for spark gap (Figure 8): if this values is higher, it could damage the system, in oppositely, the discharge can't be activated. It's important that the intensity in the output of the trigger circuit for spark gap is about 20 kV.

The LC-Inverter (Figure 6) is the most important circuit in the high voltage system to control the intensity of x-rays. In figure 9 the schematic representation of the p-spice model.

In figure 11 a typical voltage signal in output of LC inverter from P-spice simulation is shown. According to the experimental measurements (Figure 10), we obtained a similar signal with different damping due to the approximation of the model but with the same maximum value. All high voltage system is immersed in transformer oil (Figure 12) to isolate from the environment.

4. Preliminary results

From the results of p-spice model, we have tested the system by analyzing the emission of x-rays for iron target measured by photodetector from Quantrad system (diameter of 7.5 mm, absorption layer of 125 μm and dead layer of 0.3 μ). On the entrance window of photodetector, there is an aluminum foils of 12.8 μm thickness. The distance between photodector and cathode is 10.5 cm. Transmission Profile of aluminum is shown in figure 13 where is visualized one energy peak at 1.5 keV with transmission from around 20% and reach to 100% for x-rays energies which is greater than 5 keV.

15 kV of power supply is applied in two poles of spark gap (which is the input of the circuit of figure 9); in according to the high voltage system the discharge voltage is around 25 kV. System Energy is estimated to be around 240 J. Inside the interaction chamber, we exposed films by

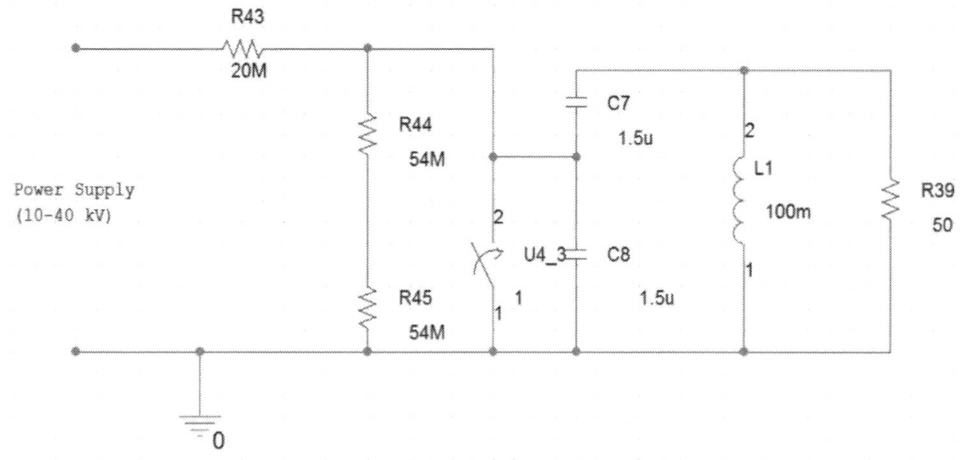

Figure 9. LC inverter - P-spice model

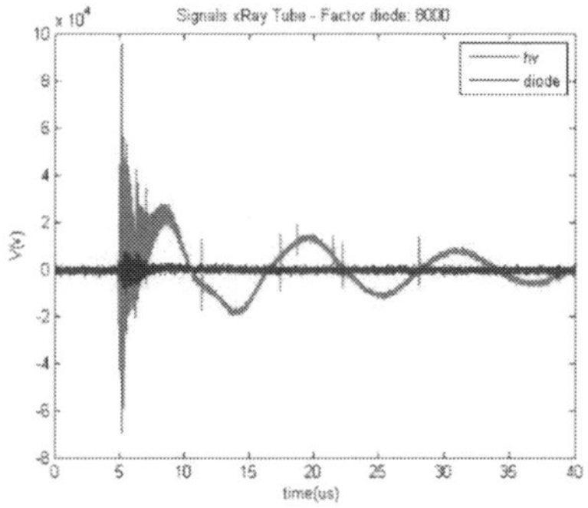

Figure 10. In red line output voltage of LC Inverter signal (Experimental)

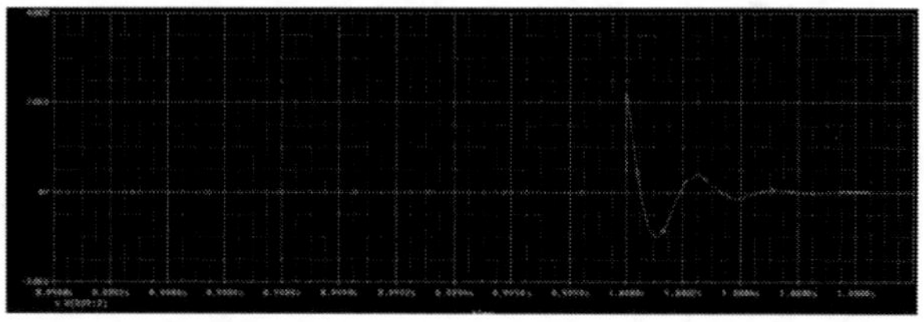

Figure 11. Output voltage of LC Inverter signal (Model)

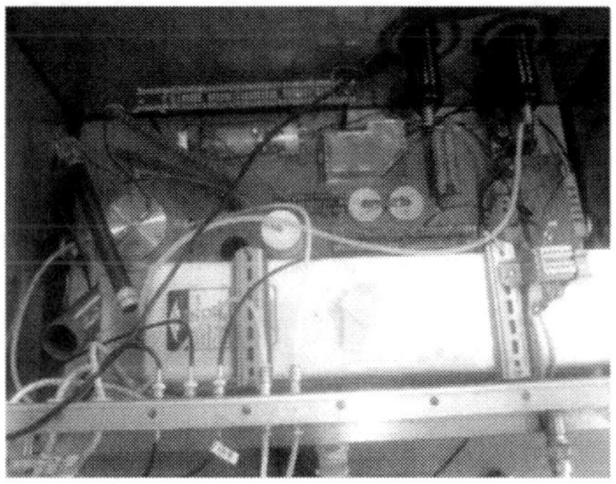

Figure 12. High Voltage system

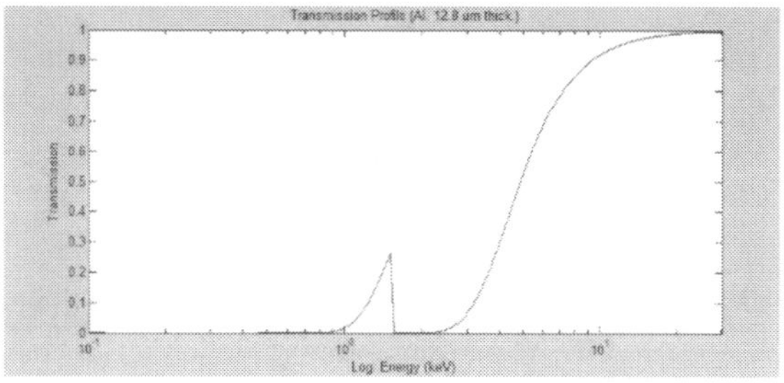

Figure 13. Profile Transmission of Aluminum (Thickness of 12.8 μm) [3]

Kodak (Film Dentaire ultra speed), to capture the samples of x-ray images (iron cross): one with aluminum foil of 12.8 μm thickness and another without aluminum foils. The x-ray images of iron cross are shown in Figure 14 where is noted a better image in the film with aluminum foil; in fact, in the film without filter, it has a blurring due to all x-rays energies components (around keV and greater than 5 keV). Since x-ray image is formed with all energies components that have a different absorption coefficient, the image of the film is not well defined. While in the film filtered is selected by a spectra where K_α and K_β energies emission of iron are more intense.

In figure 15 is shown the signal of the photodiode and current intensity. The charge is measured around 0.07 nC.

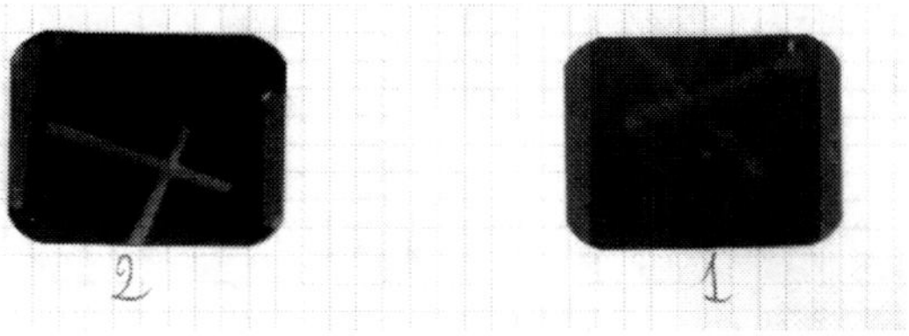

Figure 14. X-ray images captured using Films Dentaire by Kodak. Film without aluminum (1), Film with Aluminum (2).

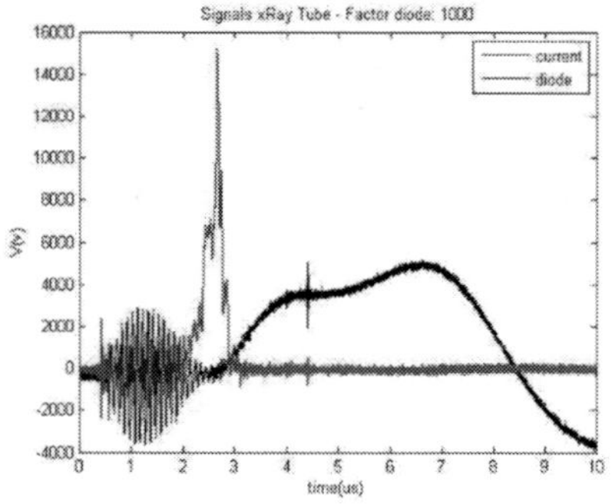

Figure 15. Output of photodiode (blue line) and current intensity (black line)

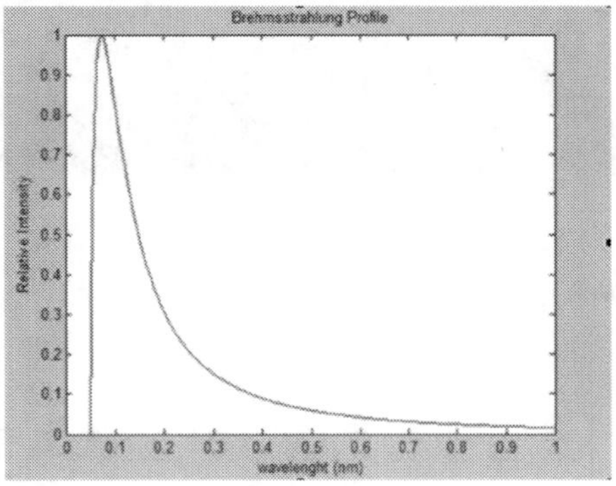

Figure 16. Standard Bremsstrahlung profile, $\lambda_{min} \cong \frac{12400}{25kV} = 0.049nm$

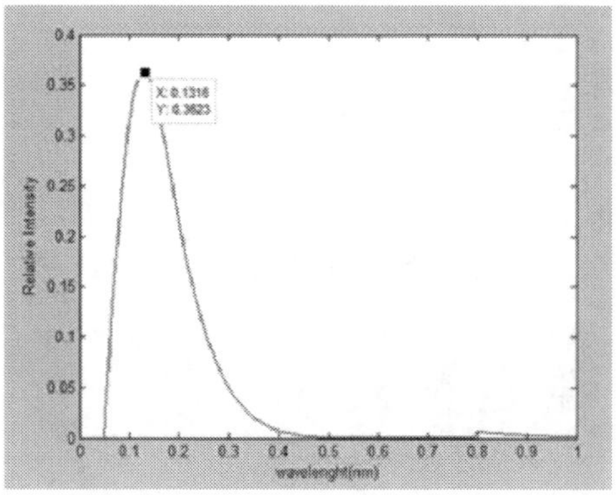

Figure 17. Complete Bremsstrahlung profile

According to the standard bremsstrahlung profile (Figure 16)

$$I \propto Z(\frac{1}{\lambda_{min}} - \frac{1}{\lambda})\frac{1}{\lambda^2} \tag{1}$$

and taking into the consideration of the detector quantum efficiency, we have calculated the complete profile (Figure 17).

From these values, we have estimated the number of photons in the input of the photodiode of $\sim 10^8$. Considering the emission of x-rays on a 4π sr solid angle, the total number of x-rays photons emitted is $\sim 10^{12}$ [4].

From these analysis we can see that the number of emitted photons is comparable with present X-ray tubes [5] but with the difference that in our source, x-rays are emitted with a time of the order of μs (figure 15), to be compared with the time of x-ray tubes traditional ranging from about 0.01 s to 10 s [5].

Another advantage of our High Voltage system is the LC-inverter; it allows to isolate the complete system during the charging phase which is essential for the safety of both people and electronic components.

5. Conclusions

The p-spice model presented is a mathematical tool to evaluate the electrical characteristics of the high voltage system to optimize the emission of x-rays. A possible future configuration will be to use an IGBT (Insulated Gate Bipolar Transistor) instead of Thyratron to compact the system and increase the energy.

The insulated-gate bipolar transistor or IGBT is a three-terminal power semiconductor device primarily used as an electronics switch and in the other devices are well-known used for combining high efficiency and fast switching.

In near future we will improve the synchronization of the system between trigger discharge with pulse laser beam. Moreover, we will examine x-rays emission by changing the distance from anode and cathode and power density of the laser [6].

6. Acknowledgement

We would like to take this opportunity to thank Prof. S. Ragazzi, Director of Gran Sasso National Laboratories of INFN for scientific and financial support with fellowship POR - LNGS.

Moreover, thank to Prof. M.G. Cifone director of MesVA Department of L'Aquila University for supporting us. This work is financially supported by INFN, PLAXA experiment.

References

[1] L. Palladino, Annual Report 2010, *INFN Laboratori Nazionali del gran sasso* - Link: http://www.lngs.infn.it/lngs_infn/contents/lngs_it/research/experiments_scientific_info/library_publications/preprint_reports/report10/lasex.pdf
[2] Habetler, T.G. et all, Design and implementation of an inverter output LC filter used for dv/dt reduction *Power Electronics, IEEE Transactions on (Volume:17 , Issue: 3)*
[3] Calculated from *http : //henke.lbl.gov/optical$_c$onstants/filter2.html*
[4] E. Bertin, Introduction to x-ray spectrometric analysis *Springer Science+Business Media, LLC*
[5] Data sheet of commercial X-ray Tube *http : //www.varian.com/media/xray/products/pdf/m105sp.pdf Varian Medical System*
[6] S.L. Gold, Thyratron-PFN, IGBT Hybrid, and Direct Switched Modulator R&D as it Effects Klystron Protection *SLAC-PUB-8475 June 2000*

Characterization of thin films for TNSA laser irradiation

M Cutroneo[1,3], L Torrisi[1], L Calcagno[2], A Torrisi[1,2]
[1)] Dottorato di Ricerca in Fisica, Università di Messina, Dipartimento di Fisica & S.d.T., V.le F.S. d'Alcontres 31, 98166 S. Agata, (ME), Italy
[2)]Dip.to di Fisica, Università di Catania, Via S. Sofia 64, 95123 Catania, Italy

E-mail: mari.cutroneo@unime.it

Abstract. Thin films of hydrogenated materials have been prepared, at Messina University, to be irradiated by high intensity lasers in TNSA conditions in order to accelerate high energetic protons with high yield and directivity. Film composition was based on polymers, metals, multilayers and nanostructures embedded in polymers. The preparation methods were different and based on different deposition techniques. High vacuum condition, nanostructures, carbon nanotubes, metal oxides and hydrates, were employed. Targets were prepared as a sheath with thicknesses ranging between 0.1 and 100 μm and surfaces of the order of some cm^2. Targets were characterized in terms of thickness, roughness, surface morphology, colours and absorption coefficient in the wavelength range 250 nm-1300 nm. Peculiar attention is given to samples with high absorption coefficient in order to improve the energy transfer from the coherent light to the generated plasma.

1. Introduction

In 1982 at Los Alamos National Laboratory have been achieved protons of 0.5MeV by using a laser at 10^{15} W/cm^2 of intensity. In the last 30 years, occurred improvement on the geometry and composition of targets, on the focalization of *fs* lasers, more knowledge on the parameters of laser-generated plasma, development of detectors with high reliability, faster, more resistant at high temperature and low background current. The state of art in 2001 evaluates the dependence of many parameters (temperature, density, electric fields, ion charge state distributions,...) on the product $I\lambda^2$, where I is the laser intensity and λ its wavelength, and shows that the ion acceleration increases with such parameter [1]. Intense charge separation in plasma and high electric fields, of the order of TV/m, has been generated inducing ions acceleration by high energy and short laser pulses [2]. Many laboratories in the world are accelerating protons up to 40 MeV/nucleon, using the TNSA (Target Normal Sheath Acceleration) approach. Recently, research has increased significantly in the direction of replacing conventional accelerators for the production and acceleration of ion beams to demonstrate the clinical applicability of laser driven-proton by irradiation of solid targets with laser pulses at high intensities. With the aim to fully utilize the advantages of protons beams laser-generated for applications, from bio-medicine to nuclear physics and from microelectronics to chemistry, the process of production of protons can be optimized using suitable hydrogenated targets. It is well-known that when an electromagnetic wave impinges a solid target, part of the radiation is adsorbed, part is scattered, part is transmitted and part is reflected. Actually, the main objective is the maximization of the absorption efficiency of the matter in order to enhance the laser energy transmitted to the target and consequently

[3] To whom correspondence should be addressed

the transfer of energy to the plasma in order to increase its temperature, density and electric field driving ion acceleration. The amount of charged particles emitted during the laser-matter interaction depends on the intensity of the absorbed radiation. In order to evaluate the amount of energy released to a solid target during the laser-matter interaction it is important to investigate on the reflectivity and absorption coefficient of the irradiated targets. The practical applications have been focused on the techniques to produce thin films to be irradiated in TNSA conditions and porous and nanostructured samples to increase the absorption effects. Thus the investigation of their optical properties becomes an important objective of the research [3]. Samples containing nanostructures, hydrogenated targets, multilayer films and microporouses targets, are able to increase the kinetic energy of the accelerated species by the non–equilibrium processes developed in laser irradiation of TNSA foils [4].

2. Materials and methods

With the aim to arrange targets with peculiar properties such as low reflectivity, high porosity and roughness of the surface, low thickness, and high absorbent nanostructures embedded, different thin film deposition techniques have been employed. In this work we will discuss only four of them that modify physical characteristics of thin films, such as the surface morphology and the inclusion of micrometric and nanometric particles. A brief description of the used four methods for thin film preparation is described in the following.

2.1. Pulse Laser Deposition (PLD)

PLD of Ti, Au, Cu and Pd thin films were obtained irradiating metallic bulk targets by laser at 0° incidence angle and positioning the substrates at different distances from the target. Substrates were polymers (mylar and polyethylene). The bulk metals were fixed on a rotating carrousel to allow a fresh surface ablation in high vacuum (10^{-6} mbar). A Nd:YAG laser at the fundamental wavelength (1064 nm), 300 mJ pulse energy, 9 ns pulse duration, 10 Hz repetition rate was used [5]. Generally Cu, Au, Pd, Ti particles have been deposited on 0.9 μm thickness mylar $(C_{10}H_8O_4)_n$ or polyethylene $(CH_2)_n$. Inside the vacuum chamber the angles in which have been fixed the substrates were 15°, 20° and 45° with respect to the normal target direction and distances ranging between 8.5 cm and 16 cm with respect to the target. Film thicknesses between 100 nm and 10 μm were obtained.

2.2. Ion Sputtering

Ion sputtering has been performed by using 1-3 keV Ar^+ ion gun with a ion current flux of 10-50 μA incident at 45° on metallic targets (Cu, Ag, Au) and on polymeric target (PMMA). Sputtered metallic atoms were deposited on polymeric substrates (mylar and polyethylene), from 0.6 to 6 μm thickness. The ion irradiated surface was 1 cm^2 and the Ar incident doses were from about 10^{15} up to about $10^{17}/cm^2$. The low ion energy induces bombardment in nuclear stopping power regime and the structural damage modifies the morphology of the target [6]. Stopping and Range of Ion in Matter (SRIM) code was employed to predict the behaviour of 2 keV Ar^+ ions interacting with the matter and to evaluate the sputtering yield (number of emitted atoms per incident ion). As for PLD, the thickness of deposited thin films has been measured using 5 MeV energy loss of alpha particles emitted form a collimated Am^{241} radioactive source. The transmitted alpha particles though the prepared thin film were detected with and without the deposited thin film. On mylar, thin films of Cu, Ag and Au with thicknesses ranging between 10 nm and 500 nm were prepared. Object of the laser TNSA investigations were both the mylar covered by thin metallic films and the same sputtered metallic targets, 10-50 micron in thickness, which show grain contours enlargement produced by the sputtering effect, as SEM image investigations have demonstrated.

2.3. Spin coating

Spin coating involves accurately dispensing a liquid onto a substrate then spinning to achieve a uniform defect-free film. It employs the use of extremely precise rotation control coupled with a closed, fully optimized process chamber. A typical spin process consists in three steps:

1) The liquid drop is deposited on (or near) the centre of the circular rotating substrate (silicon wafer or glass), generally 5cc volume of the resin fluid is employed.
2) The combination of spin speed and time selected will determine the final film thickness. This step takes some seconds and depends on the solution viscosity and temperature.
3) The time and the air permit to regulate the dry time to obtain a solidified thin film on the substrate.

Finally the thin film is detached from the substrate. In our laboratory thin film deposition was performed by using carbon nano-structures embedded in an acrylic resin in which ethanol has been added in order to reduce the viscosity. C-structures have had a size range between 10 and 100 nm which are dispersed in solution and left in a ultrasonic bath for 1 hour. 5cc volume of the resin fluid was employed. The solutions concentration was obtained using micrograms of C-powder containing carbon nanoparticles in 4 ml of resin. A 1100 rmp was the spin speed employed for 480 seconds time spinning. The obtained thin films show a thickness gradient between centre and periphery, however the use of little pieces of films, of the order of 5x5 mm^2, permits to choice the desiderate thickness, that generally ranges between 0.1 up to 10 microns or more. Thickness film measurements were performed with 5 MeV alpha particle energy loss.

2.4. Thermal Sintering

It is a technique able to embed in a matrix of Ultra High Molecular Weight Polyethylene (UHMWPE), $3x10^6$ g/mol molecular weight, 99.7% purity, 50% crystalline and 50% amorphous, micrometric and nanometric fillers, such as Fe_2O_3, $C_{16}H_{18}N_3SCl$, carbon nanotubes (CNT) and carbon nanoparticles. The fillers with a main size ranged between 50 nm and 100 nm and with a concentration ranged between 0.01 wt% , and 10 wt%, colour the polymer red for iron oxide, blue for methylene-blu and black for carbon. The procedure to realize thin targets consists in the use of ethanol mixed with carbon nanotubes in concentrations of 0.01%, 0.5%, 1%, 2% in order to facilitate their dispersion and avoid the mechanical cracking of composed final polymeric film. The powder of polyethylene at ultrahigh molecular weight was put in a mould, to form a flat and circular sample which thickness was from 10 to 100 μm and diameter of 10 mm. The mould was positioned inside the hydraulic presser at a final temperature of 200^0 C and at a final pressure of 200 bar for a time of 20 minutes. Final thicknesses were measured with surface profiler and/or mechanical micrometer.

2.5. Optical characterizations

The thin films prepared by the different techniques, having different compositions, thickness and geometry, were arranged in a target holder to perform their successive optical characterization. A Hg-Ar lamp, with characteristic lines at wavelength ranging between 200 and 1200 nm, was used as multi-monochromatic radiation source. The Fig. 1 shows the spectrum obtained by using the above-mentioned lamp that contains argon and mercury gas. It indicates the characteristic lines in UV, Visible and IR regions, and the relative intensity used for absorption measurements as a function of the wavelength. The incident radiation impinges the samples fixed into the holder-target with a normal incidence. The transmitted light from the sample is connected by an optical fibre and it is sent to a Horiba Jobin Yvon Spectrograph. The absorption coefficient of the sample was evaluated by using the Beer-Lambert' s law [7]:

$$\mu = \left(\frac{1}{x}\right)\ln\left(\frac{I_0}{I}\right) \tag{1}$$

where the absorbance μ of electromagnetic wave, is given in terms of incident light I_0, transmitted light I and thickness x of the used sample used. The experimental set-up consists of a Lynear spectroscopy software, that acquires 160 spectra per second that are stored and averaged on a PC, a suitable target holder for the lamp, the sample and the fibre and a PC for data elaboration and for the power supply of the lamp. The sample thickness was measured using the energy loss of alpha particles (coming from a radioactive source of Am-241) or the surface profiler with indeterminacy of the order of 10 nm (Tencor P-10).

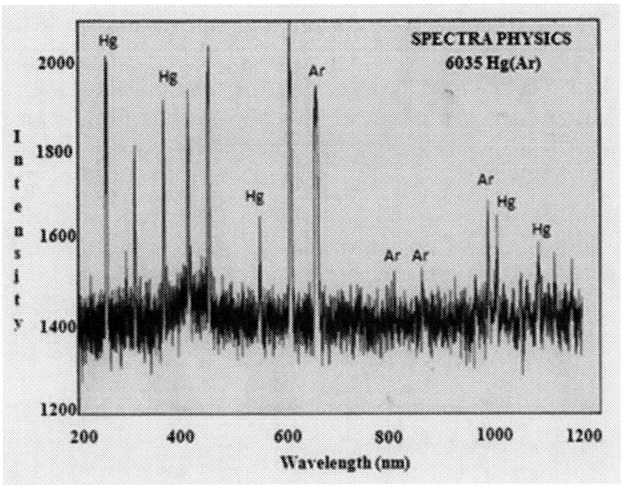

Figure 1. Hg-Ar lamp spectrum.

3. Results

The plot in Fig. 2 shows the transmission measured in a sample prepared by PLD technique depositing 52 nm of Pd on a mylar substrate 0.9 μm thickness. The points in the plot correspond to the values of the used wavelength lines. The sample, in the visible region 400 nm-700 nm, has a transparency comparable with the pristine ones. While in the infrared region, for wavelengths higher than 700 nm, it shows a higher absorption of the order of 40%. Also absorptions bands are present in the near UV region that reaches about 40% at 300 nm radiation.

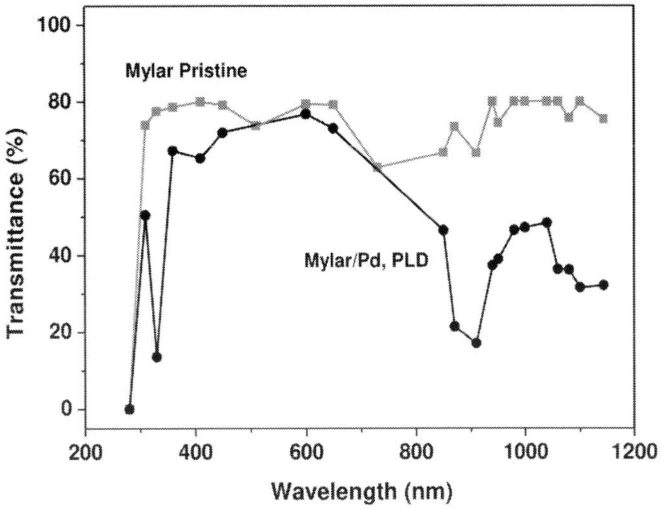

Figure 2. Transmission measurements vs. wavelengths for mylar and Pd/mylar thin film.

Polymethylmetacrylate (PMMA) targets, $(C_5O_2H_8)_n$, have been sputtered by 2.5 keV Ar ions in order to induce micrometric widening of the grain contours in the polymer surface. The evaluations of argon ions distribution in PMMA were achieved by using SRIM code [8] at the same conditions of the experimental setup. The sputtering yield of C, O and H atoms at an incidence angle of 45° is 36.44, 15.52 and 12.4 atoms/ion for C, O and H, respectively. The argon ion distribution reaches 80 Å range penetration inside the PMMA in which each ion argon removes up to 36 carbon ions producing collision, vacancies, defects and disorder in the polymer. Thus the polymer modification is extremely superficial. The samples have been treated at the same experimental conditions in order to provide ion doses at $10^{15}/cm^2$, 10^{16} $/cm^2$ and $10^{17}/cm^2$. Optical investigations performed on these modified targets show an increase of the absorbance with the increase of employed doses. Fig.3a shows a comparison between the absorbance for pristine PMMA and PMMA samples treated with a high Ar^+ absorbed dose. For the higher doses, measurements indicate a lower absorption in the UV region range and a significant increment up to 40-50% of the absorption coefficient in the IR range, above 800 nm, occurs.

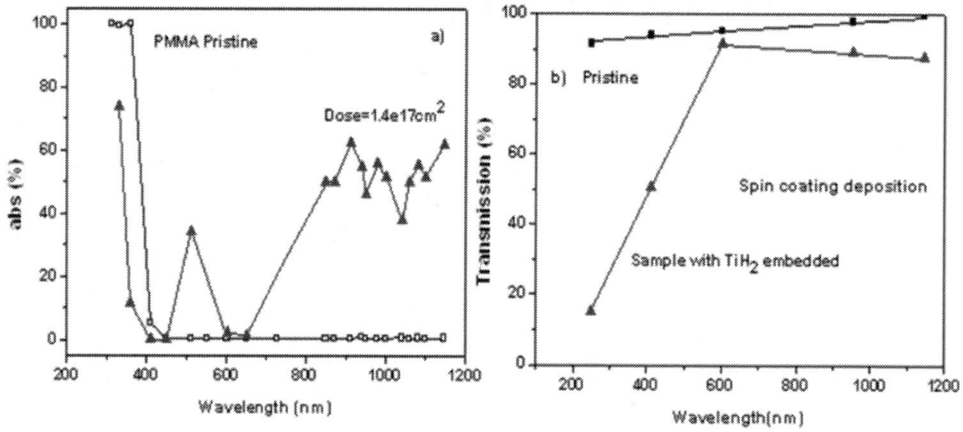

Figure 3. Absorbance vs. wavelength, for pristine and PMMA at different sputtering doses (a), and optical transmission vs. wavelength, for resin containing and not TiH_2 nanostructures (b).

Fig. 3b shows the optical transmission relating resin pristine and resin containing TiH_2 nanostructures, 50-100 nm sized, embedded by using the spinning deposition technique. This last target, has an interesting behaviour at the UV wavelength and at the IR region due to the influence of metallic component in the composite, instead in the VIS region, it presents a transmission of the order of 85% - 90% very close to the pristine sample.

The targets prepared by thermal sintering, investigated at wavelength between 250 nm and 1000 nm, show different absorption coefficients in the UV, VIS and IR regions as a function of the kind, of the thickness and of the concentration of nanoparticles embedded in the polymeric matrix.

Fig.4 reports the results of absorbance vs. wavelength measurements obtained for 100 μm thickness pure polyethylene films and for polyethylene containing different nanostructures. It can be observed as carbon nanotubes (CNT) and nanoparticles embedded into polyethylene (PE) at 0.5%-1% in weight concentration, show higher absorbance with respect to the pristine polymer of polymer containing other nanostructure species (methylene blu) in similar concentrations.

4. Discussion and conclusions

From the comparison between the absorption coefficients measured for different polymeric thin films it is possible to observe that high absorptions are obtained in the IR region for mylar films covered by thin metal films, such as Pd, Ti, Au and Cu (see Fig. 2), or for high Ar^+ ion doses sputtered PMMA (see Fig. 3.a), or for polyethylene in which peculiar nanostructures, such as carbon nanotubes and carbon nanoparticles, are embedded to concentrations of the order of 1% (see Fig. 4).

The samples prepared by spinner, show the best absorption in the IR region inserting in the solution TiH_2 nanostructures (see Fig. 3.b).

Thus radiations from Nd:YAG lasers or other IR lasers, generally employed to generate ion acceleration from TNSA plasmas, can release high energy in such thin targets placed in vacuum. The thin foils give rise to forward plasma formation and high energy ion acceleration using intensities exceeding 10^{15} W/cm^2 and sub nanosecond laser pulse duration. The effect may be more relevant if the target surface is treated to reduce the reflectivity, modifying its roughness and morphology, and if the inner target contains special absorbent species, such as micro and nanostructures absorbent the high laser light.

In order to accelerate high energy protons, very hydrogenated targets, such as polymer based foils, can be irradiated. However, in order to obtain high plasma temperature and high ion energy the thin target must contain high electron density and must have high absorption coefficient. Such requisites can be found with metallic nanostructures embedded in the thin polymeric target.
The use of polymeric structures containing CNT, with concentrations above 0.1 %, are very interesting for their capability to enhance the plasma hydrogen concentration due the CNT ability to absorb high hydrogen content [9].

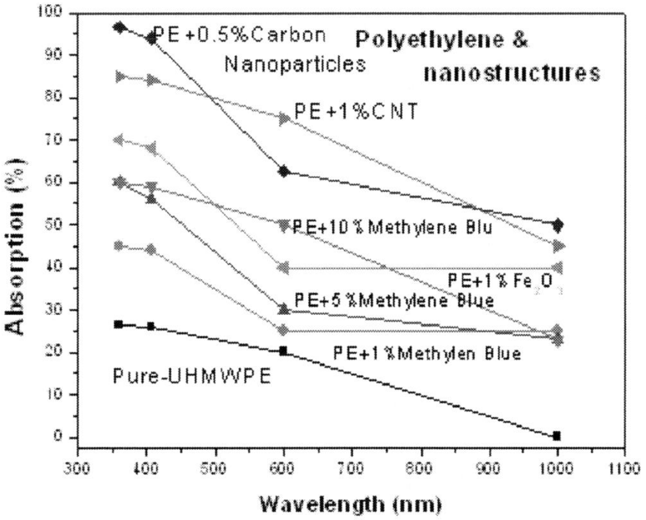

Figure 4. Absorption vs. wavelength for polyethylene pure and containing different nanostructures.

Due to the high cost of CNT multiwall nanostructures, experimental investigations have demonstrated that also the use of amorphous carbon, as nanoparticles, may induce high IR absorption if embedded in polyethylene, with the advantage of the low cost of these targets. The targets prepared as discussed previously, offer a lot of applications to be employed at medium and high laser intensities in order to obtain high hydrogenated plasmas in high vacuum and high ion acceleration [10]. These "advanced targets" can be employed to produce protons at high energy and high yield for accelerator ion sources, for chemical modifications of polymeric surfaces, for hadron-therapy accelerating protons at high energy and for nuclear physics and astrophysics, permitting to produce many events of nuclear reactions. Measurements carried out during last run at PALS laboratory in Prague, irradiating by high intensity lasers ($10^{14\text{-}16}$ W/cm^2) thin mylar foils covered by thin metals, with respect to the uncovered film, have provided an significant enhancement of the plasma density, of above 40%. Plasma shows a significant enrichment in the proton content and an increase in the mean ion velocity for thin polymeric targets irradiated in TNSA approach containing gold nanoparticles, as many detectors connected in time-of-flight configuration have demonstrated [11]. Nanoparticles and nanostrucutures, in fact, show absorption bands in the visible and IR regions due to resonant plasmon absorption effects and to vibrational states of little atom clusters .

Thus, in conclusion, the enhancement of the absorption coefficients for polymers has been used to increase the plasma temperature ad density, the kinetic proton energy and the specific ion yield (protons, carbons,...) induced in the rear side of the TNSA target by sub-nanosecond pulses, as recently reported in our literature [12-13].

References

[1] Spencer I et al., 2001 *NIM B* **183** 449
[2] Mc Kenna P et al., 2006 *Phil. Trans. R. Soc. A* **364**, 711
[3] Torrisi L, Cutroneo M, Caridi F, Gentile C, 2013 *Laser and Particle Beams*, **31**(01), 37 – 44
[4] Cutroneo M, Torrisi L, Margarone D, Picciotto A, 2013 *Applied Surface Science* **272** 50– 54
[5] Mezzasalma A M, Mondio G, Serafino T, Caridi F, Torrisi L, *Appl. Surf. Sci.*, **255** (2009) 4123
[6] Ghose D, Brinkmann U, Hippler R, 1997 *Phys. Rev. B* **55** 13989
[7] Visco A M, Campo N, Brancato N, Torrisi L, Caridi F, Cutroneo M, 2012 *APSUSC* 23023
[8] SRIM 2006, www.srim.org
[9] Rafiee Marjan A, 2012 *Journal of Computational and Theoretical Nanoscience*, **9** 12 2021-2026
[10] Torrisi L, Giuffrida L, Cutroneo M et al., 2012 *Rev Sci Instrum.* **83** Feb (2):02B315S
[11] Dhareshwar L J et al., 2010 *Journal of Physics: Conference Series* **244**, 022018
[12] Torrisi L, Cutroneo M, Andò L, Ullschmied J, 2013 *Phys. Plasmas* **20**, 023106
[13] Cutroneo M, Torrisi L, Margarone D, Picciotto A, 2013 *Applied Surface Science* **272**, 50-54

A Proton Source via Laser Ablation of Hydrogenated Targets

D Delle Side[1], V Nassisi and L Velardi
LEAS, Dipartimento di Matematica e Fisica, Università del Salento and INFN sezione di Lecce, Italy

E-mail: domenico.delleside@le.infn.it

Abstract. In this work we present results on the extraction of proton beams from a plasma generated by pulsed laser ablation of solid hydrogenated targets. The laser used was an excimer KrF operating at low irradiances (10^8–10^9 W/cm^2) and nanosecond pulse duration. The ablated targets were disks obtained by compression of TiH_2 powder. The ion emission was analyzed by the time-of-flight technique using a Faraday cup as ion collector. In order to improve the ion yield, an electrostatic extraction system was applied. Studies on the produced plasma for different laser irradiances and accelerating voltages have been performed. The results obtained show that this setup is suitable for a high yield proton source.

1. Introduction

Today it is possible to easily arrange laser beams at irradiances of the order of 10^8 – 10^{10} W/cm^2 and nanosecond pulse duration that, interacting with solid matter in vacuum, produces hot plasmas[1] by pulsed laser ablation (PLA). From these plasmas, ion beams of moderate energy can be extracted[2]. These beams have a number of interesting applications. Laser ion sources (LIS), indeed, proved over the years to be an useful tool both in the scientific and industrial fields.

Many projects across the world use them as injectors for common particle accelerators[3]. They are also successfully employed as devices for surface modification of materials[4] or for semiconductor doping[5]. Moreover, they constitute the basis for the quite ubiquitous pulsed laser deposition technique[6].

It is known in literature that hydrogenated materials are good sources of protons and heavy ions using infrared (IR) PLA setups[7-9]. IR-PLA is generally used for ion generation due to the broad use of Nd:YAG lasers in the laboratories. Moreover, at low/medium irradiances ion energy is known to scale as $I\lambda^2$, where I is the laser irradiance and λ the wavelength[10], hence the fundamental wavelength of Nd:YAG lasers is an efficient option to generate energetic ions. Furthermore, PLA performed with IR lasers generates ions with high charge states[11].

Although ultraviolet (UV) lasers induce plasmas with lower charge states, their high energy photons (4 - 5 eV) allow an efficient photoionization of the target material and direct bond breaking, leading to higher plasma densities[2]. For this reason, UV wavelengths are suitable for high flux ion and proton beams. In this work, we present the results of an UV LIS aimed at proton generation from TiH_2 targets. Our goal is to study the proton production by UV-PLA, comparing the results obtained with those already available.

[1] To whom any correspondence should be addressed.

2. Experimental setup

The experimental apparatus used was the PLATONE[12] setup. PLATONE is a LIS composed by a Compex 205 KrF excimer laser ($\lambda = 248$ nm, $\tau_{FWHM} = 23$ ns) and an electrostatic extraction system, consisting of an accelerating gap, as shown in Fig. 1. In particular, the extracting voltage (that could reach values up to 60 kV in DC mode) was applied to the first electrode, T+EC anode. In front of it, a second grounded electrode GE is placed. In this way, an intense electric field is obtained in the region EC-GE. The extraction apertures of EC and GE are coaxial and 1.5 cm in diameter. The laser beam was focused through a thin lens on the target surface. As targets we used compressed disks of TiH_2 powder, pure at 99%.

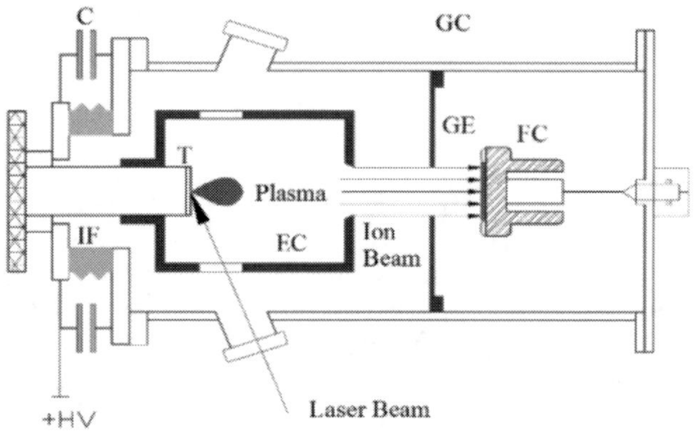

Figure1. Sketch of the experimental apparatus. IF: insulating flange, C: HV capacitors, GC: generating chamber, T: target support, EC: expansion chamber, GE: ground electrode, FC: faraday cup.

The disks were obtained by compression of 230 mg of TiH_2 powder at a pressure of 10^5 N/cm^2 for 30 minutes. This powder is relatively cheap (prices are around 1 k€ per kg) and easily available. As an example, from 1 kg of powder more than 4000 disks could be produced. Moreover, they will last for thousands of laser shots.

The targets, mounted on the support T, were irradiated in high vacuum (10^{-6} mbar) at different laser irradiances (0.6, 1.3, 2.5 and 5.1 GW/cm^2) and for different extracting voltages, ranging from 0 (free expansion) to +15 kV, in steps of 5 kV.

We characterized the resulting ion beams by means of a Faraday cup (FC), a 7.7 cm diameter aluminium disk, positioned at the right end of GC and connected to a digital oscilloscope. The FC lacked of any suppressor for secondary electrons, but the ion and proton yield obtained were corrected, taking into account secondary electrons emission (see Appendix). The total fly length available for ions, from T to FC, was 28.0 cm (17.5 cm of free expansion inside EC, 3 cm of acceleration between EC-GE and 7.5 cm between GE-FC of free drift).

3. Results

During nanosecond laser ablation, a high density plasma ($10^{17} - 10^{19}$ cm^{-3}) is obtained as a result of the laser-matter interaction. These plasmas are heated to high temperatures by the inverse bremsstrahlung and photoionization processes, expanding rapidly, perpendicularly to the target surface[13]. This expansion give rise to characteristic TOF signals when the ionic components impinge on Faraday cups. From these signals, numerous information could be obtained. During the experiments, we obtained well defined and separate TOF peaks both for protons and "Ti plasma" (Fig. 2). Applying an extraction potential, these peaks increased in amplitude, denoting a better charge extraction[14].

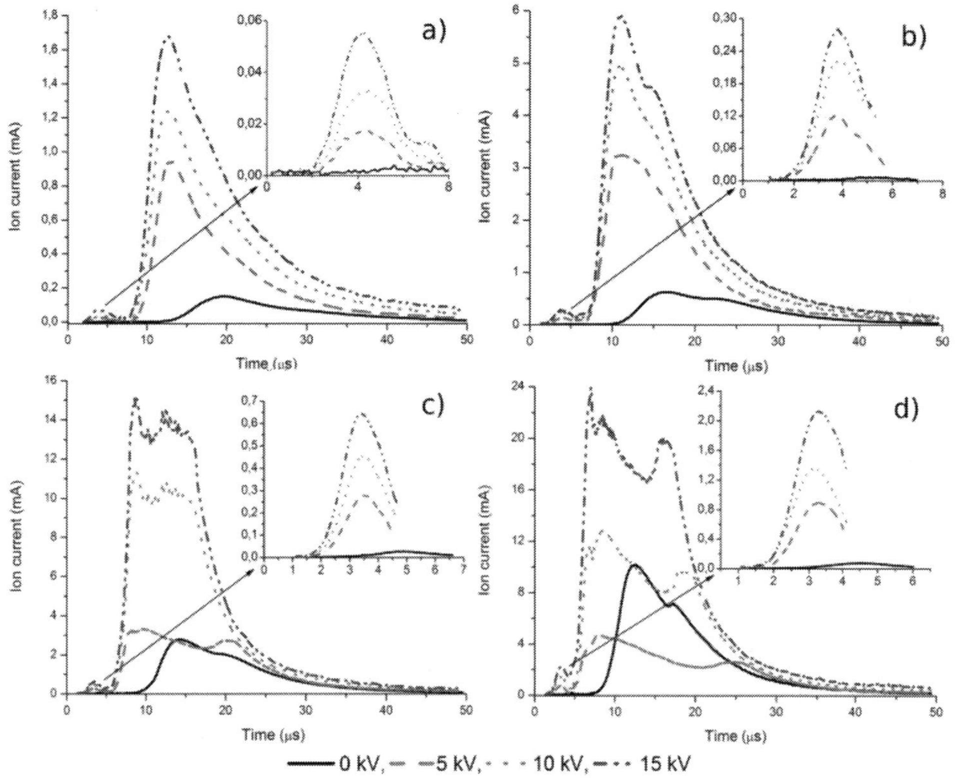

Figure 2. Ion currents recorded on FC by a digital oscilloscope at 0.6 (a), 1.3 (b), 2.5 (c) and 5.1 (d) GW/cm^2, for different accelerating voltages.

The fastest peaks (zoomed in Fig. 2) represent fast protons escaped from the main plasma plume. In order to obtain a proof of their correct identification with respect to possible impurities (Carbon, Nitrogen), we used an electrostatic barrier (EB) as particle analyzer. This device, able to select particles depending on their charge-to-mass ratio, was positioned inside GC, before FC, reducing the total fly length to 23.0 cm from T.

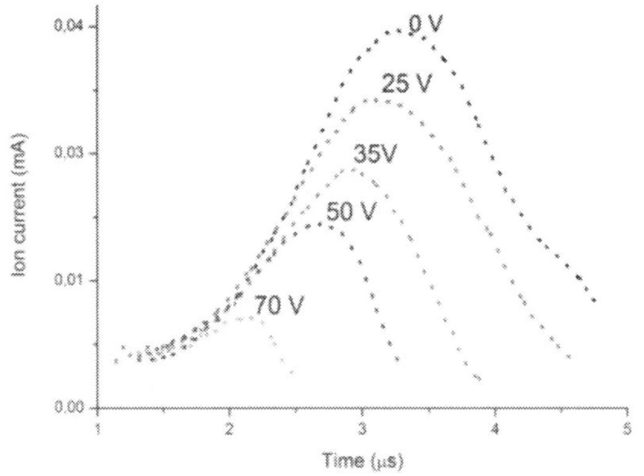

Figure 3. TOF proton currents for different stopping voltages applied to EB at 5.1 GW/cm^2 in free expansion.

In Fig. 3 are shown typical fast proton TOF signals for different stopping voltages V_b applied to EB, in order to halt particles with the charge and the mass of the proton. The precise value of V_b could be assessed using the following relation[15]

$$V_b = \frac{m L^2}{2 Z e t_{TOF}^2}, \tag{1}$$

where m is the ion mass (proton in our case), L is the fly length, Z is the ion charge state (1+ in our case), e is the elementary charge and t_{TOF} is the TOF value of the particle to be stopped. These results confirmed the correct identification of protons. For example at 5.1 GW/cm^2 in free expansion, 70 V were sufficient to reduce the major part of the proton peak, as shown in Fig. 3.

The amplitudes of the *Ti plasma* signals are sensibly higher than those of the fast protons. The former are indeed the result of the convolution of the signals of different charge states of Ti ions present in the plasma plume (in our case, the principal[16,17] is 1+) and protons trapped within the main plume. In effect, according to the target stoichiometry, one could expect a greater charge extraction for protons. Nevertheless, for example in free expansion at 0.6 GW/cm^2, the experimental results showed that the total charge obtained for the fast protons was 0.01 nC, while for Ti plasma was 2.32 nC, suggesting that a large share of protons is in the main plume. A similar behavior was observed also under the effect of the extraction potential, for all laser irradiances used. To confirm this circumstance, we performed a numerical deconvolution of the TOF signals, using the well known function introduced by Kelly and Dreyfus[18,19]

$$J(t) \propto \frac{L^2}{t^5} \exp\left\{ -\frac{m}{2kT_{KL}} \left(\frac{L}{t} - u \right)^2 \right\}, \tag{2}$$

where L is the fly length, k is the Boltzmann constant, while m, u and T_{KL} are respectively the mass, the center of mass velocity and the Knudsen layer temperature of the expanding specie. It is worth to note that we performed the deconvolutions only on the TOF signals obtained in free expansion, in this way we could safely neglect the effect of SEE, as shown in the appendix. The result obtained for free expansion at a laser irradiance of 0.6 GW/cm^2 is shown in Fig. 4.

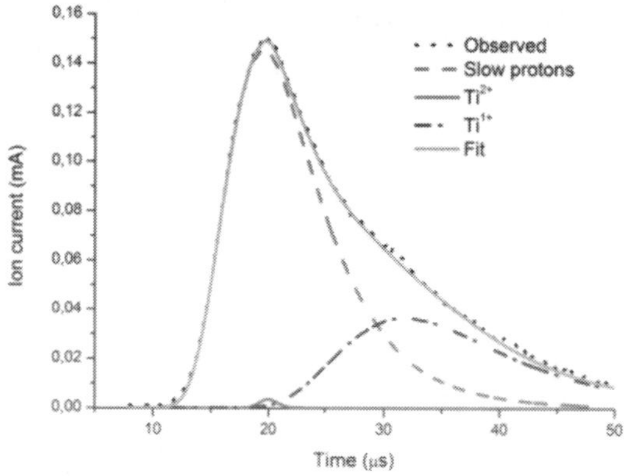

Figure 4. Deconvolution of the TOF signal obtained in free expansion at a 0.6 GW/cm^2 laser irradiance.

As it could be seen from the deconvolution of the TOF spectra, three distinct curves could be observed, corresponding to the contributions of protons, Ti^{1+} and Ti^{2+} ions. Computing their relative charges, we obtained 1.60 nC, 0.75 nC and 7 pC respectively. Using the computed values, the average charge ratio H/Ti obtained for the cases analyzed resulted 1.8 ± 0.3.

Using the TOF signals we obtained information about the fast protons kinetic energy in free expansion. Both maximum and minimum kinetic energy values were calculated using full width half maximum (FWHM) time values (Fig. 5).

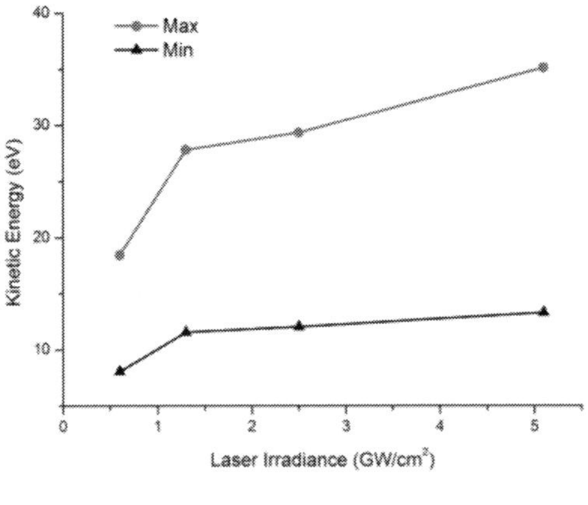

Figure 5. Maximum and minimum proton kinetic energy in free expansion mode on the laser irradiance.

The energy spread at FWHM, calculated using the relation

$$\sigma_E = \frac{E_{max} - E_{min}}{E_{max}} \times 100, \tag{3}$$

varied from 56% (at 0.6 GW/cm^2) to 62% (at 5.1 GW/cm^2).

For what concerns the charge of the fast protons, integrating the corresponding TOF signal we obtained values up to 3.52 nC (at 5.1 GW/cm^2 and 15 kV) per laser shot. Using these values we computed the proton yield per pulse, shown in Fig. 6.

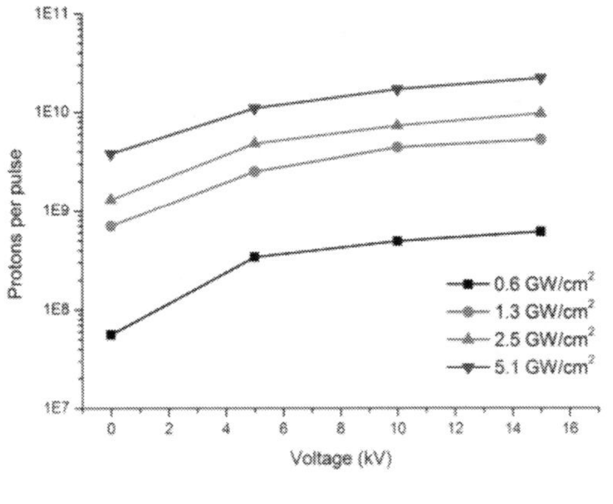

Figure 6. Protons per pulse depending on the accelerating voltage for different laser irradiances, in logarithmic scale.

The data of Fig. 6 stress a general behavior of the extraction system. At the lowest laser irradiance, the application of the extracting potential significantly increases the proton yield; while, as the irradiance increases, the effect of the potential decreases. This arise from the fact that the extracted current depends on the plasma density near the meniscus[20,21] formed at the extraction electrode (EC). The lower the plasma density, the higher is the efficiency of the extraction electrode, since the plasma is less effective in screening the extracting electric field.

The maximum extraction of protons was reached at the maximum value of laser irradiance and extracting voltage. This depends on the fact that an higher irradiance induce an higher plasma density available for extraction near the meniscus.

It is worth to note that these results were very stable and reproducible at each shot.

4. Conclusion

This work shows the potentiality of the use of UV lasers to realize high yield proton sources from TiH_2 targets. Indeed, the use of these targets is very interesting not only in IR-PLA setups, as already shown in literature[7-9], but also in the UV ones. A comparison with the results just cited is presented in Table 1, where the corresponding proton yield is shown together with some relevant experimental parameters. These results show an increased proton yield, particularly if compared to the laser irradiances used.

Table 1. Comparison of proton yield with those available in literature; in parenthesis is shown the value for the protons trapped within the main plume

	Target	Laser wavelength (nm)	Laser irradiance (W/cm^2)	protons/pulse
Current work	TiH_2	248	5×10^9	3.8×10^9 (3.8×10^{11})
Sekine et al.[8]	MgH_2	1064	2×10^9	4.0×10^8
Sekine et al.[8]	ZrH_2	1064	1×10^9	7.1×10^8
Torrisi et al.[9]	TiH_2	1064	2×10^{10}	2.8×10^9

As shown, it is possible to enhance the proton yield both increasing the laser irradiance and the extracting voltage. Indeed, we obtained fast proton bunches with fluxes up to 10^{10} proton/pulse and this represent an enhancement with respect to the literature. Moreover, applying a magnetic particle filter, it is possible to extract also the protons trapped in the main plasma plume. Their yield, shown in Table I within parentheses, is notable and could be even increased by means of the extracting potential. Further work will deserve more attention to the main plume composition and to comparative analysis with other hydrogen-rich targets.

As a general conclusion, we can assert that LIS aimed at low energy proton production using TiH_2 are more effective using lasers in the UV range. Indeed, these sources are intended to provide protons to devices that will accelerate them to substantially higher energies. In this context it is important to obtain high fluxes, in order to reduce the effects of charge losses during the beam transport. As shown, high fluxes can be obtained by taking advantage of the higher ionization fraction induced by the UV radiation[2].

Appendix

It is worth to note that the FC used was lacking of any suppressor to prevent secondary electron emission (SEE) due to ion impact. We know that this could lead to misleading estimation of quantities. It is known that the main parameters that determines the yield of SEE are the ion energy and its charge state[22]. In the experiments under exam we dealt with low ion energies, obtained as the result of the applied accelerating voltage (for a maximum of 15 kV). Moreover, as stated above, the UV laser is known to induce ion with low charge states; consequently due both to the laser used and to the target composition we dealt mainly with singly charged particles.

Another key parameter in the yield of SEE is the presence of oxides on the surface of the FC. These impurities, indeed, are known to greatly increase this undesired effect. For this reason, before any measurement we performed a prior degassing together with a cleaning process through sputtering of ions on the FC surface.

In these conditions it is reasonable to expect that SEE is low, although not negligible. Consequently, using a suppressor, we estimated the SEE yield. The effect of SEE yield on the charge Q collected on FC could be represented by the equation

$$Q_{measured} = (1+\gamma)Q_{ions}, \tag{A.1}$$

where γ is the SEE coefficient. In particular, restricting the measures to the fast protons, we found that coefficient is lower with respect to the main plasma, owing to the higher kinetic energy of Ti ions. In Table A1 are presented the values obtained for γ. The values of γ have been used to correct the yields that we deduced from the experimental data.

Table A1. Values obtained for the SEE coefficient.

Accelerating Voltage (kV)	γ for fast protons	γ for main plasma
0	0.01	0.01
5	0.06	0.25
10	0.10	0.41
15	0.13	0.50

References

[1] Belloni F, Doria D, Lorusso A, and Nassisi V 2004 *Rev. Sci. Instrum.* **75** 4763
[2] Torrisi L, Gammino S, Andò L, Nassisi V, Doria D, Pedone A 2003 *Appl. Surf. Sci.* **210** 262
[3] Okamura M, Pikin A, Zajic V, Kanesue T and Tamura J 2009 *Nucl. Inst. and Meth. in Phys. Res. A* **606** 94
[4] Noli F, Lagoyannis A and Misaelides P 2008 *Nucl. Inst. and Meth. in Phys. Res. B* **266** 2437
[5] Rosinski M, Badziak B, Parys P, Wołowski J and Pisarek M 2009 *Appl. Surf. Sci.* **255** 5418
[6] Chrisey D B and Hubler G K 1994 *Pulsed Laser Deposition of Thin Films* (John Wiley and Sons, New York)
[7] Torrisi L, Caridi F and Giuffrida L 2011 *Laser Part. Beams* **29** 29
[8] Sekine M, Kondo K, Okamura M and Hayashizaki 2012 N *Rev. Sci. Instrum.* **83** 02B318
[9] Torrisi L, Cavallaro S, Cutroneo M, Margarone D, and Gammino S 2012 *Rev. Sci. Instrum.* **83** 02B310
[10] Láska L, Jungwirth K, Krása J, Pfeifer M, Rohlena K, Ullschmied J, Badziak J, Parys P, Wolowski J, Gammino S, Torrisi L and Boody F P 2005 *Appl. Phys. Lett.* **86** 081502
[11] Láska L, Jungwirth K, Králiková B, Krása J, Pfeifer M, Rohlena K, Skála J, Ullschmied J, Badziak J, Parys P, Wolowski J, Woryna E, Gammino S, Torrisi L, Boody F P and Hora H 2003 *Plasma Phys. Control. Fusion* **45** 585
[12] Nassisi V, Delle Side D and Velardi L 2013 *Appl. Surf. Sci.* **272** 114
[13] Capitelli M, Casavola A, Colonna G and De Giacomo A 2004 *Spectrochim. Acta B* **59** 271
[14] Lorusso A, Siciliano M V, Velardi L and Nassisi V 2010 *Appl. Phys. A* **101** 179
[15] Nassisi V and Pedone A 2003 *Rev. Sci. Instrum.* **74** 68
[16] Velardi L, Siciliano M V, Delle Side D, and Nassisi V 2012 *Rev. Sci. Instrum.* **83** 02B717
[17] Belloni F, Doria D, Lorusso A, Nassisi V and Krása J 2006 *Rev. Sci. Instrum.* **77** 03B301
[18] Kelly R and Dreyfuss R W 1988 *Surf. Sci.* **198** 263
[19] Krása J, Jungwirth K, Krouský E, Láska L, Rohlena K, Pfeifer M, Ullschmied J and Velyhan A 2007 *Plasma Phys. Control. Fusion* **49** 1649
[20] Brown I G (editor) 2004 *The Physics and Technology of Ion Sources*, 2nd Edition (Wiley-VCH Verlag GmbH & Co. Weinheim)
[21] Wilson R G and Brewer G R 1973 *Ion Beams* (John Wiley & Sons, New York)
[22] Sternglass E J 1957 *Phys. Rev.* **108** 1

Intraocular lens employed for cataract surgery

A M Roszkowska[1] and L Torrisi[23]

[1]Corneal Physiopathology and Refractive Surgery Service, U.O.C. di Oftalmologia, Azienda Ospedaliera Universitaria, Messina, Italy
[2]PhysicsDepartment and S.d.T., Messina University, V.le F.S. D'Alcontres 31, 98166 S. Agata, Messina, Italy

E-mail: Lorenzo.Torrisi@unime.it

Abstract. The aim of this paper is to illustrate the techniques of cataract surgery with implantation of intraocular lenses and some physical properties of the used materials. The new technology, coupled with extensive experience and the studied cases, permits to increase the standardization and accuracy of the engravings, by reducing the use and handling of surgical instruments inside the eye. At present it is possible to replace the cataract with crystalline lenses based on biopolymers such as PMMA, silicone, acrylic hydrophilic and hydrophobic acrylic. These materials are increasingly able to replace the natural lens and to ensure the fully functional of the eye. The role of femtosecond lasers in cataract surgery, to assist or replace several aspects of the manual cataract surgery, are discussed.

1. Introduction

The crystalline lens is a convex lens, flattened anterior-posterior direction, located behind the iris, suspended with a round ligament complex. Along with the cornea, it composes the main part of the ocular dioptre. Fundamental properties of the lens are transparency and dioptre power. Exclusive feature is the ability to change its dioptre power in order to focus on the retinal surface images from nearby objects. This ability called "accommodating" occurs with the change of curvature of the lens, especially, of the front face, and is a reflection of the focus-induced retinal images that determines the ciliary muscle contraction with the simultaneous convergence of bulbs and miosis.

The term is defined as any cataract lens opacity can result in a loss of transparency and a diffraction and scattering of light. Cataract is the most common among ocular pathologies. Its symptoms include a gradual reduction of visual acuity, altered colour perception, emergence of refractive defects and glare [1]. Fig. 1a shows a photo of an eye with cataract. The formation of cataracts is a para-physiological process that determines a slow and progressive crystalline opacity. There is no medical treatment to stop this deterioration and the only possible therapy is surgical. The technique involves the removal of the crystalline lens and replacing it with an artificial one that can correct visual defects as well as pre-existing myopia, hyperopia, astigmatism and presbyopia.

The intervention is completely painless, lasts approximately twenty minutes, in not complicated cases, and runs with a relatively simple and painless procedure. Actually, the technology presents in

[3] To whom correspondence should be addressed

the operating room, and microsurgical operations for the surgery, requires precision and grace in the order of tens of microns for maximum post-operative result [2].

The growing development of artificial crystalline increasingly reliable, careful patient selection and the prior study of the individual case allow us to get a high reliability of the post-operative results, managing to simultaneously fix most defects existing visual as: myopia, astigmatism and hyperopia. Fig. 1b shows an artificial lens implanted in eye.

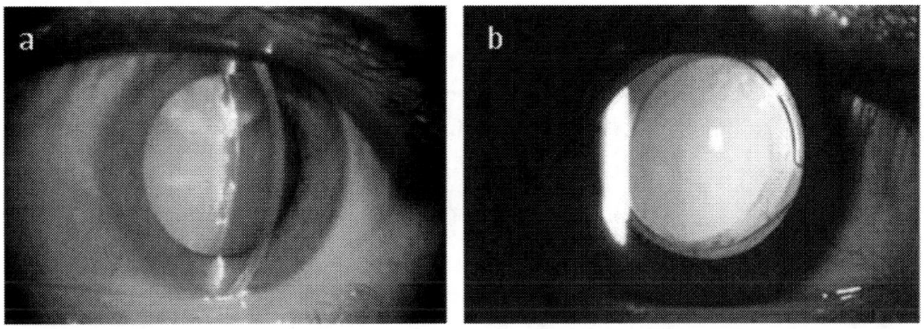

Figure 1. Eye with cataract (a) and eye with artificial crystalline implant (b).

2. Materials and methods

The cataract surgery is preceded by specialists visit and functional tests to ensure preventive health and the characteristics of the eye, to assess their capabilities and determine the exact conditions for deciding the most appropriate intervention. Some of the preliminary examinations required are:

- *Examination of visual acuity*: it is used to determine how many tenths you can see with best optical correction. Statistically a person in full welfare reaches 10/10 (ten tenths) of visual acuity.
- *Bio-microscopic examination* with a slit lamp: high magnification to examine the anatomy of the eye and cataract characteristics.
- *Evaluation of the fundus oculi*: specific examination of the retina and the optic nerve.
- *Numerical count and morphology of corneal endothelial cells*: when there are doubts about the number and viability of cells that make up the posterior layer of the cornea.
- *Tonometry:* measurement of rare eye pressure.
- *Ultrasonic/laser biometry,* to measure the length of the eyeball. It permits to calculate and customize the template and the optical power of the artificial lens to be implanted during surgery.
- *Stratigraphy OCT (Optical Coherence Tomography)* of the macular region: indicated for the presence of retinal and macular diseases (only in selected patients.)

Currently there are two basic categories of intraocular lenses (IOL), rigid and soft ones, made of various types of materials and with various shapes and sizes with different optical characteristics.

The **rigid lenses** are made of poly-methyl-methacrylate based on PMMA monomer $(C_5O_2H_8)_n$, a special plastic similar to perplex. They have the advantage of being easy to build and to have a low cost. They have a big drawback of requiring a large incision to be introduced into the eye and to apply several stitches, healing times become longer and, therefore, also the period for full recovery.

The **foldable lenses** made of silicone or acrylic material, based on dimethyldioxane monomer $(C_2H_6OSi)_n$, which can be more hydrophilic or hydrophobic requiring more complex techniques and expensive but they have the great advantage to be inserted through tiny incisions that do not require stitches and allowing the patient anatomic and visual rehabilitation very fast.

Modern cataract surgery requires, without doubt, the implantation of an artificial crystalline lens. PMMA is a material that at body temperature shows a glass-like appearance and is stiff and brittle, based on a polymer structure to individual chains closely joined together. It has a perfect transparency to visible light and a low optical dispersion index. It is hydrophobic and this can result in unwanted tack with endothelial cells and inflammatory cells in aqueous reducing the corneal transparency. A typical transmission spectrum measured in folding IOL is shown in Fig. 2 for PMMA (a) and (b) Acrylic [3]. The modern IOL have about 5 mm optic diameter and have a smooth texture to be folded and inserted into the eye through an incision of only 2-2.5 mm that normally it does not require stitches. Silicone has the advantages of having low costs of production, to be sterilized in autoclaves, high biocompatibility and it is very soft. Silicone doesn't produce endothelial damage but it has the disadvantages to decentralize the lens, to have a low refractive index, to be not very thin and to be chemically sensitive to IR radiation. Instead, acrylic based lens have the advantages to have high refractive index and thus to have little thickness, to be absorbent in the UV region, to have a wetting angle variable and to have high flexibility to be easily implanted. Acrylic lens has the disadvantages to be more expensive.

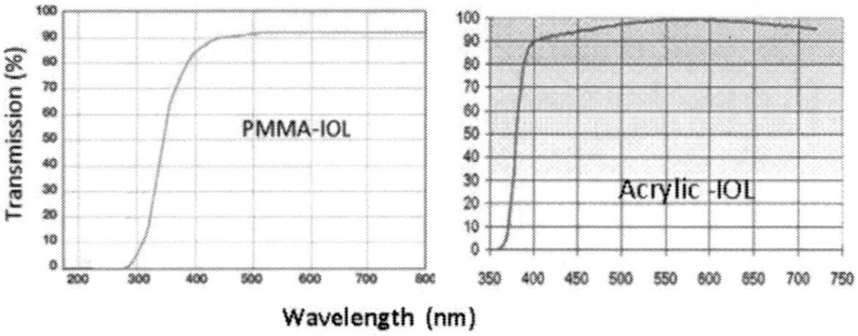

Figure 2. Optical transmission of a typical PMMA intraocular lens (IOL) (a) and of an Acrylic based IOL (b).

There are different kinds of IOL with design and different physical and optical characteristics: monofocal, multifocal and accommodative, toric, aspheric, apodizzated and UV filters. Fig. 3 shows some typical design of IOL lenses. Routinely incisions are created manually by introducing a sharp, sterile blade into the cornea. The use of a manual blade makes it difficult to control the length and architecture of the incision, which may affect the stability of the wound under pressure. Following the surgery this could manifest by the corneal wound leaking increasing the potential risk of infection.

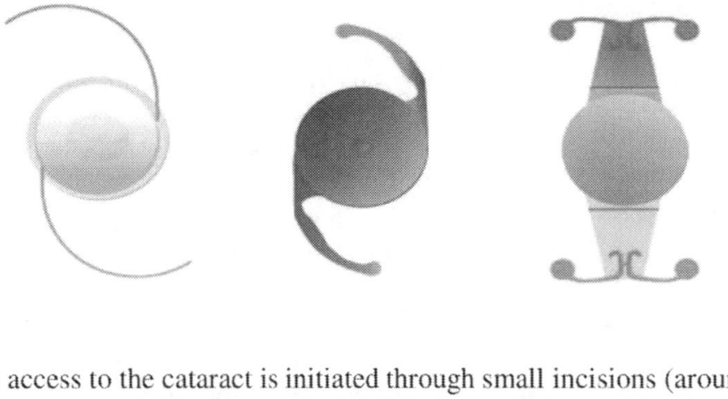

Figure 3. Three typical designs of IOL elastic lens based on silicone or acrylic.

The access to the cataract is initiated through small incisions (around 2mm in length) made into the peripheral cornea zone. Known as Clear Corneal Incisions (CCI) they remain the preferred method for surgeons accessing the anterior chamber of the eye during cataract surgery. Previously large incisions (approximately 5-7mm in length) were made into the sclera (white part of the eye). This is a highly vascularized part of the eye and which required particular consideration for older patients on blood thinning therapy. To ensure that the incisions were adequately sealed, sutures were required following surgery. The need for sutures prolongs the eye recovery process. The CCI surgery is micro invasive and can be considered to be self-sealing (that means no sutures are required).

3. Femtosecond lasers

A femtosecond laser is a laser which emits optical pulses with a duration well below 1 ps (*ultrashort pulses*), i.e., in the domain of femtoseconds (1 fs = 10^{-15} s). It thus also belongs to the category of ultrafast lasers or ultrashort pulse lasers working in mode locked in near infrared light to create precise subsurface cut (photodisruption of deeper tissues).

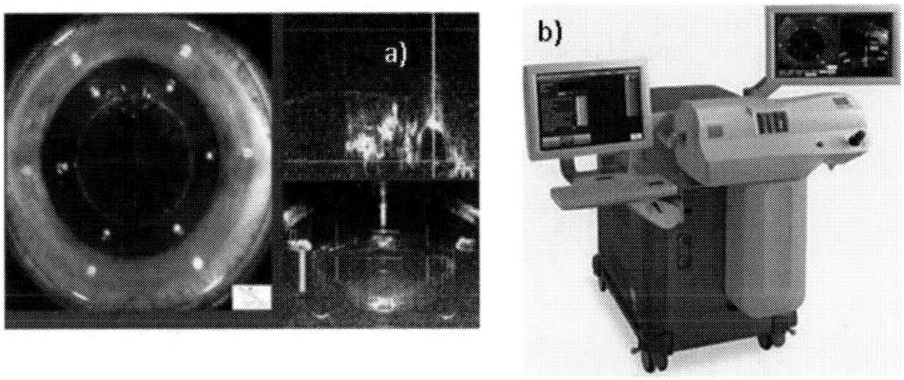

Figure 4. Photo of a phase during cataract procedure laser surgery (a) and femtosecond laser with OCT eye imaging (b).

The role of femtosecond lasers in cataract surgery is to assist or replace several aspects of the manual cataract surgery. These include the creation of the initial surgical incisions in the cornea, the creation of the capsulotomy, and the initial fragmenting (breaking up) of the lens, as indicated in Fig. 4a for a new generation surgery. The femtosecond laser may also produce incisions within the peripheral cornea to aid the correction of pre-existing astigmatism. The preliminary results are very promising. In cataract surgery *fs* laser cutting is performed inside and on the surface of the crystalline lens. Since lens position and orientation differs from eye to eye, precise 3-dimensional imaging is required to define location of the lens capsule in order to properly apply the laser cutting patterns. Image-guided laser cataract surgery is based on the use of integrated Optical Coherence Tomography (OCT). A typical instrumentation is reported in Fig. 4b. Laboratory studies have shown that the femtosecond laser produces consistent and stable incisions in the eye. This has been attributed both to the controlled, reproducible generation of the incisions and the configuration of the corneal wound created. It would suggest that laser-cut, clear corneal incisions offer the potential of less risk of wound leak thereby further reducing the risk of infection following surgery. In addition to creating the main cataract surgical wound, femtosecond lasers are used to create limbal relaxing incisions (LRIs). LRI microsurgery can reduce the amount of astigmatism presents in the eye and often improves the post-operative uncorrected visual acuity. Typically, corneal incisions to control astigmatism have been performed by hand held blades. The major factors determining their effectiveness are the incision length, depth, and uniformity, the location relative to the centre of the cornea, the incision geometry, the patient's age, and the amount of astigmatism. The precision of the laser created incisions should allow the surgeon greater control over the final refractive endpoint possibly leading to improved visual outcomes. Again further data is required at this point to confirm these assertions.

4. Results and discussion

The treatment of cataracts is surgical and it involves only the removal of the lens replacement and implantation of artificial crystalline lens. Today it runs in surgery, under local anesthesia, no stitches and it allows a good visual recovery as early as next day. It is a technologically advanced intervention, which requires an optimal experience of the surgeon and his team but also of a modern and perfect instrumentation. The latest techniques for micro-surgical cataract surgery are using mini-invasive tools that help the surgeon to break up and remove the opaque lens nucleus without damaging the delicate structures inside the eye. With the aid of a microscope, the operator inserts a small tube with a diameter of 0.8 mm. which emitting ultrasound, sputter crystalline fibres allowing their full suction. This technique is called "ultrasonic phacoemulsification" [4]. Fig. 5 shows a scheme of phacoemulsification with the implantation of an intraocular lens. This technique allows you to remove the cataracts through an opening of only 2.75 mm. Through the same aperture is then injected a foldable artificial lens, which once positioned opens like a flower definitively at the headquarters levels.

a)

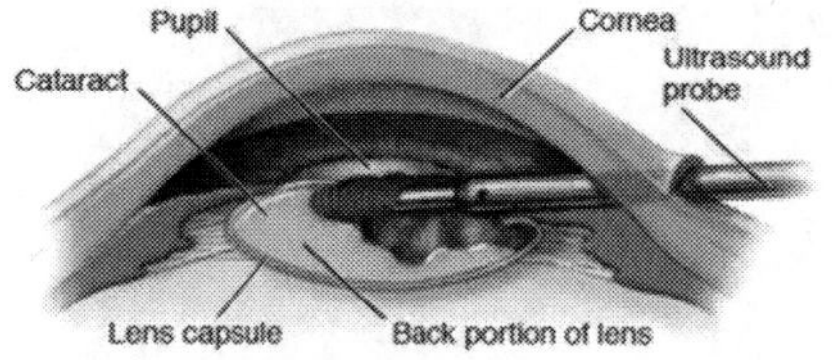

b)

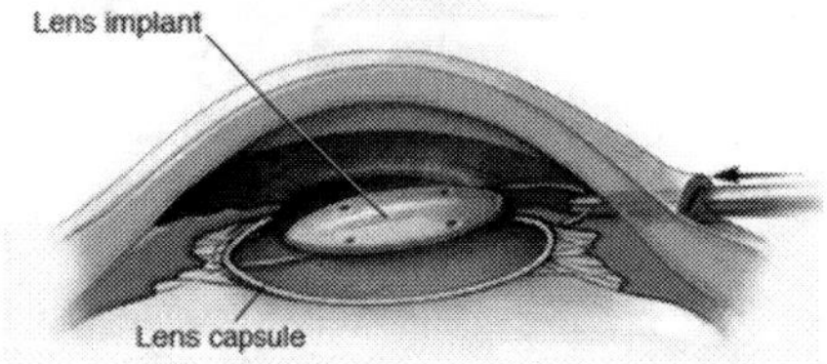

Figure 5. Scheme of phacoemulsification (a) with implant of an intraocular lens (b).

The phacoemulsification aims to respect the natural anatomy of the eye, even with a valve incision, which in most cases does not require suturing minimizing astigmatism and inconvenience to the patient. In some phacoemusificators it is used a ultrasound probe that revolves on its axis both longitudinal and torsional (OZIL), it helps the surgeon in the delicate fragmentation of the nucleus before his aspiration.

The spread in cataract surgery with micro necessitated the use of IOL with optical big dish, can be implanted through the tunnel after being bent and relax once placed inside the eye. Silicone and acrylic based IOLs can be folded and implanted through a cut of between 1.8 mm and 3.2 mm. The acrylic IOL can be hydrophobic and hydrophilic, in relation to the amount of water present in their structure [5].

The hydrophobic IOL showed no tendency to become cloudy, although fingerprints have been described left on the optical plate by bending and insertion tool, and the possibility of formation of bright particles inside the optical plate. The silicone IOL showed the possibility of yellowing and hydrophilic acrylic IOL to become cloudy for the accumulation of calcium and phosphates on the surface and in the substance of the optics of the lens.

When the cataract is removed, it reappears in 20% of cases within a range going from few months to about one year after surgery and the vision may return to decrease for the emergence of so-called secondary cataract. Secondary cataracts is caused by posterior capsule opacity that is claiming support the lens and now it keeps in place the new artificial crystalline lens. This fabric over time can lose its natural transparency decreasing the vision. Several years ago it was necessary to return to the operating room. Today this can be easily treated on as outpatients, with no hassle and permanently with the help of Yag laser to perform so called posterior capsulotomy by disruption of the masses. This laser emits bursts of 100 mJ at 1064 nm IR field with duration of nanosecond that can be focused on the opaque posterior capsule to vaporize when giving their energy. However, radiation is transparent in thin IOL. The use of laser light in cataract surgery, for a variety of purposes, it is becoming increasingly widespread. Next to the worldwide used ultrasonic phacoemulsification in cataract surgery, actually the application of the femtosecond lasers become true. Such devices use energetic pulses with duration of about 10 fs, used in repetition rate, to perform the anterior

capsulorhexis and to crush the cataract material. Near infrared impulses "ablate the tissue without heat which makes it even less invasive ocular surgery. The cataract operation is completed by removing the masses by a surgeon. After the cataract operation eyesight recovery is usually very fast. This method simplifies the cataract surgery in some selected cases however it cannot be used in advanced cataracts or in presence of corneal opacity.
After the cataract surgery can use a laser light for shaping the customizable or intraocular lenses hi-tech. In this case the laser light model intraocular lenses to meet the Visual requirements of the patient and correct vision defects remains after cataract surgery. From this point of view in physics research relating to the field of laser-tissue interaction is helping the clinic to improve more and more results. Studies on laser ablation of IOL lenses made of PMMA, Silicone and Acrylic resins are increasing in actual literature.

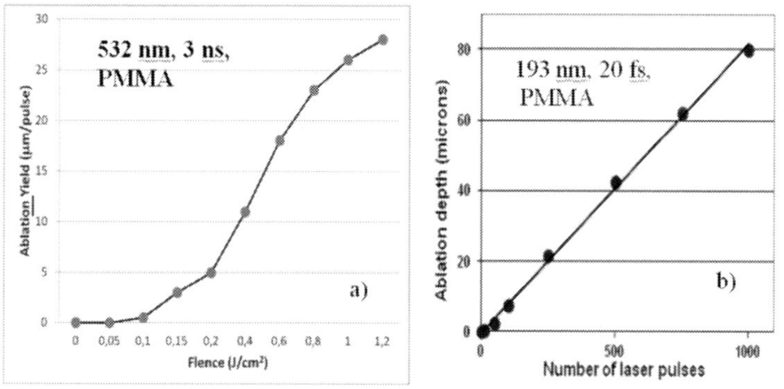

Figure 6. Ablation yield of IOL-PMMA as a function of the laser fluence at 532 nm wavelength and for 100 laser shots (a) and Ablation depth in PMMA produced by fs laser pulses at 193 nm wavelength and at 18 J/cm^2 fluence (b).

Fig. 6a shows an example of the result obtained at the Physics Department of the University of Messina for the ablation of PMMA as a function of fluence with a Nd: Yag laser at 532 nm second-harmonic wave-length and 3 ns pulse. Measurements were performed by measuring the ablation mass per 100 laser shots. The ablation shows a threshold value, of about 100 mJ/cm^2, beyond which the ablation increases nearly linearly with the laser fluence up to values of about 1 J/cm^2. After this value the ablation yield tends to saturate [6]. Fig. 6b shows the ablation yield in PMMA-IOL, in terms of removed depth vs. number of laser pulses, produced by *fs* laser shots at 18 mJ pulse energy, 193 nm wavelength, 0.1 mm^2 spot, and 18 J/cm^2 laser fluence (b).

The new technology, coupled with extensive experience operating casuistry and of more than 24,000 patients operated in the last 20 years, allows us today to further increases the standardization and accuracy of engravings, 40% reduction of the use and handling of surgical instruments inside the eye. In summary the new cataract surgery involves less trauma, more precision, more standardization and faster visual recovery and predictable.

The femtosecond laser is able to create a near perfect, round opening in the anterior capsule by dissecting it with a spiral laser pattern crossing the anterior capsule. To avoid distortion of the incoming laser beam on gas bubbles and tissue fragments, the spiral pattern is applied first posterior to the capsule, and advances anteriorly. The surgeon then is able to simply remove the capsule with surgical forceps. This has several potential benefits over the radially-created manual procedure. In conventional cataract surgery the incidence of anterior capsular tears has been documented from 0.79% in very experienced hands to 5.3% within teaching institutions.

The capsulotomy may also impact on the visual outcome. An irregularly-shaped capsulotomy may influence the position of the implanted IOL leading to decentration and tilt which may cause a decrease in the patient's quality of vision. The ability to create a precise, well-centered capsulotomy should therefore optimize the surgeon's ability to achieve the patient's anticipated visual outcome. Measurements have shown that in comparison to manually created capsulotomies those created with a laser have significantly lower internal aberrations following surgery. The femtosecond laser has the capability to assist the fragmentation (breaking up) of the cataract. The laser applies a number of pulses to the lens in a pre-designed pattern which then allows the surgeon to use current technology to remove the lens matter. This additional step has been shown to reduce the average time and energy required to break up and remove the lens by approximately 50%. Inherently this should make the overall procedure safer and less traumatic to the eye, which may further reduce the risk of postoperative swelling and lead also to a faster visual recovery. The ability of surgeons to provide significant advantages to patients at risk of major intraoperative and postoperative complications will continue to increase and ultimately it may represent a substantial benefit of femtosecond technology. This may prove to be the greatest potential advantage of laser-assisted cataract surgery although data significant to provide a clinical benefit is not available at this time in point.

5. Conclusions

For the cataract surgery a key parameter is represented by a careful selection of the most suitable intraocular lens. Usually you run several tests before you can locate intraocular lenses for each patient. Thanks to scientific research much boost in this area to usable materials, the means and the technologies available today to the ophthalmologist, there are now several types of intraocular lenses and research and innovation continue to make great strides in this area.

The Standard IOL can be employed for near and far view. However, under conditions of low contrast or night vision the vision quality is not optimal.

The Aspheric IOLs allow a better quality of night vision compared with traditional intraocular lenses. The Toric IOL can correct astigmatism.

The hi-tech IOL can be customized with light in selected cases that provide the best view possible without glasses. The Multifocal IOL that have two or more focal points allows you to see well from far and from near after cataract operation.

The Accommodative IOL through a special form of the lens restores the natural mechanism of focus to close that generally is lost with the age.

The IOL for macular holes (OriLens and LMI Lipshitz Macular Implant), ideal for those suffering from cataracts and macular holes, it's a telescopic lens system (intraocular telescope) that allows you to enlarge the image.

The use of last generation *fs* lasers permits to resolve the cataract problems using a less invasive surgery a very accurate instrumentation which permits clear and very controllable corneal incisions, precise capsulotomy, accurate phacofragmentation of the cataract and to resolve complications reducing the pressure on the eye and the risk for the patient.

In conclusion, the cataract surgery, thanks to the new instrumentation development, has become a clinical procedure less invasive and always with appreciable results that significantly improves the quality of life of patients.

References

[1] R. F. Steinert, "Cataract Surgery", 3rd Edition, Expert Consult - Department of Ophthalmology, University of California, Irvine CA, by Roger F. Steinert, MD, 2013
[2] J.P. Gills, R. Fenzl, R.G. Martin, "Cataract surgery: the state of the art", SLACK, 1998
[3] L. Torrisi, A. Italiano, E. Amato, F. Caridi, M. Cutroneo, C. A. Squeri, G. Squeri & A. M. Roszkowska, Rad. Eff. And Def. In Solids 167(9), 2012, 1-10
[4] P.G. Ursell, D.J. Spalton, M.V. Pande, E.J. Hollick, S. Barman, J- Boyce, K. Tilling Journal of Cataract & Refractive Surgery 24 (3), 1998, 352–360
[5] C. Scolaro, L. Torrisi, M. Cutroneo, F. Caridi, A.M. Roszkowska and E. Pedullà "A Liquid contact angles on biocompatible Surfaces" Proc. III Work. On Plasma, Sorgenti, Biofisica e Applicazioni, Lecce 19 ottobre 2012, 38-42
[6] L. Torrisi, A. Borrielli and D. Margarone Nucl. Instr. And Meth. In Phys. Res. B 255, 373-379, 2007

Research on laser-removal of a deuterium deposit from a graphite sample

M Kubkowska[1], E Skladnik-Sadowska[2], K Malinowski[2], M J Sadowski[1,2], M Rosinski[1] and P Gasior[1]

[1] Institute of Plasma Physics and Laser Microfusion (IPPLM), 01-497 Warsaw, Poland
[2] National Centre for Nuclear Research (NCBJ), 05-400 Otwock, Poland

E-mail: Elzbieta.Skladnik-Sadowska@ncbj.gov.pl

Abstract. The paper presents experimental results of investigation of a removal of deuterium deposits from a graphite target by means of pulsed laser beams. The sample was a part of the TEXTOR limiter with a deuterium-deposited layer. That target was located in the vacuum chamber, pumped out to 5×10^{-5} Torr, and it was irradiated with a Nd:YAG laser, which generated 3.5-ns pulses of energy of 0.5 J at $\lambda_1 = 1063$ nm, or 0.1 J at $\lambda_3 = 355$ nm.

PACS: 52.38.-r, 52.38.Mf, 52.70.Nc, 81.05.Uw, 82.50.Bc, 52.40.Hf

1. Introduction

It is well known that graphite tiles or carbon fibre composites (CFC) are often used in various tokamak facilities [1-3]. During plasma discharges performed with the pure deuterium-gas filling, some portion of deuterium is deposited and occluded upon surfaces of the inner armour of the tokamak chamber. This deuterium is partially removed during successive tokamak discharge, but some portion remains in the exposited parts for a long time. In fact, in addition to deuterium, there are deposited also some heavy impurities (e.g. Si, B, Ni, Cr, Fe and W species), which originate from eroded surfaces of the tokamak armour [4].

In order to induce a decomposition and removal of the deposited layers, some efforts were undertaken to use repetitive laser pulses [5-7]. Some preliminary studies with the use of a Nd:YAG laser, which were undertaken at the IPPLM, gave positive results [8], but information about efficiency of the deuterium removal was rather scarce. On the other hand, the massive accumulation of fusion fuel (i.e. deuterium and tritium) in carbon-containing inner parts of the tokamak chamber might have a serious impact on the safety and economy of the tokamak operation, particularly in future ITER experiments [9]. Therefore, the main aim of the reported research was to investigate the efficiency of a laser-removal of deuterium from a graphite sample, which was earlier exposed to deuterium discharges in a tokamak.

2. Experimental setup

The investigated target constituted a deuterium-deposited graphite sample (Toyo Tanso, IG-430U) cut out from a TEXTOR toroidal limiter, which was earlier exposed to a series of discharges performed with the pure deuterium filling. That target was placed in the centre of the quasi-spherical vacuum chamber equipped with a pumping system and several operational or diagnostic ports. The vacuum chamber was initially pumped out to the final pressure of 5×10^{-5} Torr.

One of the side ports, oriented at an angle of about 60^0 to the target surface, was equipped with an optical window used to let in pulsed laser beams from a Nd:YAG laser, which could generate 3.5-ns pulses of energy of 0.5 J at λ_1 = 1063 nm, or 0.1 J at λ_3 = 355 nm. The diagnostic port oriented perpendicular to the target surface was connected with an electrostatic ion energy analyzer equipped with the deflection system consisted of two coaxial metallic cylinders. Ions, which reach the detector (photomultiplier), were separated on the path of flight due to their energy and mass spread. Another diagnostic port, oriented at an angle of about 60^0 to the target surface in the opposite direction than the laser beam axis, was used for measurements with an ion collector coupled with a fast oscilloscope. The detector constituted a Faraday cup which was applied for measurements of ion current in a stream of charged particles. These two ion diagnostic tools were similar to that used during the previous preliminary studies [8], but in the reported study the use was also made of a Mechelle®900 spectrometer in order to apply the optical emission spectroscopy (OES) technique. The scheme of an experimental setup is presented in Fig 1.

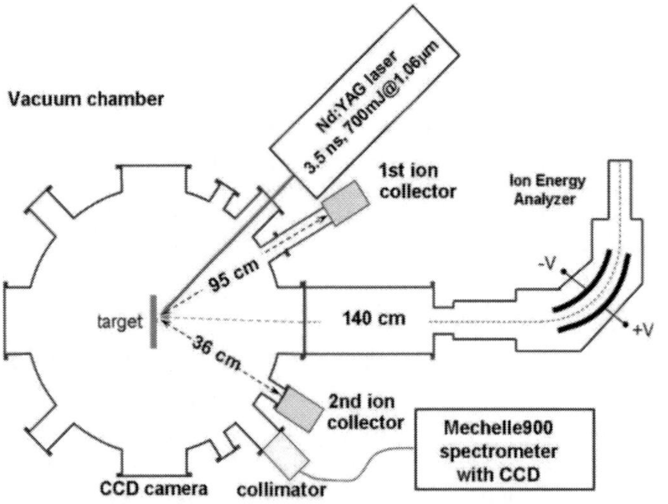

Figure 1. Experimental setup used for studies of C-targets irradiated by laser pulses.

As shown in Figure 1, in order to record optical emission spectra the use was made of another diagnostic port. It was equipped with a quartz window and an optical collimator, which was coupled with the Mechelle®900 spectrometer through a quartz-fibre cable. The spectrometer recorded optical spectra in the wavelength range of 300–900 nm by means of a PCO-SensiCam camera coupled with a computer and an appropriate spectroscopic software [10-11].

3. Experimental results

The reported experiments were started by taking time-integrated pictures of a plasma plume produced from the target irradiated by a single laser pulse. Conditions of the laser focusing were changed by applying lenses of f = 10-70 cm focal length, placed at a distance l_f = 10-75 cm from the target. Corresponding time-integrated pictures were compared. Examples of such pictures are shown in Fig. 2. The applied f = 10 cm lens with a protecting glass can be seen in Fig. 2a.

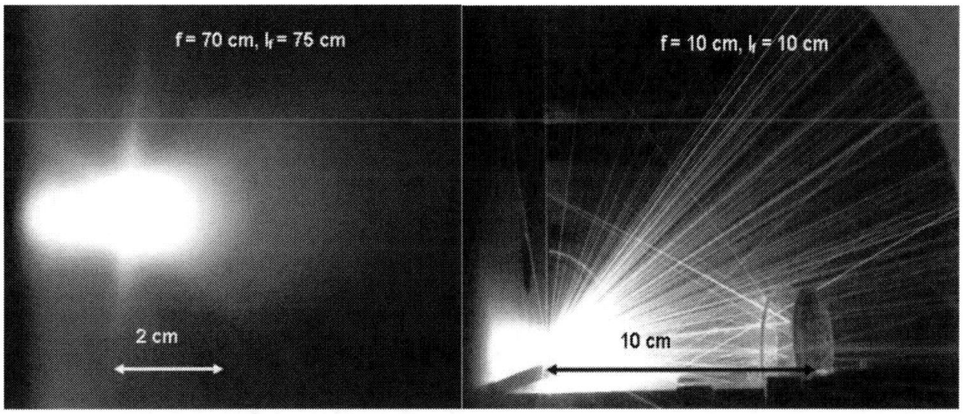

Figure 2. Time-integrated pictures of plasma produced by a single laser pulse focussed upon a deuterium-saturated carbon-target: a) at f = 70 cm, l_f = 75 cm; b) at f = 10 cm, l_f = 10 cm.

From the time-integrated picture presented in Fig. 2a one can easily see that the plasma plume, produced during irradiation of the investigated target by a pulsed laser beam at f = 70 cm and l_f = 75 cm, was quasi-symmetrical to the axis perpendicular to the target surface. When the lens of a f = 10 cm was situated near the target, at a distance l_f = 10 cm form its centre, the time-integrated picture shown in Fig. 2b demonstrated distinct traces of ion species (mainly carbon ions) emitted from the eroded target.

The optical emission spectrum, as emitted from the laser-produced plasma plume, was recorded by means of a Mechelle®900 spectrometer for different numbers of laser pulses. A comparison of a chosen part of such a spectrum, which was time-integrated over each 5 successive shots (generated with a frequency 1 Hz) during a series of 50 laser pulses, is presented in Fig. 3.

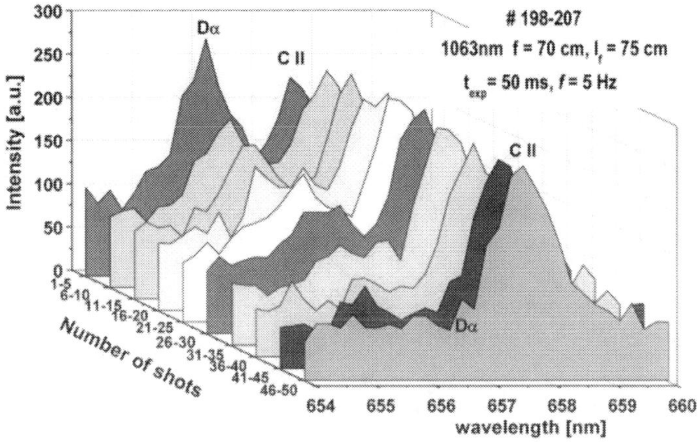

Figure 4. The chosen part of the optical emission spectrum (ranging from 654 nm to 660 nm), which was recorded for each successive 5 shots during a series of 50 laser pulses (1063 nm, 0.5 J) focussed at l_f = 75 cm, with lenses of f = 70 cm, upon the deuterium-saturated graphite target ($C + D_2$).

From the spectra shown in Fig. 3 one could easily see that the intensity of spectral lines corresponding to C II ions did not change noticeably during the whole series of the applied laser pulses, while the intensity of spectral lines from deuterium (the D_α line in that case) decreased considerably. Hence, one could deduce that the erosion of the graphite substrate was comparable for each laser shot, while the successive laser shots caused a removal of deuterium from the irradiated target.

These statements were confirmed by an analysis of the recorded spectra, and in particular by the measured intensity of the D_α line ($I_{D\alpha}$) versus the number of laser shots and the histogram showing changes of the $I_{D\alpha}/ I_{CII}$ as a function of the laser shots, as presented in Fig. 4.

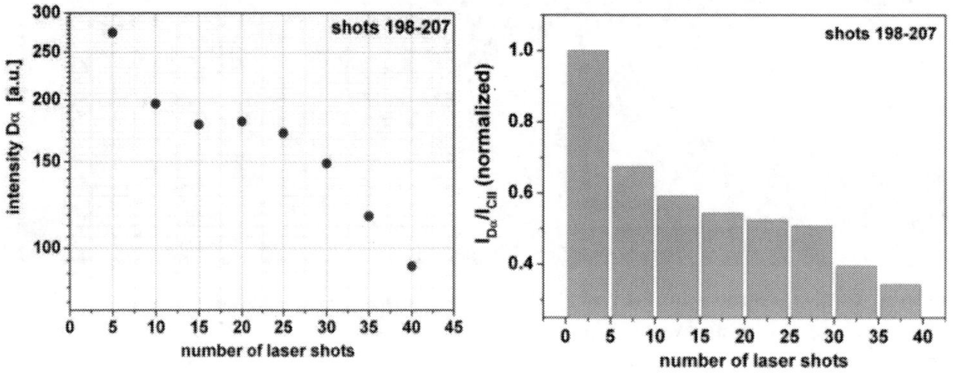

Figure 4. Decrease in the D_α intensity and in the $I_{D\alpha}/ I_{CII}$ ratio, as observed with an increase in the number of the laser pulses.

It should be noted that spectroscopic measurements analogous to those presented in Figs. 3 and 4 were also carried out for other focusing conditions (i.e., l = 10 cm and l_f = 10 cm), and the obtained results were very similar when the laser was operated at the 1st harmonic λ_1 = 1063 nm, at energy 0.5J.

In contrary, when the laser wavelength and energy was changed, i.e. when the 3rd harmonic λ_3 = 355 nm was used and its energy was decreased to 0.1 J, there were recorded carbon lines of lower intensity and no deuterium lines. Hence, one could deduce that 0.1 J pulses were insufficient to induce a removal of the deposited deuterium.

During the detailed analysis of the OES measurements particular attention was paid to changes in profiles of the D_α and C II lines in order to determine parameters of the investigated ion species. Some results are presented in Fig. 5.

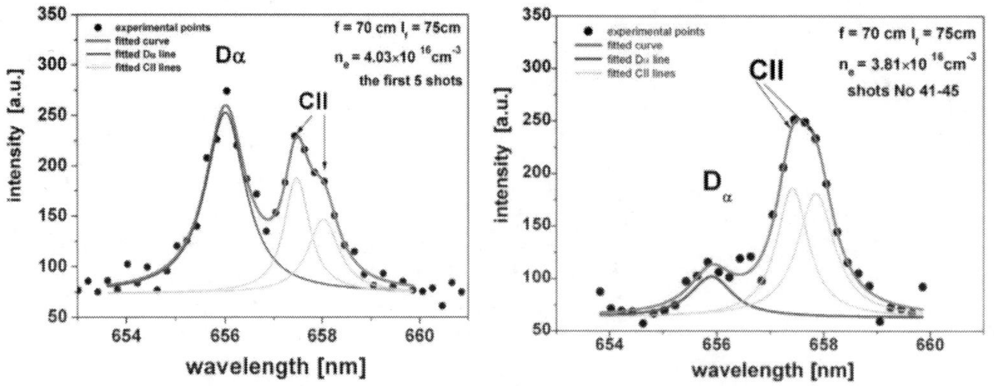

Figure 5. Fitting of profiles of the D_α and C II spectral lines for the spectrum recorded for the first 5 laser pulses (left) and for that observed after 41-45 shots (right).

On the basis of the profiles presented in Fig. 5 it was estimated that the deuterium population after 40 laser pulses decreased very strongly. In contrary, after such a treatment the population of C II ions did not change considerably. This conclusion was consistent with the previous statements. The electron concentration (N_e), which was determined from the Stark broadening of the D_α line profile recorded for the first 5 laser pulses, amounted to about 4 x 10^{16} cm^{-3}, while this parameter estimated from the D_α profile recorded during shots nos. 41-45 was equal to about 3.8 x 10^{16} cm^{-3}. It should be noted that in the last case intensity of the analyzed line was considerably lower. After changes of the focusing conditions (i.e., changes of a distance between the lens and the target, at the constant focal length f) the N_e changed from about 4 x 10^{16} cm^{-3} (see above) to about 7 x 10^{16} cm^{-3}.

In order to investigate the emission of ions from the laser-irradiated target the use was made of the ion collector described above. Some examples of ion signals, which were recorded with that collector, are presented in Fig. 6.

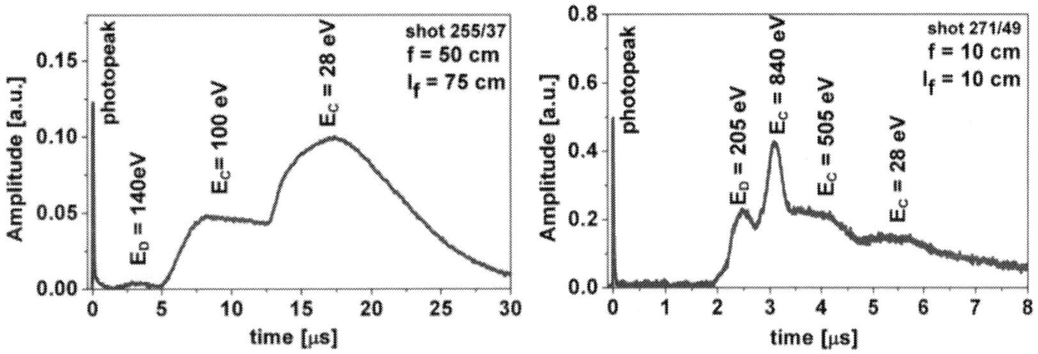

Figure 6. Time-resolved ion signals recorded with the ion collector situated at a distance of 36 cm from the target irradiated by 5 laser pulses (λ_1 = 1063 nm, 0.5 J) at different focusing conditions: with the lens of f = 50 cm placed at a distance l_f = 75 cm (left), and with the lens of f = 10 cm placed at a distance l_f = 10 cm (right)

Basing on the well know time-of-flight (TOF) technique it was possible to estimate ion energies corresponding to the recorded ion peaks. The results of simple computations showed that (depending on the laser focussing conditions) energies of deuterons ranged from about 50 eV to about 200 eV, while energies of carbon ions (considering mainly C II) ranged from about 30 eV even to about 850 eV.

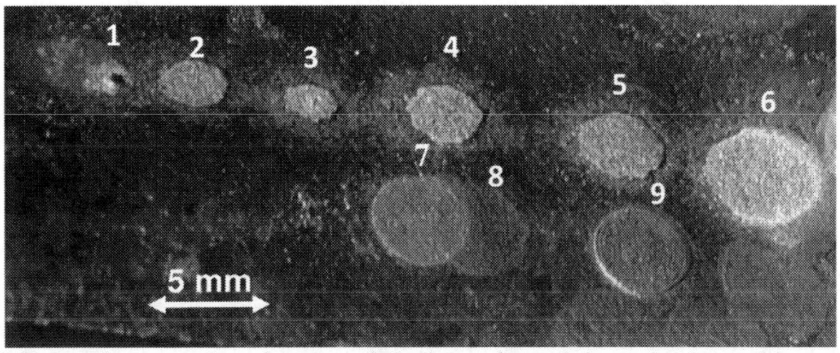

Figure 7. Macro-photography of a part of the investigated deuterized-graphite target, which shows micro-craters formed after the irradiation at different focusing length (f = 50 cm and 70 cm) and different distances of the lenses from the target (l_f = 70 cm, 75 cm and 80 cm). All craters were produced by the 1063-nm laser pulses. The crater no. 1 was produced at f = 70 cm and l_f = 75 cm; the crater no. 2 was produced by at f = 70 cm and l_f = 80 cm; the crater no. 3 - at f = 70 cm and l_f = 85 cm; the craters nos. 4 and 5 – at f = 70 cm and l_f = 70 cm; the crater no. 6 - at f = 50 cm and l_f = 70 cm; the crater no. 7 - at f = 50 cm and l_f = 80 cm; and craters nos. 8 and 9 were produced at f = 50 cm and l_f = 75 cm.

The laser-erosion of the investigated target was also investigated by macro-photography of the irradiated surface. An example is shown in Fig. 7.

Pictures of the micro-craters produced during the reported experiment are indicated by successive numbers in Figure 7. Measurements performed with an optical microscope showed that these micro-craters were of 1200-3000 μm in diameter and 30-150 μm in depth.

4. Conclusions

The main results of the reported study can be summarized as follows: On the basis of the recorded optical spectra it was found that the removal of deuterium from the surface layer of the target depends strongly on the laser wavelength and energy. Considerable differences in the intensities of the investigated spectral lines (mainly D_α and C II) were observed after the applied series of the laser pulses of λ_1 = 1063 nm and energy 0.5 J. This observation showed a high efficiency of the deuterium removal. In contrary, the use of laser pulses of λ_3 = 355 nm 0.1 J led to very weak erosion of the substrate material (carbon), but no removal of deuterium was achieved. More accurate analysis of spectroscopic data confirmed the above statements.

Measurements performed with an ion collector did also confirm the emission of deuterons and carbon ions. Signals obtained from the ion energy analyzer (at the highest applied power flux density) showed that the plasma included carbon ions with an electric charge ranging up to 4+. The signals from the ion collector allowed to estimate that in that case the peak energy of the carbon ions amounted to about 1 keV.

In conclusion it might be stated that to ensure an efficient removal of deuterium deposits one should apply laser pulses of λ_1 = 1063 nm and energy above 0.5 J focused upon a spot of 1-3 mm in diameter, ensuring power density of $0.7\text{-}1\times10^9$ W/cm^2. In the future it is planned to perform experiments with the 3rd harmonic of Nd:YAG laser to reach a higher power density and to observe eventual differences in the deuterium removal process.

Acknowledgement

This work, partly supported by the European Communities under the contract of Association between EURATOM and IPPLM (No. FU07-CT-2007-00061), was carried out within the framework of the European Fusion Development Agreement. This scientific work was also partly supported by Polish Ministry of Science and Higher Education within the framework of the scientific financial resources in the year 2013, allocated for the realization of the international co-financed project.

References

[1] Skinner, C. H., Gentile, C. A., Ascione, G. Carpe, A., Causey, R. A., Hayachi, T., Hogan, J., Langish, S., Nishi, M., Shu, W. M., Wampler, W. R. and Young, K. M., *J. Nucl. Mater.* **290-293**, 486 (2001).
[2] Andrew, P., et al., Stork, D. (Ed.) Technical Aspects of Deuterium Tritium Operation at JET (special issue), .*Fusion Eng. Des.* **47**, 211 (1999).
[3] Fortuna, E., Rubel, M. J., Philipps, V., Kurzydlowski, K. J., Martens, Ph., Miskiewicz, M., Pisarek, M., Van Oost, G. and Zieliński, W., *J. Nucl. Mater.* **367-370**, 1507 (2007)
[4] Rubel, M. et al., *J. Nucl. Mater.* **307-311**, 111 (2002).
[5] Shu, W.M., Kawakubo, Masaki, K., Y. and Nishi, M. F., *J. Nucl. Mater.* **313–316**, 584 (2003).
[6] Skinner, C. H., Bekris, N., Coad, J. P., Gentile, C. A., Hassenein, A., Reisdwig, R. and Willms, S., *Phys. Scripta* **T103**, 34 (2003).
[7] Ivanova, D., Rubel, M., Philipps, V., Schweer, B., Freisinger, M., Gierse, N., Penkalla, H., Petersson, P., Dittmar, T., J. Nucl. Mater. 415, S801 (2011).
[8] Gasior, P., Badziak, J., Czarnecka, A., Parys, P., Wolowski, J., Rosinski, M., Rubel, M. and Philipps' V., *Phys. Scripta* **T123**, 99 (2006).
[9] Loarer, T., Corre, Y., Delpech, L., Devynck, P., Douai, D., Ekedahl, A., Guilhem, D., Gunn, J. P., Klepper, C. C., Marandet, Y. and Vartanian, S., *Nucl. Fusion* **53**, 033003 (2013).
[10] Kurucz-database: http://www.pmp.uni-hannower.de/cgi-bin/ssi/test/kurucz/sekur.html
[11] NIST database: http://physics.nist.gov

High harmonic generation from atom clusters

M Aladi and I B Földes

Wigner Research Centre for Physics, Hungarian Academy of Sciences, Association EURATOM HAS, Budapest, Hungary

E-mail: aladi.mark@wigner.mta.hu

Abstract. High harmonics generation driven by 40 fs laser pulses is investigated in three different noble gases, both in gas and cluster regime depending on gas type and backing pressure. Pressure dependence of the harmonic intensity shows different scaling laws in case of cluster formation compared with monomer gas. Both the higher harmonic yield and the stronger spectral broadening of harmonics in xenon gas show the presence of clusters in the jet. The central wavelength of harmonics has a blue shift increasing with increasing laser intensity which is a consequence of free electron contribution to the phase matching.

1. Introduction

One of the main objects in the process of high-order harmonic generation (HHG) is to increase the conversion efficiency. In most of the cases atomic gas is used as target in form of gas jet or static gas cell. Gas jets can be generated by commercial solenoid valves in which case clusters may be formed in this high backing pressure jet due to the rapid cooling by the adiabatic expansion into vacuum. Clusters are promising targets for HHG and in our experiments we investigate characteristics of cluster harmonics compared with harmonics from monomer gases. In the case of gas harmonics the laser intensity limit is a considerable problem, because ionization mitigates the HHG process. Clusters are considered as suitable medium for HHG, because this medium forms a transition between gas and solid and it gives a higher conversion efficiency to harmonics. Ionization of the clusters however sets a limit to their application, too. The purpose of the present work is to determine the limits of applicability of cluster targets compared with atomic medium.

2. Experimental condition

High harmonics can be produced simply by focusing laser light into a gas target. We investigate the intensity of harmonics depending on laser intensity and backing pressure in the case of three noble gases.

2.1. Experimental setup

In the experiment a Ti:sapphire laser beam was used with 800 nm central wavelength, 1 kHz repetition rate. The pulse duration is 40 fs with 4 mJ pulse energy. The gas jet target used in the present experiment was characterized earlier [1]. The generated harmonic beam is characterized using a toroidal grating, which images the target onto a microchannel plate detector (MCP). The spectrum appears on the phosphor screen of the MCP and it is imaged onto a normal CCD.

2.2. Noble gas as cluster source

Noble gases as helium, argon and xenon were used as targets. In a gas-cluster medium the average $\langle N \rangle$ number of atoms in one cluster can be featured by means of the semi-empirical Γ^* Hagena-parameter:

$$\Gamma^* = k_h \cdot P_0 \cdot (0.74 \cdot d)^{0.85} \cdot T_0^{-2.29} \text{ with } \langle N \rangle = 33 \cdot \left(\frac{\Gamma^*}{1000}\right)^{2.35},$$

where k_h is the empirical condensation parameter depending on the gas, P_0 is the backing pressure, d is the nozzle diameter and T_0 is the initial gas temperature. For xenon $k_h = 5500$, for argon $k_h = 1650$ and for helium $k_h = 4$. Therefore xenon is the most convenient gas for cluster formation, whereas in helium no clusters are expected and argon is an intermediate gas for cluster presence.

3. Observations

3.1. Intensity dependence

Odd harmonics were observed up to the 45th order. The intensity of harmonics was investigated depending on the laser intensity whereas the laser energy was kept constant and the position of the gas jet was varied relative to the focal plane. In Figure 1 we can see, that the 15ω harmonic yield in xenon is nearly an order of magnitude higher than that in the other gases at 6 bar backing pressure. It can also be seen that the conversion to harmonics is stronger in argon than in helium. Increasing the intensity above 10^{14} W/cm^2 the conversion saturates and then it decreases. It is similar for each gas, independently from the eventual cluster formation. Although the conversion efficiency is the highest for xenon where clusters are formed, on the other hand due to the lower ionization potential for xenon - independently from the cluster formation - the conversion starts to decrease even at lower laser intensity than for argon. Thus it is confirmed that ionization limit of harmonics generation is independent from cluster formation.

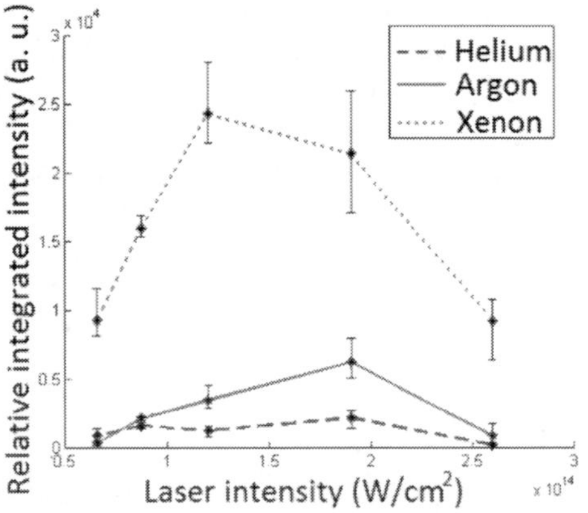

Figure 1. Laser intensity dependence of 15ω harmonic signal at 6 bar backing pressure.

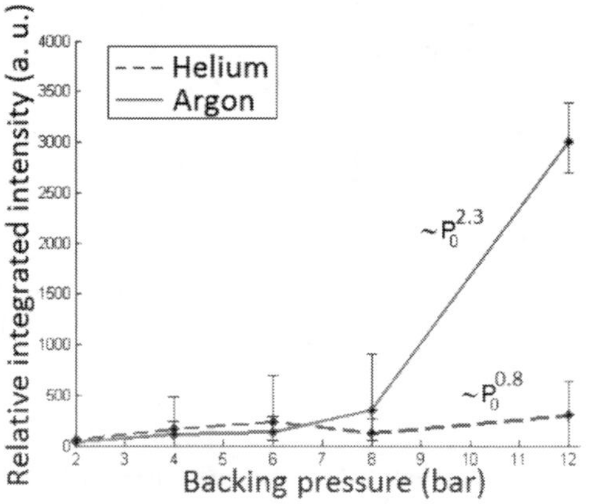

Figure 2. Backing pressure dependence of 27ω harmonic signal at $6.5 \cdot 10^{13}$ W/cm^2 laser intensity.

3.2. Backing pressure dependence

Backing pressure dependence of the harmonic intensity at fixed laser intensity shows the effect of cluster formation. Figure 2 illustrates it for argon for a special example of the 27th harmonics, in which case at lower pressures when no clusters are formed the conversion efficiency is low. A different scaling law can be seen above 6 bar due to cluster formation. For helium the conversion efficiency slowly increases with pressure, whereas for argon, where clusters are expected an approximately square-law dependence can be observed. This dependence is different from some earlier observations, where P_0^2 dependence was observed for atomic gas and P_0^3 dependence for clustered gas [2]. In the present experiment just cluster targets show a P_0^2 dependence. Note that in the early experiments of Donelly et al. [2] a longer, 160 fs pulse duration was used, which duration exceeds the characteristic time for Coulomb explosion of the cluster.

3.3. Spectral broadening

Our observation confirms the earlier results that the spectral widths of harmonics are larger for clusters than for monomer gases. Figure 3 illustrates that at 12 bar backing pressure harmonic components are significantly broader in xenon than in helium. This spectral broadening is increasing with increasing pressure.

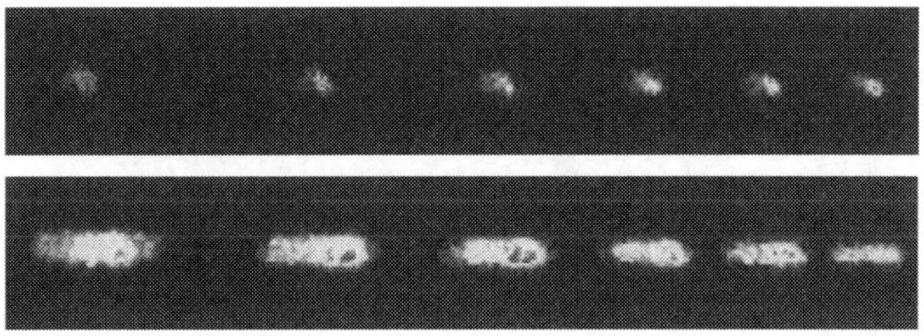

Figure 3. Spectral broadening of harmonic spectrum at 12 bar backing pressure in the case of helium (top) and xenon (bottom).

3.4. Spectral blueshift

Harmonics spectra are illustrated at three different laser intensities for helium and xenon. In Figures 4 and 5 the backing pressure was 2 bar. Clearly the harmonic yield is much higher in xenon (Figure 5) than in helium (Figure 4). Argon with 12 bar backing pressure shows similar spectra than xenon at 2 bar.

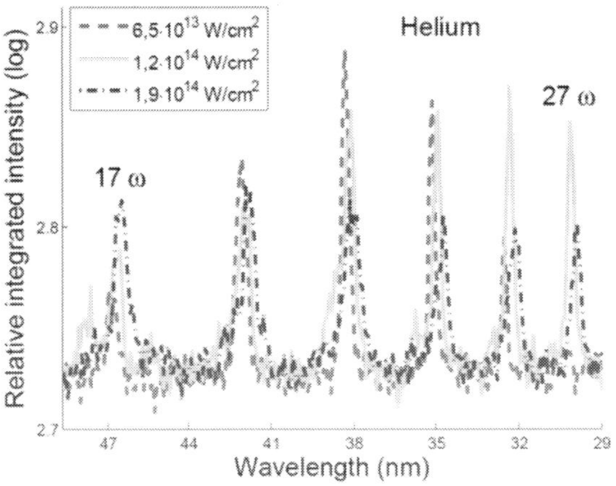

Figure 4. Integrated spectrum at 2 bar backing pressure for helium target at three different laser intensities.

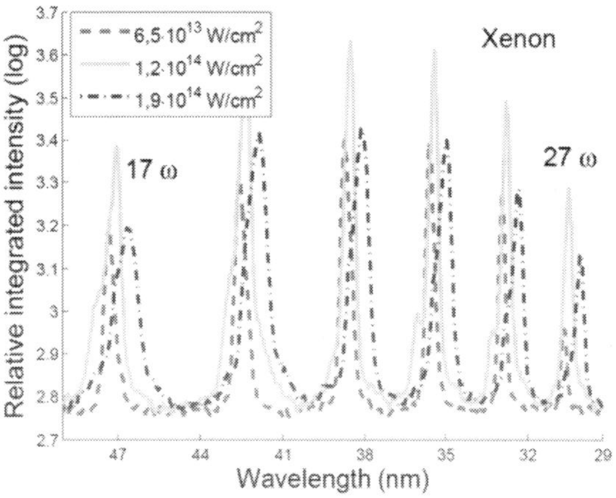

Figure 5. Integrated spectrum at 2 bar backing pressure for xenon target at three different laser intensities.

It is evident from these spectra that the central wavelength of harmonics has a frequency shift toward shorter wavelengths. This ionization effect is known as frequency blueshift [3], which is effect of the free electrons onto the propagation and thus onto the phase matching of the harmonics with the incoming beam. The blueshift is illustrated quantitatively in case for the 21ω radiation in Figure 6.

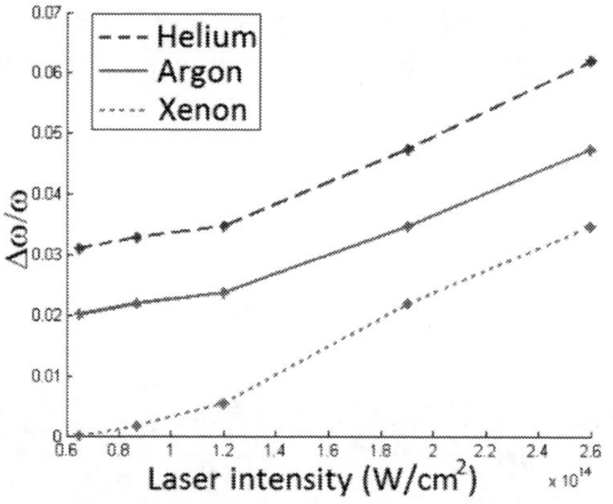

Figure 6. Laser intensity dependence of spectral blueshift for 21ω at 6 bar backing pressure.

4. Conclusion

It was demonstrated through a series of experiments that clusters may enhance the conversion efficiency to high harmonics in a limited range of laser intensities, below the ionization limit. On the other hand separating the contributions from monomers and clusters is not simple [4] because the order of magnitude difference between the clustered xenon and the monomer helium is only partly caused by cluster formation, partly by the generally higher efficiency for xenon. On the other hand the steep increase of harmonics conversion efficiency in argon when cluster formation starts at higher pressures gives a clear evidence for the beneficial effect of clusters on harmonics generation. Additionally the spectral broadening of harmonic spectrum can refer to the existence of clusters in the gas jet as well. On the other hand the observed spectral blueshift is caused by the effect of free electrons onto the phase matching, i.e. not a cluster-effect. This blueshift gives a possibility for tuning the wavelengths of harmonics with the laser intensity.

Acknowledgement

The authors acknowledge the help in the laser operation and the valuable discussions with A. Barna, P. Dombi, I. Márton and P. Rácz. This work, supported by the European Communities under the contract of the Association between EURATOM and the Hungarian Academy of Sciences, was carried out within the framework of the European Fusion Development Agreement. The views and opinions expressed herein do not necessarily reflect those of the European Commission. It was also supported by the Hungarian Nr. ELI 09-1-2010-0010 hELIos-ELI contract and the COST MP1208 and MP1203 activities.

References

[1] Rakowski R *et al* 2005 *Nucl. Instrum. Methods Phys. Res.* A **551**, 139
[2] Donelly T D *et al* 1996 *Phys. Rev. Lett.* **76**, 2472
[3] Shin H J *et al* 2001 *Phys. Rev.* A **63**, 053407
[4] Ruf H *et al* 2013 *Phys. Rev. Lett.* **110**, 083902

Hall magnetohydrodynamic model of the quasi-single helicity self-organization of RFP

A Gurin

Institute for nuclear research NAS of Ukraine, Prospect Nauky, 47, Kiev, 03680, Ukraine

E-mail: aagurin@ukr.net

Abstract. In the framework of two-fluids Hall magnetohydronynamic description of a paramagnetic z-pinch with the low safety factor (q<<1 on the pinch axis), the theoretical model of kink mode stabilization is developed. It is shown that global kink oscillations of paramagnetic pinch, whose instability had been found in the single-fluid MHD theory, generate due to the Hall effects a rotation of plasma, which, in turn, stabilizes the kinks of the plasma column. The fluctuations are considered in terms of the cylindrical boundary problem for the pinch equilibrium with arbitrary components of the magnetic field, $(0,B_\theta,B_z)$, hydrodynamic velocity, $(0,V_\theta,V_z)$, and the Hall parameter. The consideration is fulfilled in approximation of "cold" plasma when the Alfven magnetic activity plays the key role. The hydrodynamic velocities V_θ, V_z are determined numerically from momentum balance collision-free equations averaged over the helical fluctuations. The developed self-consistent quasi-linear magnetohydronynamic model of the marginally stable kink spectrum reflects some physical features of the quasi-single helicity regime in reversed field pinches.

1. Introduction

The last decade has seen substantial progress in understanding the physics of toroidal discharges, qualified as a reversed field pinches (RFP) with improved confinement. The new paradigm has established, under which a current maintenance in RFP is characterized by laminar quasi-single-helical (QSH) oscillations - the kinks with numbers $m = 1$ and $n >> 1$ [1-3]. Two questions are of current interest: the nature of the QSH wave spectrum and its mechanism of self-organization.

Magnetohydronynamics (MHD) spectra are the most extensively discussed subject in the pinch physics. Linearized version of the MHD equations was first formulated as a differential boundary value problem of order 2 for linear pinches as far back as before tokamak epoch [4]. The Hain-Lüst equation [4] allows determine the non-local frequency dispersion and radial structure of helical pinch oscillations for an arbitrary magnetic fields and parameters of plasma without flows. Many authors have rediscovered the equation in various forms. In references [5,6] the most usable view presented, and in [7] in the same terms hydrodynamic flows of the plasma, $(0,V_\theta(r),V_z(r))$, are taken into consideration. Disclaimer of immobility plasma in the spectral theory is very important in view of the extensive evidence that any equilibrium plasma is accompanied by its differential rotation, and the relatively weak low-frequency fluctuations are associated with plasma flows. However, inclusion of plasma velocities in the analysis requires the explicit knowledge of the radial distributions of $V_\theta(r)$ and

$V_z\,(r)$. Thus, in each case, an appropriate model matching fluctuation spectra and plasma equilibriums is in a need.

In this paper, with reference to the RFP, the model proposed, in which the boundary value problem includes Hall effects of two-fluid dynamics in addition to plasma flows. While in the standard one-fluid MHD theory of ideal plasma the zero real frequency is characteristic for unstable wave oscillations (Re$\{\omega\}$=0 whereas Im$\{\omega\}$>0), in the Hall MHD theory non-zero real frequencies arise. Therefore it turns out, the balance equation averaged over fluctuations become non-trivial and can be satisfied only taking into account of plasma flows. Below the plasma velocities are determined numerically from the momentum balance equations presented as quadratic forms averaged over fluctuations. Substituting these velocities in the computational code for the boundary value problem reveals the stabilizing effect of the induced plasma flows on each of the unstable modes. The iterative procedure minimizing increments permits us to call the amplitudes that define a marginal stable kink spectrum.

The entire treatment is performed in the cold plasma approximation, in which thermal effects are omitted. Moreover, the model considers an ideal plasma without any collisions, whereby only detects a tendency towards self-stabilization of resonant modes, while the limit of strict stabilization is unattainable in the iterative procedure because of their singularity. This significantly limits the physical relevance of the model but provides a basis for a broader description of the Hall mechanism of the RFP dynamic self-organization.

2. Basic equations.

The Hall version of MHD theory introduces the current velocity **U** into description as follows:

$$\frac{\partial}{\partial t}\mathbf{B} = rot(\mathbf{V}_e \times \mathbf{B} - \eta\, rot\mathbf{B})$$

$$\mathbf{V}_e = \mathbf{V} - \mathbf{U}, \quad \mathbf{U} = \frac{rot\mathbf{B}}{\sqrt{\Pi}N}$$

$$N(\frac{\partial}{\partial t}\mathbf{V} + (\mathbf{V}\nabla)\mathbf{V}) + \nabla \mathrm{P} = rot\,\mathbf{B}\times\mathbf{B} \qquad (1)$$

$$\frac{\partial}{\partial t}N + \nabla N\mathbf{V} = 0, \quad P = \beta_0 N^{5/3}$$

$$\eta = \frac{c^2}{4\pi\sigma}\frac{1}{aV_A} \propto 10^{-6} - 10^{-7}$$

$$\Pi = \frac{4\pi N_0 e^2 a^2}{Mc^2} = \frac{\omega_{ci}^2}{\omega_A^2} \propto 40 - 50$$

Equations.(1) are written in the dimentionless form using the Alfven units: V_A=$B_0(4\pi\rho_0)^{-1/2}$ for velocities, the plasma column radius a and the frequency ω_A= V_A/a for all sizes and frequencies, where Alfven velocity V_A is expressed through physical parameters on the pinch axis. The density N and magnetic field B are reduced to 1 on the axis: $N = B = 1$ at $r = 0$ in cylinder coordinates (r,θ,z). In (1), the constant Hall-Braginskii parameter Π and resistivity η are estimated for real RFPs.

The linear version of (1) for all fluctuating fields represented as, e.g., $\mathbf{B}(r)$+ $\delta\mathbf{B}(r)\exp(i\zeta)$+c.c. where $\zeta = k_z z - m\theta - \omega t$, is suitably to express in terms of the electron displacement ξ_{er}: $\delta V_{re}(r) = -\,i(\varpi + \mathbf{k}\cdot\mathbf{U})\xi_{re}(r)$, where $\varpi = \omega - \mathbf{k}\cdot\mathbf{V}(r)$, $\mathbf{k} = (0, k_z, -m/r)$. Useful variables are also $F = \mathbf{k}\cdot\mathbf{B}$ and $G = r[\mathbf{B}\times\mathbf{k}]_r$. With assumption $\beta_0 = 4\pi P_0/B_0^2 = 0$, i.e. neglecting thermal pressure effects, and only in this case if $1/\Pi > 0$, one can obtain a Hain-Lust-like boundary problem of order 2:

$$\frac{d}{dr}Ar\frac{d}{dr}r\xi_{re} + Q\xi_{re} = 0, \qquad \xi_{re}(0) \neq 0, \;\; \xi_{re}(1) = 0 \qquad (2)$$

in which A and Q are expressed trough ϖ, Hall contributions and the frequency of θ-rotation, $\Omega = V_\theta/r$, in explicit form:

$$A = \frac{1}{k^2r^2}[F^2 + \frac{\varpi^2 N}{D}\tilde{G}^2], \qquad F = k_z B_z - \frac{mB_\theta}{r}, \qquad G = k_z r B_\theta + mB_z, \qquad \tilde{G} = G + \frac{\Omega^2 k^2 r}{\sqrt{\Pi}\varpi}$$

$$D = \varpi^2 r^2 N - k^2 r^2 B^2 - \frac{1}{\sqrt{\Pi}}(B_\theta V_z' - \frac{B_z}{r^3}\frac{d}{dr}r^4\Omega) - \frac{1}{\Pi}(\varpi^2 - 2\Omega\frac{1}{r}\frac{d}{dr}r^2\Omega)],$$

$$Q = N\varpi^2 - F^2 + 4(\frac{k_z^2 B_\theta^2}{k^2r^2} - N\Omega^2 + mN\frac{\Omega^3}{\varpi}) - k^2r^2(\frac{N\Omega^4}{\varpi^2} + \frac{NU^2}{D}) - \\ - r\frac{d}{dr}\left[\frac{B_\theta^2}{r^2} - 2\frac{k_z G B_\theta}{k^2 r^3} + \frac{\varpi N \tilde{G} U}{D}\right] \qquad (3)$$

$$U = 2\varpi\frac{k_z r B_\theta}{k^2r^2} + 2\Omega B_z - \frac{\Omega^2 G}{\varpi} - \frac{1}{\sqrt{\Pi}}(\varpi^2 - 4\Omega^2 + 2m\frac{\Omega^3}{\varpi})$$

Equations (2) and (3) in the case $\Pi \to \infty$ and similar formulas [4] in the case $\beta_0 = 0$ (i.e. $P = 0$) are equivalent.

The mean fields $B_r(r)$, $B_\theta(r)$ can be substituted into (3) as a solution of equations that define in a sufficiently general form the decomposition of the rot**B** along the directions **b=B**/B and **b**^= **b**×$\mathbf{e}_r$ with two coefficients $\alpha(r)$ and $\beta(r)$:

$$.\frac{1}{r}\frac{d}{dr}rB_\vartheta = \frac{\alpha B_z - \beta B_\vartheta}{B^2}, \qquad \frac{d}{dr}B_z = -\frac{\alpha B_\vartheta + \beta B_z}{B^2} \qquad (4)$$

If $\alpha = \alpha_0 B_z$, $\beta = 0$, we have the force-free paramagnetic model [8]. A more general model that takes into account the fluctuations follows from the non-inertia equation of electron motion which takes role of the Ohm's law in the Hall MHD theory

$$\mathbf{E} + \mathbf{V}_e \times \mathbf{B} = \eta \mathrm{rot}\mathbf{B} \qquad (5)$$

Equation (5) does not contain thermal terms in our case. In steady state, $\mathbf{E} = E_0\mathbf{e}_z - \Phi'(r)\mathbf{e}_r$, so the **b**-projection of (5) gives in our dimentionless description: $E_0 = \alpha_0\eta$, where $\alpha_0 = 4\pi\sigma aE/cB$ is a constant for the paramagnetic model expressed through driving electrical field E and magnetic field B at the pinch axis in physical units. In our calculations α_0 is assumed equal 4 in accordance with data [1-3].

To study the effect of fluctuations on the pinch equilibrium, should be averaged the Ohm's law components along directions **b** and **b**^:

$$\alpha = \alpha_0 B_z + \frac{B}{\eta}[\langle B_r(\mathbf{b}^\wedge, \mathbf{V}_e)\rangle - \langle V_{re}(\mathbf{b}^\wedge, \mathbf{B})\rangle] \\ \frac{V_{er}}{\eta} = -\frac{\alpha_0 B_\theta + \beta}{B^2} + \frac{1}{\eta B}[\langle B_r(\mathbf{b}, \mathbf{V}_e)\rangle - \langle V_{re}(\mathbf{b}, \mathbf{B})\rangle] \qquad (6)$$

The brackets ‹…› denotes the averaging over the wave phase ζ in the single-mode approximation (recall that phase ζ is determined above by the complex presentation of perturbations, e.g., $\delta N(r)e^{i\zeta}$ + c.c).

Smallness of parameter η allows estimating the amplitude of fluctuations with the low scale of $\eta^{1/2}$ value to give sufficient contribution into parameters of configuration. Indeed, the non-radial flow perturbations and the mean radial transport are very slow compared to the Alfven velocity in reality. Transition to rescaled values, $\xi_{re} = \eta^{1/2}\tilde{\xi}_{re}$, $V_{re} = \eta\tilde{V}_{re}$, excludes the parameter η in the formulas (6) and below for values with tilde. The η-analysis defines the parameter β from the radial momentum balance (1) as a zero-order value, $\beta = -rN\Omega^2$ under condition $\beta_0 = 0$, giving the diamagnetic rotational correction to the paramagnetic model.

To calculate the flows, it is enough to use the procedure ‹...› for non-radial components of the momentum balance (1) where terms with time derivates can be omitted if the quasi-steady states with rotation exist. It is easy to see that under the single-mode consideration following equations take place:

$$\frac{1}{r^2}\frac{d}{dr}r^2\langle NV_rV_\theta\rangle = \left\langle B_r\frac{1}{r}\frac{\partial}{\partial r}rB_\theta + \mathbf{k}\cdot\mathbf{B}\frac{\partial}{\partial\varsigma}B_\theta\right\rangle = \frac{1}{r^2}\frac{d}{dr}r^2\langle B_rB_\theta\rangle$$
$$\frac{1}{r}\frac{d}{dr}r\langle NV_rV_z\rangle = \left\langle B_r\frac{\partial}{\partial r}B_z + \mathbf{k}\cdot\mathbf{B}\frac{\partial}{\partial\varsigma}B_z\right\rangle = \frac{1}{r}\frac{d}{dr}r\langle B_rB_z\rangle \tag{7}$$

Integrating equations (7) and using the η-analysis one can obtain the simple local presentation for non-radial velocities:

$$V_{\theta,z} = \frac{1}{\langle\Gamma_r\rangle}2\,\mathrm{Re}\{\delta\tilde{B}_r\delta\tilde{B}^*_{\theta,z} - N\delta\tilde{V}_r\delta\tilde{V}^*_{\theta,z}\}, \qquad \langle\Gamma_r\rangle = N\tilde{V}_{er} + 2\,\mathrm{Re}\{\delta\tilde{N}\delta\tilde{V}^*_{er}\} \tag{8}$$

3. Main results

The boundary eigen-value problem (2) for modes $m = \pm1$ was solved by the shooting method initially in the framework of the paramagnetic model of motionless cold plasma. It is found the kink m=1 is instable within an interval of k_z in accordance with known results [6]. Figure 1 shows the growth rates (increments) within interval $1.66 < k_z < 2.37$ (upper curve) and the occurrence of the oscillation frequencies, Re{ω}, to be low in the Alfven scale (lower curves - note the factor 10^2!). Low frequencies are result of the weak influence of the Hall effects on magnetosonic wave dispersion in the case of $\Pi >> 1$, as can be clearly argued in terms of oscillations in a uniform magnetic field. They noticeably depend on the Hall parameter within the interval $10<\Pi<50$ of practical interest. The increments are saturated for these Π because influence of the Hall parameter on Im{ω} manifests itself if $\Pi<<1$. In our case, the subinterval $k_z > 2$ corresponds to resonant "internal" modes, for which the condition $F = \mathbf{k}\cdot\mathbf{B} = 0$ introduces a singularity of solution (2) for stable modes. However, equation (2) is valid for the analysis and numerical solution while Im{ω}>0 for all k_z.

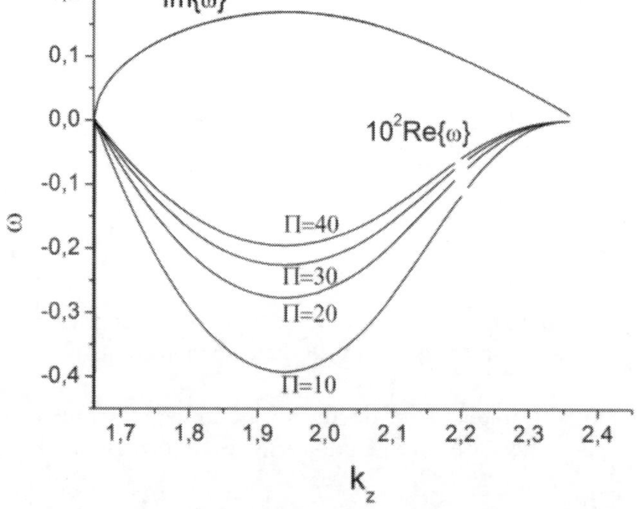

Figure1. Dependence of the kink increments and frequencies on k_z and Hall parameter Π in cold plasma of the paramagnetic pinch without flows.

The appearance of the real valued frequencies of unstable kinks if $V_\theta = V_z = 0$ but Π is finite makes their amplitudes to be complex values. It is important under calculation of the velocities (8) because an additional contributions into mean values ‹…› appear due to Π. The diagram on figure 2 illustrates the process of stabilization of the kink resonant modes under the influence of plasma rotations generated by theirs own field oscillations calculated in accordance with the expressions (8). The growth rates are plotted against the rescaled amplitude $\tilde{\xi}_{re}(0)$ rising from zero when they start at corresponding values in figure 1. The set of internal kinks m=1, 2.1< k_z < 2.3 are chosen for which the greatest amplitudes are required and a "dominant mode" can be called. The data are obtained in iterative procedure substituting all field amplitudes into equations (8) and (6), and then into (4) and (3), step by step increasing amplitude. As a result, the limiting values of amplitudes can be found although the singularity of stable modes breaks off the iterations when increments become very close to zero.

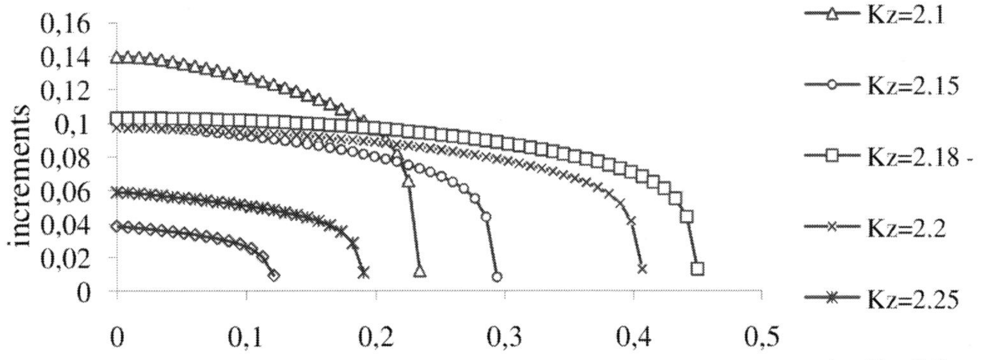

Figure 2. Stabilization of different kinks under influence of flows with increasing amplitude $\tilde{\xi}_{re}(0)$ (Π=40).

4. Conclusions

It is shown that the non-local kink-modes are unstable in the cold plasma of paramagnetic pinches and can be stabilized by the flows generated by own oscillations under influence of the Hall effects. The one mode may be dominant in the spectrum of wave numbers k_z that may be compared with some toroidal number n. Theoretical quasi-linear model determining the rotational velocities and the algorithm taking into account the plasma rotation are proposed.

However the presented model don't able to describe a static maintenance of saturated amplitudes as well to find the values of real frequencies of resonant modes in the strict limit Im{ω}=0 due to their singularity. More dissipative processes as well as all β-effects of non-scalar pressure must be involved into the model. Some convergence with the theory of tearing modes taking into consideration real low-frequency fluctuations of the pinch is to be achieved.

References

[1] Escande D E, Martin P, Ortolani S et al 2000 *Phys. Rev. Let.* **85** 1662
[2] Martin P, Marreli L, Spizzo G et al 2003 *Nucl. Fusion* **43** 1855
[3] Frassinetti L, Brunsell P R, Drake J R et al 2007 *Phys. Plasmas* **14** 112510
[4] Hain K and Lüst R. 1958 *Z. Naturforsch* **13a** 936
[5] Goedbloed J P 1971 *Physica* **53** 501
[6] Goedbloed J P and Hagebeuk H J L 1972 *Phys. Fluids* **15** 1090
[7] Wang C, Blockland J W S, Keppens R and Goedbloed J P 2004 *J. Plasma Phys* **70** 651
[8] Kadomtsev B B 1962 *Nucl.Fusion (Suplement Part 3)* 969

Plasma Physics by Laser and Applications 2013 Conference (PPLA2013) IOP Publishing
Journal of Physics: Conference Series **508** (2014) 012018 doi:10.1088/1742-6596/508/1/012018

Laser Ablation Characterization in Laboratori Nazionali di Legnaro

D Scarpa[1] , P Nicolosi[2,3], A Franci[2], A Tomaselli[6], M Manzolaro[1], S Corradetti[1,4], J Vasquez[1,5], M Rossignoli[1], M Calderolla[1], A Monetti[1], A Andrighetto[1] and G Prete[1].

[1]INFN, Laboratori Nazionali di Legnaro, Legnaro (Padova), Italy.
[2]Dipartimento di Ingegneria dell'Informazione,Universit degli Studi di Padova.
[3]IFN-CNR UOS Padova
[4]Dipartimento di Scienze Chimiche dellUniversit di Padova, Padova, Italy.
[5]Dipartimento di Ingegneria dell'informazione dellUniversit di Padova, Padova, Italy.
[6]Universit degli Studi di Pavia, Dipartimento di Ingegneria Industriale e dell'Informazione, Via Ferrata 1, Pavia, Italy.

E-mail: daniele.scarpa@lnl.infn.it

Abstract. Using high power laser focalized into a target material generates plasma and it has the consequence to immediately extract and ionize atoms from the target itself. This process is the starting point to have a simple and compact ion source, usually named Laser Ion Source (LIS). This kind of sources are arising in the scenario of ion sources especially in refractory elements ions production, where the atomization of the material to ionize is the main issue because of its high evaporating temperature. These considerations and the fact that ion sources are becoming nowadays more and more important to several fields of science and technology, open an interesting line of research that our group at Laboratori Nazionali di Legnaro wants to investigate. Experiment involves characterization of produced ions by measuring charge state and amount of ions created. Measurements will be performed with several power densities and varying ions collector distance and potential respect to the target. These simple experiments are the necessary preliminary steps to characterize the system and to start a solid future development onto possible different and effective ways to perform ions sources using laser.

1. Introduction

To fully understand a Laser Ion Source (LIS) [1] [2] behavior it is mandatory to familiarize and to deeply analyze laser ablation processes onto target materials. This is the aim of the new laser setup implemented at Laboratori Nazionali di Legnaro, where we start to study laser ablation processes. A Q-switched Nd:YAG is used to ablate target placed in front of an ions collector inside an high vacuum chamber. Thanks to this simple configuration, inspired by [3], it will be possible to test the system configuration, the signal and data collecting procedure in order to move a secure first step into this new line of research for our group. This work will synthesize the firsts experiments and results with this simple configuration and it will show possible future steps for the experimental setup and measurements.

Journal of Physics: Conference Series **508** (2014) 012018 doi:10.1088/1742-6596/508/1/012018

2. Experimental setup

The experiment involves several systems and can be resembled in figure 1. A laser beam is focalized onto a target inside a vacuum chamber and a collector, placed in front of the target, collects the ion signal from the expanding plasma plume.

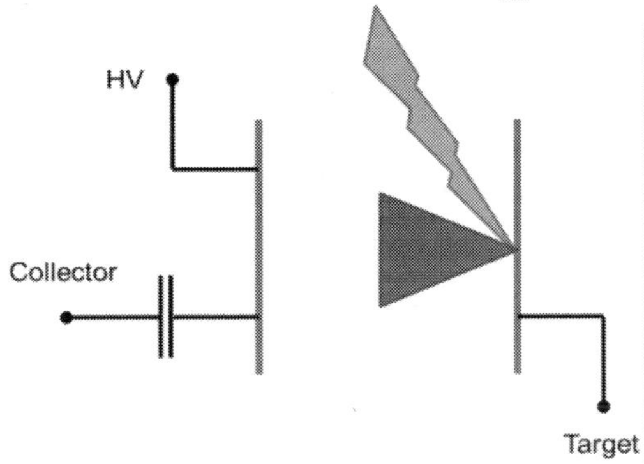

Figure 1. Simplified experimental setup

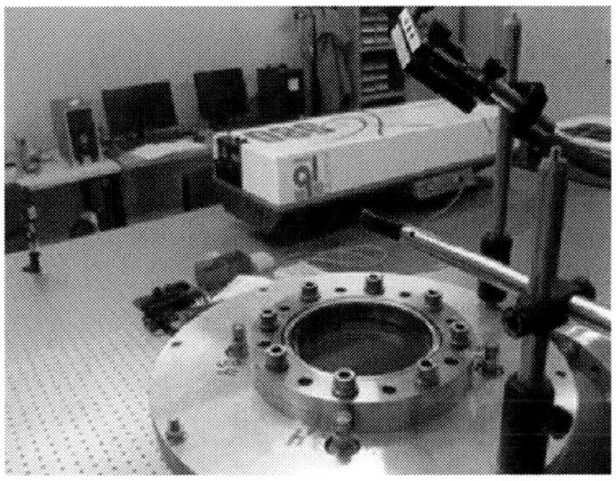

Figure 2. View of vacuum chamber

2.1. Laser and focalization system

The used laser is a Quantel YG980 Q-switched Nd:YAG laser system, operating at its fundamental wavelength (1064 nm), capable to deliver up to 2 J energy 10 ns pulses at maximum repetition rate of 10 Hz. Laser beam, entering in the vacuum chamber with an angle between 20 and 45 degrees, is focalized onto the target by a 20 cm focal length lens.

2.2. Vacuum Target Chamber

The vacuum chamber (figure 2) was entirely designed and assembled at LNL. The high vacuum is ensured by a diffusive oil pump (Edwards DIFFSTAK 63/150) capable of 135 liter/sec. The final vacuum value measured after 2 hours pumping is $2{,}5 \cdot 10^{-5}$ mbar.

Inside the chamber is placed the target-collector system. Target is made of a Copper plate, electrically isolated by a plastic support. It can be rotated to expose different area to each shot of the laser beam and it can be electrically polarized positive or negative respect to the collector and the chamber by an external high voltage power supply. Distance between plates can be varied from 2 to 5 cm. Signals are collected by electrical connection of the plates directly to the oscilloscope (Tektronix TDS 3034C) channels with 50 Ohm terminations. A 3D section of the chamber is showed in figure 3.

2.3. Data collection

In order to collect data from the system an automatic procedure was implemented using MATLAB® as command interpreter between the PC and the oscilloscope. Data of interest, such as laser pulse duration and plates voltages, are automatically recorded and plotted in the PC monitor for immediate evaluation.

3. Results

We measured several parameters to evaluate laser ablation processes and ions generation. Distance between target and collector plate was fixed to 4,5 cm, polarization voltage between plates and laser pulse energy were variable parameters. Figure 4 shows an example of data

Figure 3. Section view of the chamber and target.

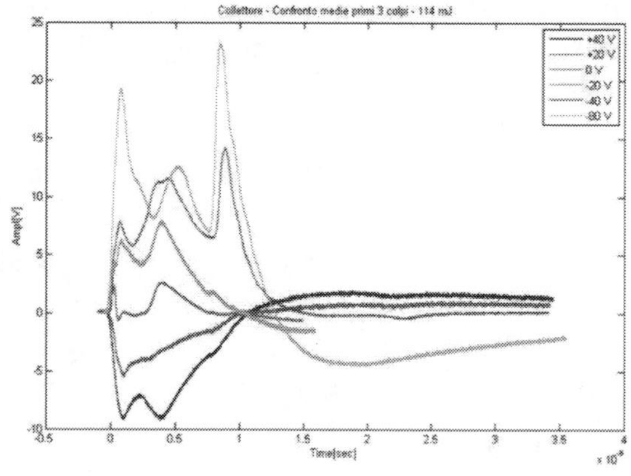

Figure 4. Collector plate signals at different polarization voltages (blue - 40V; red 0V; ocher 80V.

collected representing signals generated by extracted charges at different polarization voltages of the collector plate respect to the target plate. Signals are positive or negative due to the type of particle (ions or electrons) attracted by the plate polarization, peaks analysis are under development. Results are consistent with the existing literature confirming that the apparatus is ready to a more intense measurement campaign. Furthermore these processes could be used for a first evaluation of the produced charge state for ions extracted from the plasma.

The high speed of the material, ions and neutral, expelled by the laser ablation process combined to the reduced distance between the target and the collector, does not allow an easy analysis of the charges expelled. Furthermore the absence of some kind of suppressor system on the collector side, allows generation of secondary emission electron even from the neutrals atoms impinging on the plate.

4. Future Developments

As described in the previous section, the distance between the plates is the main illness of this kind of experimental setup. Since the obvious solution is to increase the plates distance, we decided to develop a new setup, including a 1,5 m home made Time of Flight mass spectrometer. In this solution the increased distance between the plates is used to perform a powerful analysis instrument capable to obtain information on the ions charge state production and distribution. Works are in progress using PROE® to design the mechanical setup (figure 5) and SIMION® to study ions trajectories (figure 6) and to define plates voltages.

5. Conclusions

Experiments of ions generation by laser plasma were performed in LNL. An easy setup composed by a target-collector system was assembled to start studies on plasma characterization and observed results are consistent with the existing literature. Future developments, involving the assembly of an home made Time of Flight mass spectrometer, are planned to better characterize these processes of ions formation in term of charge state distributions and efficiency of the generation process.

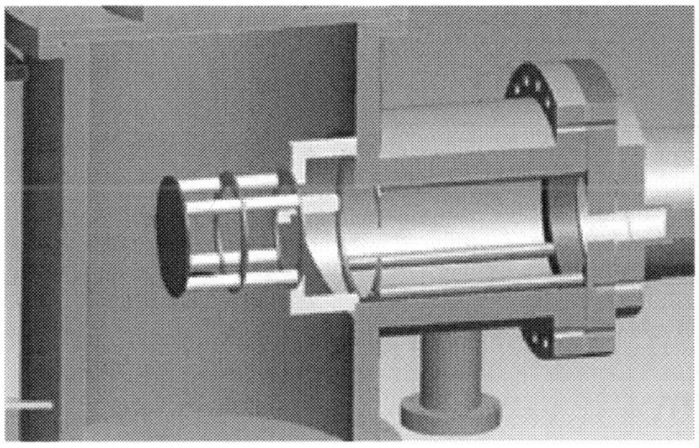

Figure 5. Mechanical design of the ToF.

Figure 6. SIMION® ions trajectories simulation

Acknowledgments

Authors want to thank Michele Lollo and Edoardo Visentin for their precious support and mechanical constructions.

References

[1] Sharkov and Scrivens 2005 *IEEE TRANSACTIONS ON PLASMA SCIENCE* **33-6**
[2] Belloni et al. 2006 *Proceedings of EPAC 2006, Edinburgh, Scotland* THPCH**139** 3319-3321
[3] T Yoshitake 2000 *J.of Material Science* **35** 1469-1474

Plasma Physics by Laser and Applications 2013 Conference (PPLA2013) IOP Publishing
Journal of Physics: Conference Series **508** (2014) 012019 doi:10.1088/1742-6596/508/1/012019

High harmonics generation at the interaction of an intense *fs* laser pulse with an overdense plasma layer

A Mihailescu, V Stancalie and V Pais
National Institute for Laser, Plasma and Radiation Physics, P.O.Box MG-36, Magurele, 077125 ROMANIA

E-mail: andreea.mihailescu@inflpr.ro

Abstract. Harmonics generated with *fs* laser pulses are not only among the most intense sources in the wavelength range of a few tens of nanometers but also a far better source than X-ray lasers from the coherence point of view. It is anticipated, that in the near future, the harmonics issued from femtosecond laser-produced plasmas will represent a coherent source much brighter than all other XUV sources apart from the XFEL. These particular aspects, correlated with the fact that high order harmonics produced by reflection of an ultrashort and intense laser pulse on an overdense plasma layer impose virtually no restrictions on the intensity of the laser employed, lead the way to a series of potential applications in plasma diagnostics and nonlinear optics. This work presents a theoretical study on the high harmonics generation mechanism occurring at the reflection of a femtosecond Ti-Sapphire terawatt laser pulse ($\lambda_0 = 0.8\mu m$ and pulse durations around 100-150 *fs*) on an overdense steep Al plasma layer. Insights on how different laser parameters influence this generation mechanism and its efficiency are provided.

1. Introduction

Among the non-linear effects that have attracted the attention of the scientific community is high harmonics generation. This is primarily due to their numerous potential applications, like obtaining attosecond or zeptosecond pulses, gaining insight on various plasma parameters such as local electron density and plasma expansion velocity or strong electric and magnetic fields as well as electron transport within a target.

High harmonics generation (HHG) by means of laser interaction with laser-produced overdense plasma layers is an alternative to the same mechanism that occurs in rare gases, proving itself a far more reliable approach in terms of conversion efficiency rate. While HHG in gases is related to atomic nonlinearities, exhibiting increased inefficiency at higher laser intensities due to the ionization occurring in the gaseous medium, in case of plasma targets there is no limitation when it comes to the incident laser intensity, a higher intensity strongly enhancing the production of harmonics. Unlike laser interaction with gaseous mediums, which provide rather low harmonics intensity, harmonic radiation generated by relativistic lasers reflecting on overdense plasma layers may even exceed the intensity of the focused laser by several orders of magnitude. This consequence leads the way to a wide variety of potential applications such as opacity measurements of warm and high density matter with increased temporal and spatial resolution.

When talking about HHG, it is important to consider the type of laser used as the properties of the high order harmonics are strongly influenced by the incident radiation. Furthermore, the conversion efficiency of the laser energy into harmonics is also dependent on the incident pulse. For a *p*-polarized femtosecond laser pulse, Tarasevitch et al. [1] established a conversion efficiency comparable to that of gas harmonics, its order being of 10^{-6} for the 10th order harmonic in case of an incident laser wavelength $\lambda_0 = 0.8\mu m$ with an intensity ranging within $10^{17} - 10^{18} Wcm^{-2}$.

Using a femtosecond laser pulse imposes several restrictions such as a pre-pulse free laser and a high contrast pulse. A pre-pulse leads to significantly less intense harmonics and decreases the highest order of observable peaks while a poor pulse contrast may lead to the suppression of the generation of high harmonics.

Other notable factors that have been accounted for, are the steep density gradient and the polarization of the laser beam. Up to present, HHG has been achieved almost exclusively with *p*-polarized incident laser pulses. However, recently, Tarasevitch et al. [2] have obtained high order harmonics with an *s*-polarized relativistic intensity laser pulse.

Taking into consideration that high harmonics generated with a femtosecond laser lead to a power per solid angle of at least an order of magnitude higher than the one resulted from those generated with a picosecond laser, this paper aims to provide insights on the influence of different parameters of a *fs* Ti-sapphire terawatt laser ($\lambda_0 = 0.8\mu m$ and pulse durations around 100-150 *fs*) on the high harmonics generation mechanism and its efficiency.

2. Using the Particle-in-Cell approach to model high harmonics generation. Results.

Theoretical or experimental investigations of high harmonics generation have reported considerable improvement by using particle-in-cell simulations [3,4,5,6].

This present work uses the LPIC++, a 1D3V fully relativistic code with high spectral resolution in order to describe the laser-plasma interaction. The plasma- placed in the centre of the simulation box- is considered as a steep layer, plane, preionized and collision-less in a 1D description. The laser is interacting either at normal, either at oblique incidence, tapping on the plasma layer from the left side. Practically, this code treats the plasma kinetically, solving the relativistic equations of motion for a set of electrons and ions incorporated within a macroparticle. Each macroparticle obeys a distribution law that satisfies a kinetic Vlasov equation. The 1D3V PIC code accounts for the three components of the particle velocities. Densities, currents and electromagnetic fields are defined on a 1D grid and updated after each particle push. The electromagnetic fields are calculated using Maxwell's equations. The longitudinal electric field is obtained from Poisson's equation while the transverse electromagnetic fields yield from the corresponding current density after having solved the wave equation. The basic algorithm for solving these equations is that of Birdsall and Langdon [7] and it includes the "leap-frog" scheme for resolving particle pushing.

The code further defines the laser pulse by introducing a time-dependent condition for the transverse fields at the front side of the simulation box. The oblique incidence is treated by using the Bourdier method [8].

This study assumes a Ti-sapphire femtosecond laser system. Its intensity is one of the parameters that were adjusted, going as far as the relativistic regime. In other words this intensity is varied between $10^{17} W/cm^2$ and $10^{19} W/cm^2$. The pulse shape was chosen a sine function and the durations of the pulse are 100, respectively 150 *fs*. The emission of the reflected light and, therefore, of the harmonics is restricted to the specular direction, all other effects leading to an emission into a broader angular space being neglected. However, it is known that, in the case of tightly focused pulses, on near-critical plasmas, the electron density is pushed with tremendous force and fast, so that the reflection surface changes orientation during each optical cycle, leading to emission away from specular direction [9].

The plasma layer was chosen to be aluminium, one time ionized. The ions are considered as an immobile background in which electrons move. Femtosecond laser-produced plasmas have little or no time to expand. During the interaction, the plasma surface can thus be represented with a good approximation by a simple step profile, overdense, so that it acts like a mirror reflecting incident light specularly. This is the oscillating mirror model introduced by Bulanov et al. [10].

The density of the plasma layer was varied in the range between $n_e = 2n_c$ and $n_e = 15 \cdot n_c$.

Although the 1D approach of this study might seem simplistic and limited, it has the advantage that it allows for a large number of time steps per laser cycle, along with more particles per cell such that high orders of harmonics can be resolved. Realistic plasma densities demand a high number of particles per cell in order to tackle all important non-linear surface dynamics. Relativistic surface oscillations are responsible for high harmonics generation along with resonance absorption and accounting for the dynamics of this surface is challenging, numerically as well as in terms of computational resources and simulation time.

The number of macroparticles per cell was set to 200 and the temperature adjusted accordingly, taking into consideration both the real physical conditions as well as the numerical ones. The total number of cells in the simulation box is 4096 out of which the plasma occupies 1500 cells, right in the middle of the box. A total of 1500 of vacuum cells can be found in front of the plasma layer in order to avoid the escape of particles. The number of cells per laser wavelength is 1000 and the number of simulation steps is 100 steps per laser cycle.

Simulation results reveal strong oscillations of the critical plasma surface driven by the normal component of the laser field and by the ponderomotive force. Figure 1 and 2 depict such oscillating plasma surfaces, driven by field strengths of $a_0 = 0.7$ (corresponding to $I\lambda^2 = 6.76 \cdot 10^{17} Wcm^{-2} \mu m^2$), respectively by $a_0 = 1$ ($I\lambda^2 = 1.38 \cdot 10^{18} Wcm^{-2} \mu m^2$), with the laser either normally incident, either reaching the surface at an incidence angle of 30° relative to the normal. The laser is considered p polarized.

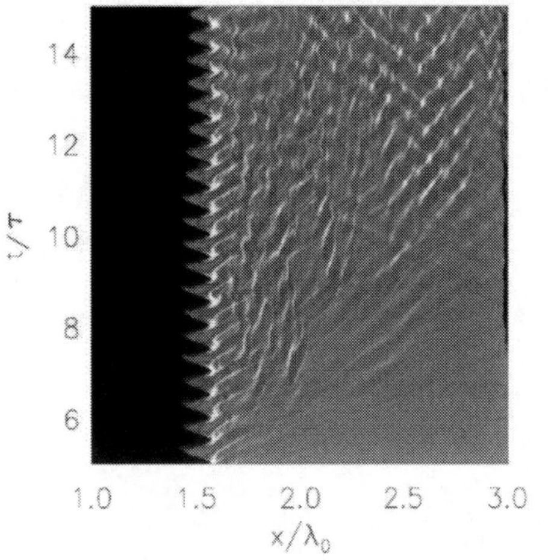

Figure 1 a). Normal incidence, laser field strength $a_0 = 0.7$, pulse duration 150fs, electron density relative to the critical plasma density $n / n_c = 4$.

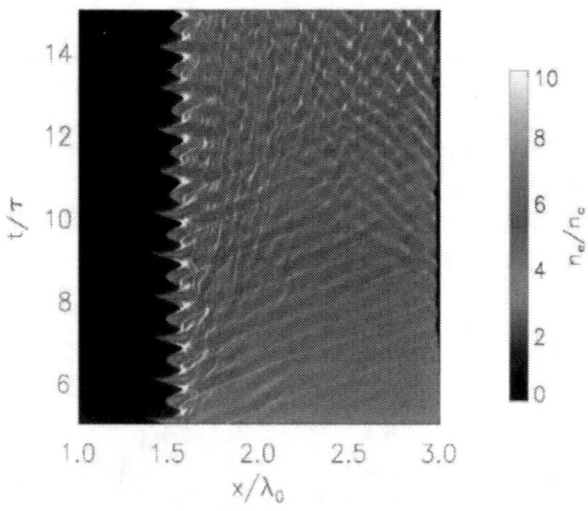

Figure 1 b). Oblique incidence with $\alpha = 30°$, laser field strength $a_0 = 0.7$, pulse duration 150fs, electron density relative to the critical plasma density $n / n_c = 4$, polarization p.

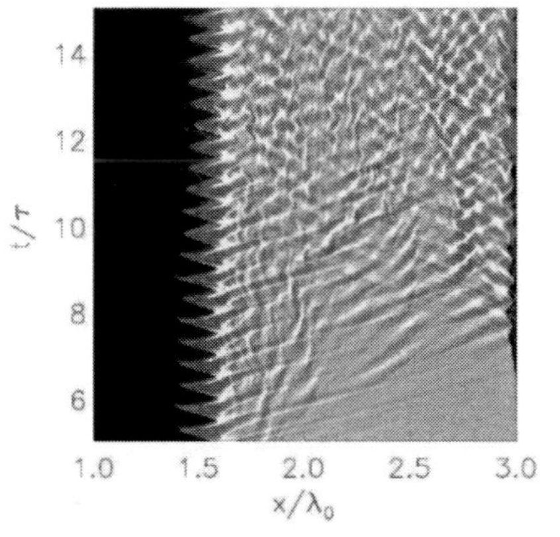

Figure 2 a). Normal incidence, laser field strength $a_0 = 1$, pulse duration 100fs, electron density relative to the critical plasma density $n/n_c = 5$.

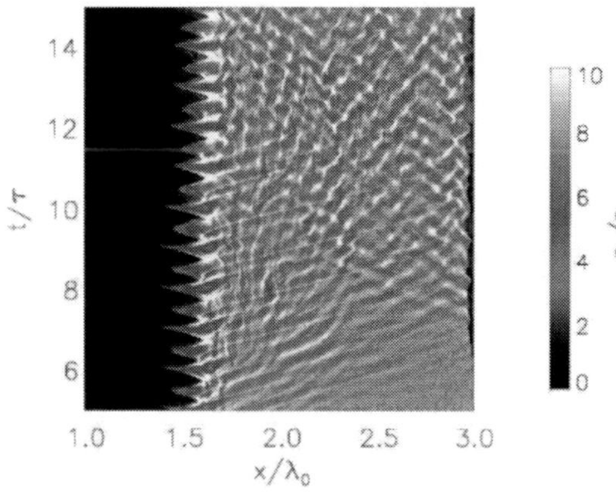

Figure 2 b). Oblique incidence with $\alpha = 30°$, laser field strength $a_0 = 1$, pulse duration 100fs, electron density relative to the critical plasma density $n/n_c = 5$, polarization p.

At intensities above 10^{18} W/cm^2, electrons swing in the laser pulse with relativistic energies. A significant change in the interaction between the electrons and the electromagnetic field occurs in the relativistic regime. The dipole approximation is no longer valid as the amplitude of the electron oscillations is not small with respect to the laser wavelength. Even a harmonically oscillating electron can be a source of harmonics [11]. The plasma electron dynamics does not depend on a_0 and the plasma electron density n_e separately but rather both factors influence its behaviour. This can be expressed by introducing a similarity parameter $S = (n_e / n_c) / a_0$. By observing the yield spectra and the behaviour of plasma in certain density and laser pulse characteristics, one can derive assumptions about the desired plasma density when having a particular laser intensity or about the required laser intensity when knowing the density ratio in order to achieve a desired harmonics spectra.

Figures 3 and 4 illustrate the harmonics spectra obtained by reflection on the oscillating critical plasma surfaces pictured in Figures 1 a) and 2 b). The harmonics spectra obtained here are power spectra of the electric field.

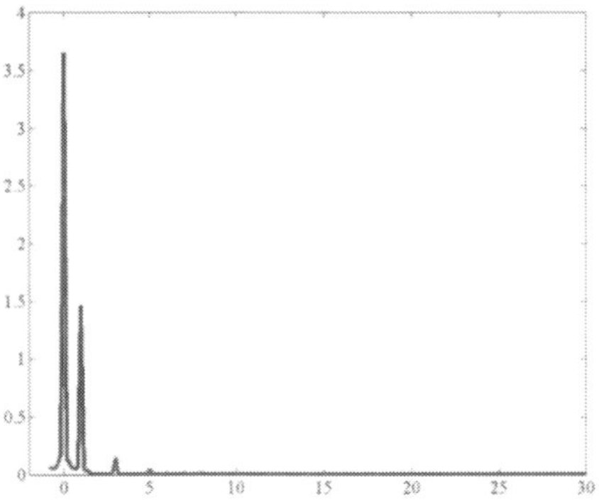

Figure 3 a). Harmonics spectra. Oblique incidence $\alpha = 30°$, $a_0 = 1$, pulse duration 100fs, $n/n_c = 5$, polarization p.

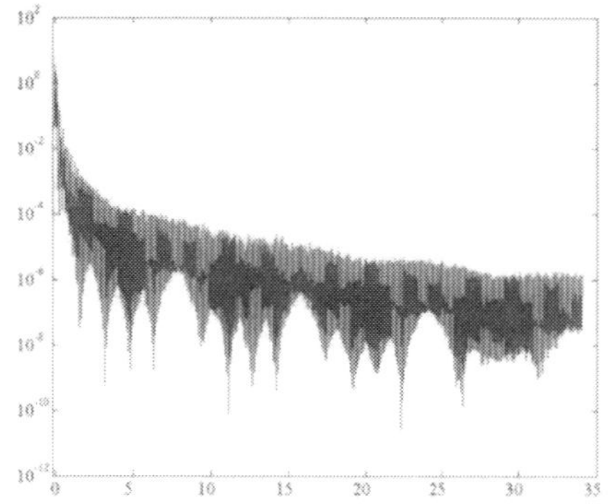

Figure 3 b). Harmonics spectra. The axes have been set logarithmic scales.

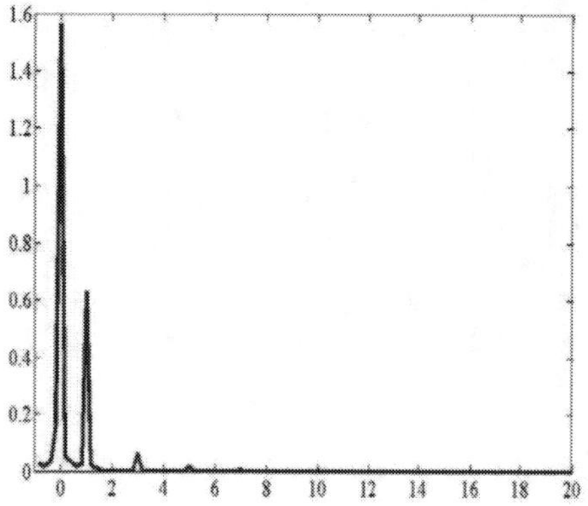

Figure 4 a). Harmonics spectra. Normal incidence. $a_0 = 0.7$, pulse duration 150fs, $n/n_c = 4$.

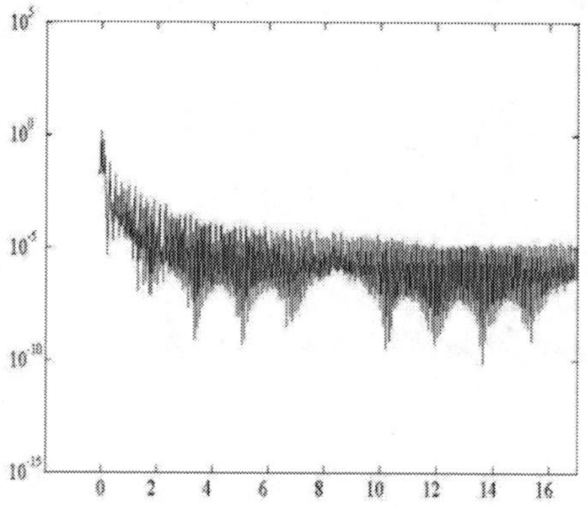

Figure 4 b). Harmonics spectra. The axes have been set logarithmic scales.

3. Conclusion

Harmonic emission increases with the laser intensity but it also depends on the plasma density. A high ration between the electron density and the critical density leads to a nearly flat plasma surface, thus affecting the harmonics yield. A too low ratio causes an extreme disturbance to the plasma layer, consequently completely altering the nature of the reflected pulse. At sufficiently high intensities and moderate densities, surface deformities can be caused by Rayleigh-Taylor like instabilities driven by the imbalance between the ponderomotive laser pressure and the thermal plasma pressure, resulting in spectra rich in high order harmonics.
Shortening the pulse duration as well as using pulses interacting with plasma at oblique incidence were found to produce more harmonics and of higher amplitudes.

References

[1] Tarasevitch A., Orisch A., von der Linde D., Balcou Ph., Rey G., Chambaret J.P, Teubner U., Klopfel D. and Theobald W. 2000, *Phys. Rev. A* **62**, 023816.
[2] Tarasevitch A., Wunsche C. and von der Linde D. 2008, Strong Field Laser Physics, edited by T. Brabec, *Springer Series in Optical Sciences* **134**, 301-319.
[3] Norreys P.A., Zepf M., Moustaizis S., Fews A.P., Zhang J., Lee P., Bakarezos M., Danson C. N., Dyson A., Gibbon P., Loukakos P., Neely D., Walsh F. N., Wark J. S. and Dangor A. E. 1996, *Phys Rev. Lett.* **76**, 1832.
[4] Zepf M. et al. 1996 *Phys. Plasmas* **3**, 3242.
[5] Lichters R., Meyer-ter-Vehn J., Pukhov A. 1996 *Phys. Plasmas* **3**, 3425.
[6] Tarasevitch A., Lobov K., Wunsche C. and von der Linde D. 2007 *Phys. Rev. Lett.* **98**, 103902.
[7] Birdsall C.K. and Langdon A.B. 1991 *"Plasma physics via computer simulations", [Series on Plasma Physics]*, Taylor & Francis Group, LLC, New York.
[8] Bourdier A. 1983 *Phys. Fluids* **26**, 1804.
[9] Naumova N. M., Nees J.A., Hou B., Mourou G.A. and Sokolov I.V. 2004 *Opt. Lett.* **29**, 778.
[10] Bulanov S. V., Naumova N. M. and Pegoraro F. 1994 *Phys. Plasmas* **1**, 745.
[11] Landau E., Lifshitz 1975 *The classical Theory of Fields* (Pergamon, Oxford, 1975) 4th Ed.

Thomson parabola spectrometry of laser generated plasma at PALS laboratory

M Cutroneo[1], L Torrisi[1,2], S Cavallaro[2], L Ando'[2], A Velyhan[3]
[1] Dip.to di Fisica, Università di Messina, V.le F.S. D'Alcontres 31, 98166 S. Agata, Messina, Italy
[2] INFN-Laboratori Nazionali del Sud, V. S. Sofia 62, 95123 Catania, Italy
[3] ASCR, PALS, Na Slovance 2, 18221 Prague 8, Czech Republic

E-mail: mari.cutroneo@gmail.com

Abstract. Laser generated Plasma has been obtained at PALS laboratory in Prague irradiating thin films by Target Normal Sheath Acceleration (TNSA) regime. The irradiated targets were polymers and metals with embedded nanostructures and different thicknesses. In the present work, plasma has been characterized by using Thomson Parabola Spectrometer placed in forward direction. The regime of laser intensity was of the order of 10^{16}W/cm^2 at 1.3 μm wavelength. Simulations performed by TOSCA code have been employed to compare theoretical prevision with experimental data. This approach permitted the recognition of parabolas and the evaluations of ion charge, energy and mass-to-charge ratio. Results revealed that the maximum ion acceleration is obtained n metallic foils for optimal thickness of the order of 10 μm and for target containing nanostructures responsible for the increase of the plasma electron density and resonant absorption effect, as will be presented and discussed.

1. Introduction

In the last years many improvements occurred in the field of laser-matter interaction and in many laboratories of the world have been achieved accelerated ion beams of multi-MeV energy [1]. Much kind of detectors have been employed to analyse the ions beams plasma generated, such as ion collectors, semiconductor and track detectors.

A useful instrument able to detect energy spectra of all ions species in a single laser shot, in a well defined solid angle, is the Thomson Parabola Spectrometer (TPS) [2]. This detector can be available for high laser power density and sub-nanosecond laser pulses irradiating solid matter and producing hot plasmas expanding in vacuum at supersonic velocity. The non-equilibrium charge distribution of the plasma generates high electric fields driving ion acceleration at energies exceeding 1 MeV per charge state [3]. The laser-accelerated ion streams can be characterized in term of energy and charge distributions, maximum ion energy, angular divergence and emission yield, as well as shot-to-shot reproducibility. The Thomson Parabola Spectrometer deflecting a little sampling of ion emission is fast and shows high sensitivity in mass-to-charge ratio. However comparative techniques of ion analysis are need to avoid possible errors of interpretation in parabola recognition, energy measurements and yield emission. To this TPS spectra generally are compared with time-of-flight (TOF) spectra obtained using semiconductor or ion collector detectors, track detectors and scintillators.

2. Experimental setup

At PALS laboratory in Prague an Iodine laser system with 10^{16} W/cm^2 laser intensity, 1.315 μm fundamental wavelengths, 300 ps pulse duration and a beam focus diameter of 70 μm, operating in single shot mode, was employed [4]. Laser irradiates solid thin films in Target Normal Sheath Acceleration (TNSA) regime in order to generate non-equilibrium plasma in forward direction i.e. in the rear side of the target. The irradiation produces a high electric field driving the ion acceleration. Irradiated targets were thin polymers, such as polyethylene (PE) and mylar and metallic foils of Al and Au.

Ion acceleration emitted from plasma was monitored employing different detectors based on time-of-flight (TOF) configuration, such as ion collectors (IC) and SiC semiconductor detectors, track detectors (CR39) and TPS. Generally SiC detectors were placed at different angles and at 60 cm from the target.

SiC semiconductor detectors were employed efficiently because of their fast response, not sensitivity to the visible light (energy gap 3.26 eV) and proportionality to the energy of detected ions. Their current density signals, J_{SiC} depends on [5]:

$$J_{SiC} = e\,(n_i\,E_i/\varepsilon)\,\mu_{eff}(U_d/d) \qquad (1)$$

where e is the electron charge, n_i is the ion density, E_i is the ion energy, ε is the energy necessary for electron-hole pair creation, μ_{eff} is the electron mobility in SiC, U_d is the bias applied to the semiconductor detector (~ 600 V) and d is the thickness of the semiconductor sensitive layer (~ 100 μm).

A Thomson Parabola Spectrometer has been placed along the normal to the target surface in a forward direction at about 1 m from the target. A sketch of TPS structure is reported in figure 1.

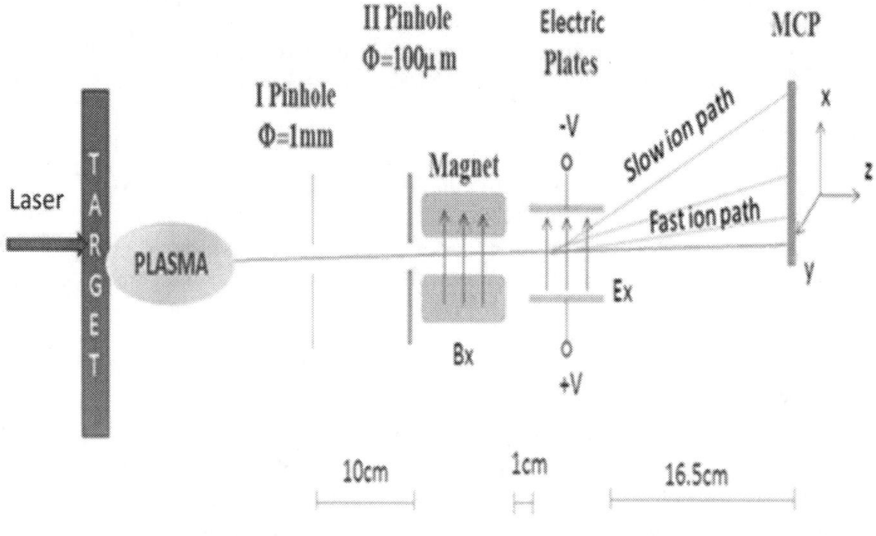

Figure 1. Sketch of Thomson Parabola utilized at the PALS laboratory in Prague.

Two pinholes collimate the ions species accelerated along the normal to the target surface; the first has 1 mm diameter and the second, laced at 10 cm distance from the first, has 100 μm diameter. This second pinhole is located at a distance of 5mm with respect to the magnet. A magnetic field of 0.06-0.12 T and a parallel electric field at 0.5-1.4 kV/cm have been applied orthogonally to the direction of the incident ions. Charged particles are deflected by electrostatic and magnetic fields towards the multi-channel plate (MCP) fixed at a distance of 16.5 cm from the electrostatic plates. The deflection of ions produces a good mass/charge and charge state separation by the image parabolas on a MCP coupled to a phosphorus screen and a fast CCD Camera [6, 7]. By the comparison between the experimental parabola images and the simulation parabolas achieved by Opera 3D/ TOSCA code and Mat-LAB software [8] it is possible to measure the mass per charge state, the charge state and the energies of the detected ions. Moreover, assuming the MCP/phosphorus light signal to be proportional to the number of detected ions, it is possible to have preliminary information about the ion energy and charge state distributions. The Simulations permit the parabola recognition using the real geometry of the used TPS and the real values and shapes of the used magnetic and electric fields.

3. Results

Measurements performed at 10^{16} W/cm^2 laser intensities, irradiating thin foils in TNSA configuration, permitted to evidence that ions are emitted at energies up to about 4 MeV per charge state. Fig. 2a reports a typical TPS spectrum obtained by irradiating 6 μm thin PE target with 590 J pulse energy. The spectrum shows a circular zone where photons and neutral particles arrive on MCP and a lot of parabolas outgoing from this circle. Fig. 2b shows the conversion of the experimental spectrum in gray scale levels and the simulation data (lines) overlapped to the experimental one, as obtained by Opera 3D/ TOSCA code and Mat-LAB software.

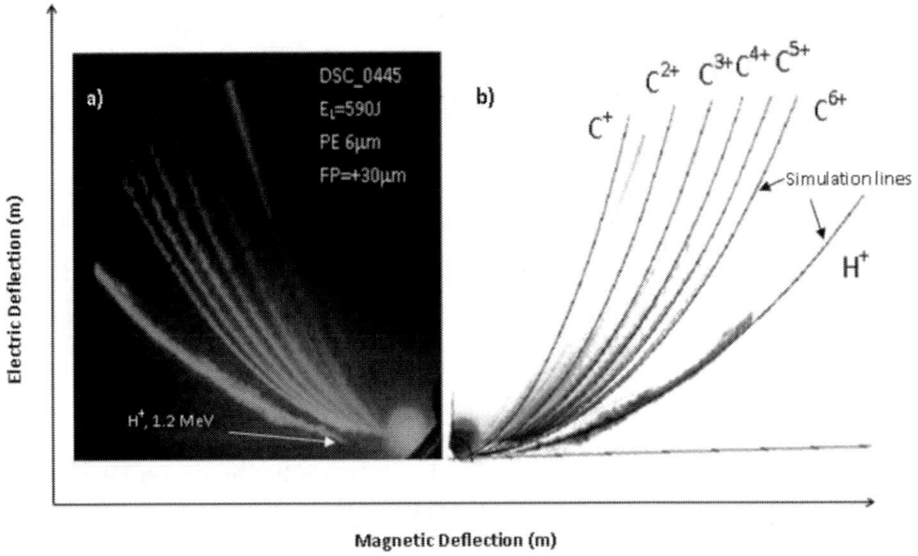

Figure 2. Experimental parabolas from a 6 μm PE irradiation at 590 J (a), and spectra transformation in gray scale with identification of the different parabolas (b).

The lower parabola is due to proton deflections and the others to C ions with all the six charge states. The parabola points nearest to the circular zone are due to the detection of highest energy ions. The distance between the protons parabola and the centre circle is compatible with a proton maximum energy of 1.2 MeV.

In order to confirm the agreement with experimental TPS results and TOSCA simulations, measurements were compared with those obtained using SiC detector spectra.

Figure 3 shows a typical spectrum of a SiC detector obtained during the same laser irradiation of 6 μm PE giving the spectrum of Fig. 2. SiC is placed in forward direction at 30° and 60 cm flight distance. Spectrum shows a photopeak, which represents the start of the TOF measurements, a background due to electron and/or electron Bremsstrahlung detection and a large peak due to the ion detection. The faster ions are due to protons located at 35 ns and corresponding to a kinetic energy of 1.5 MeV, in good agreement with TPS spectrum.

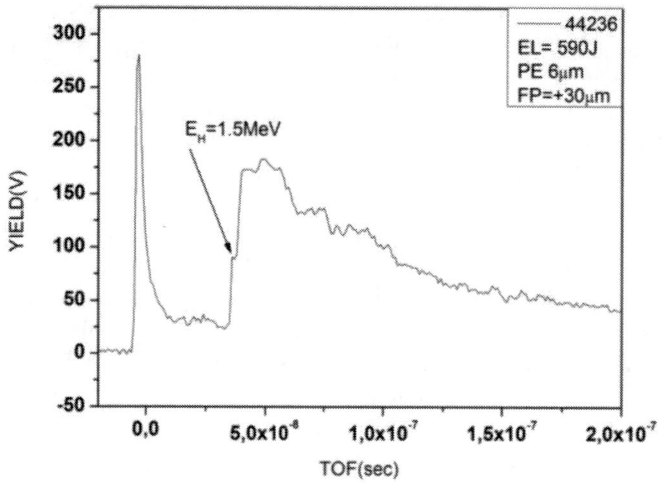

Figure 3. SiC spectrum achieved irradiating PE-TNSA target 6 μm thick.

In order to confirm the agreement with experimental TPS results and TOSCA simulations, measurements were compared with those obtained using SiC detector spectra.

Figure 3 shows a typical spectrum of a SiC detector obtained during the same laser irradiation of 6 μm PE giving the spectrum of Fig. 2. SiC is placed in forward direction at 30° and 60 cm flight distance. Spectrum shows a photopeak, which represents the start of the TOF measurements, a background due to electron and/or electron Bremsstrahlung detection and a large peak due to the ion detection. The faster ions are due to protons located at 35 ns and corresponding to a kinetic energy of 1.5 MeV, in good agreement with TPS spectrum.

The relative light intensity of the TPS parabolas permits to elaborate the ion energy and charge state distributions. For the 6 μm PE TNSA irradiation hydrogen and carbon atoms are all ionized.

Figure 4a shows the ion energy distributions with, as a first approximation, Boltzmann-like shapes. Protons reach the energy of 1.5 MeV while carbon ions reach the energies of 900 eV, 1.8 MeV, 2.7 MeV, 3.6 MeV, 5 MeV and 6 MeV for the charge states C^{1+}, C^{2+}, C^{3+}, C^{4+}, C^{5+} and C^{6+}, respectively. Thus in average an acceleration of about 1 MeV/charge sate is obtained. Fig. 4b shows the charge distribution indicating that proton charge is higher with respect to the carbon one, due to the higher hydrogen content for stoichiometry and for absorbed gas in the target.

Irradiating polymer in which metallic nanostructures are embedded, resonant absorption effects have been observed due to plasmon resonances induced by the laser light. Irradiating thin metallic foils with the optimal thickness of 10 μm the plasma electron energy increases. Using advanced targets in which reflection coefficient is reduced, due to peculiar surface treatments, the absorption coefficient of the laser light increases. In these three cases the electric field driving ion acceleration generated in the rear side of the TNSA target increases and, with the used laser intensity, protons with a kinetic energy above 4 MeV have been measured.

4. Discussion and conclusions

Measurements performed with a Thomson Parabola Spectrometer give information about ion species, energy distribution and charge states distribution (that can be given by the intensity light of the CCD images) in a single laser shot.

An interesting obtained result is that the proton energy achieved by using TPS is lower than the value obtained by SiC detector. This behaviour is explained considering the higher sensitivity of the semiconductor device, with respect the TPS one's. The plasma temperature, evaluated as a first approximation from the maximum charge states measured through TPS, increases with the electron density of the target. This means that high plasma temperatures and high acceleration of ions can be obtained in heavy metallic targets with respect light metals or polymeric ones.

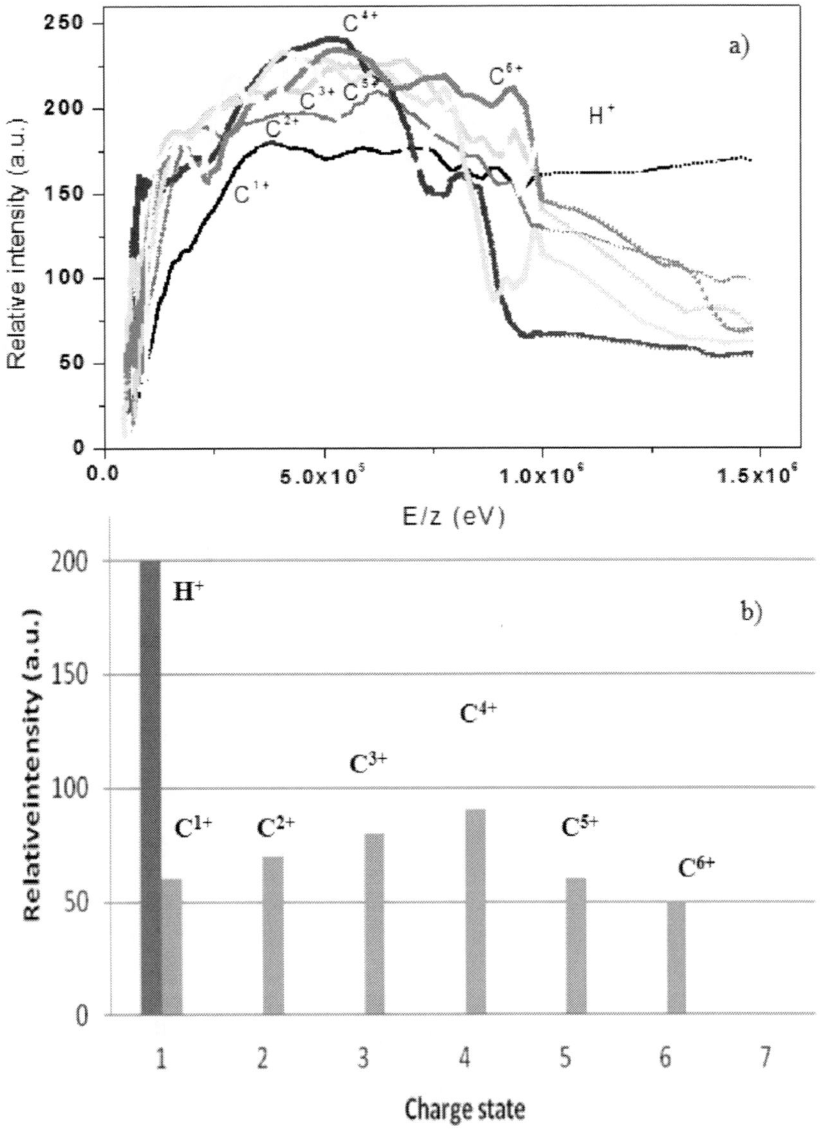

Figure 4. distributions (a) and charge state distributions (b) for 0.6 μm polyethylene target irradiated at 590 J.

References

[1] M. Hegelich et al., 2002 *Phys. Rev. Lett.* **89**, 085002

[2] W. Mrozet al., 2000 *Rev. Sci. Instr.***7**, 1417

[3] M. Borghesi, J. Fuchs, S.V. Bulanov et Al., 2006 *Fusion Science and Tech.* **49**, 412

[4] L. Torrisi, M. Cutroneo, S.Cavallaro et al., 2011 ISBN:978-83- 926290-1-6

[5] M. Cutroneo, P. Musumeci, M. Zimbone, L. Torrisi et Al., 2012 *Journal Material Research*, pp 87 – 93

[6] E. Woryna, P. Parys, J. Wolowski, and W. Mroz, 1996 *Laser Part. Beams* **14**, 293

[7] L. Torrisi, A. Borrielli, F. Caridi et al.,2008 *ECA* Vol. **32D** P-2.144

[8] www.technosoft.biz/ns/tosca.php

Journal of Physics: Conference Series **508** (2014) 012021 doi:10.1088/1742-6596/508/1/012021

A study of the secondary electrons yield produced by laser induced ions on Faraday cups as a function of beam angle

V Nassisi[1,2,3], D Delle Side[1,2], L Velardi[1,2], G Buccolieri[1,2], A De Benedittis[1] and F Paladini[1]

[1] LEAS - Dipartimento di Matematica e Fisica "*Ennio De Giorgi*", Università del Salento, Via per Arnesano, 73100, Lecce - Italy

[2] INFN Sezione di Lecce, Via per Arnesano, 73100, Lecce - Italy

E-mail: vincenzo.nassisi@le.infn.it

Abstract. We analyze the dependence of the Secondary Electron Emission yield on the beam incidence angle (θ) in the case of Faraday cups used for diagnosis of laser-plasma. We show that this dependence follows a $\sec^{\alpha}(\theta)$ law, consequently $\theta \neq 0$ situations (i.e. non-normal incidence) should be treated with care to avoid wrong measurements.

1. Introduction

Faraday cups (FCs) are widely used as charged particle detectors for characterizing ion and electron beams[1]. In such devices, the charged particles reaching the conductive (metal) surface of the detector induce a current, generally recorded on oscilloscopes or other digital devices. Although the working principle is very simple, the impact of energetic charged particles on the conductive surface is a rather complicated process. In particular, the interaction of the charged particles with the electrons of the surface could causes excitation of electrons and ionization both in the incident and the target species. When this excitation happens near the surface it mainly results in the emission of electrons. This process is called Secondary Electrons Emission (SEE) and could potentially lead to wrong estimation of currents in FC detectors. Indeed, secondary electrons leaving the surface of the detector could be erroneously interpreted as a net positive charge, overestimating ion currents or underestimating electron ones. To overcome this problem, polarized grids in front of the detector surface are generally used[2], in order to suppress electron escape. The aim of this work is to study how SEE depends on the angle of incidence of the projectiles, in order to explore if new geometrical configurations of FCs are applicable.

2. Theory

FCs' theory is quite simple. Charged particles impinging on the detector surface generate a proportional current that is transmitted to the recording electronics by means of a transmission line. This line has generally a R = 50 Ω characteristic impedance, to avoid any reflections during the propagation of the signal. At the same time, also FCs should exhibit the same impedance. For example, a simple FC housed within a vacuum pipe constitutes a capacitor together with the housing. In particular, one should take care of reducing the FC capacitance (C_c) as much as possible trough an

[3] To whom correspondence should be addressed

ad hoc design. In effect, this parameter influences the damping time of the transmission line (TL), given by $R \times C_c$, that has to be small with respect to the duration of the beam under exam, otherwise it would adds to it, compromising it diagnosis. In figure 1 two circuits used for FC are presented.

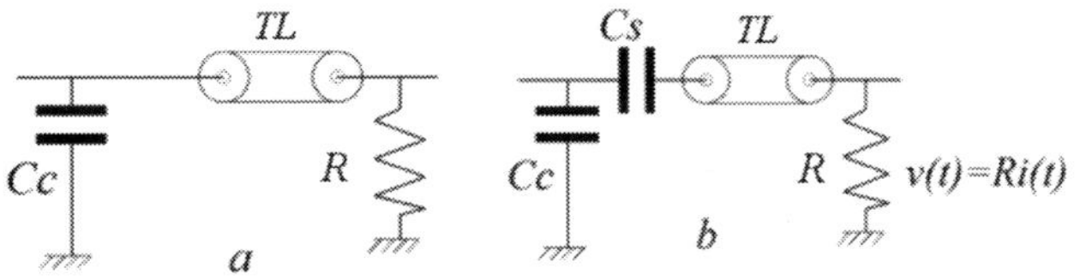

Figure 1. Circuit schematics of Faraday cups for: a) charged beams; b) plasma beams. These devices measure the current *i(t)* generated by the charges impinging on the detector plate.

Using FCs for diagnosis of a plasma flux requires polarization of the surface of the detector in order to ensure that only (for example) ions reach the FC. This imposes to use of a second capacitor, C_s, in order to separate the polarizing voltage from the signal to be diagnosed, as shown in figure 1b.

Let's suppose that we want to diagnose an ion beam (the same applies also to electrons). If the beam ions have sufficient energy, the interaction with the surface of the detector could lead to SEE. This process is the result of the excitation of the electrons of the external shells both of the detectors and of the ions. If the excited electrons are produced near the surface of the detector, depending on their energy, they could escape from the surface, leading to wrong current and total charge measurements. In particular, the total charge registered will be given by

$$Q = (1+\gamma)Q', \tag{1}$$

where Q represents the charge registered, Q' the real charge of the species under exam and γ is the SEE coefficient (in the case of ions $\gamma > 0$, while $\gamma < 0$ for electrons). For this reason, FCs generally use almost transparent electrodes in order suppress SEE, as shown in figure 2.

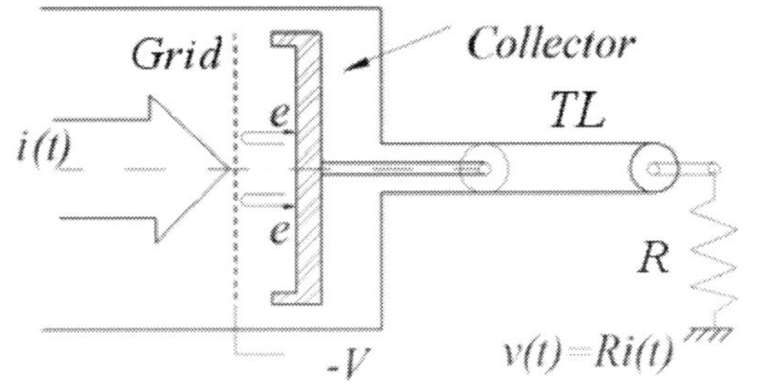

Figure 2. Example of a FC with a grid that acts as SEE suppressor. In the case of plasma ions diagnosis, for example, the suppressor is put at a negative voltage, greater (in absolute value) than the one applied to the metal plate of the FC detector.

The setup illustrated in figure 2 allows to safely consider $\gamma = 0$ in equation (1) at the cost of lower recorded charge/current. Indeed, the suppressor electrodes are generally realized with grids with meshes that stop a share of the charged particles directed towards the FC detector.

An important parameter that has to be considered when dealing with SEE induced by ion beams, is the orientation of the surface with respect to the propagating beam. If we indicate with θ the angle between the beam and the normal to the plate (therefore, $\theta = 0$ represents the case of normal incidence), it could be shown that SEE coefficient depends on θ as

$$\gamma(\theta) = \gamma(0)\sec^{\alpha}(\theta), \tag{2}$$

with α representing a characteristic of both the incident ions and of the plate material[3]. It is worth noticing, that this dependence works only when SEE is given to the kinetic energy of the beam (Kinetic Electron Emission). When the energy of the beam is low, another phenomenon took place, Potential Electron Emission and it behaves differently[3].

3. Materials and methods

The apparatus used during our measurements is the PLATONE setup[4,5]. PLATONE is an electrostatic linear accelerator coupled to a laser ion source (LIS). It uses a Compex 205 excimer laser (λ = 248 nm and τ_{FWHM} = 23 ns) to generate a plasma inside a vacuum chamber. In particular, the plasma is generated within an expansion cylinder (EC, 18 cm long) isolated from ground and connected to an high voltage generator. In this way, plasma expands freely in EC until it reaches a 1.5 cm diameter extraction hole where it starts being electrostatically accelerated by the voltage applied on EC. A pierce ground electrode was placed at a distance of 3 cm from the extraction hole. After this electrode, at a distance of 2 cm, we placed a FC, as the one described in figure 2.

The target used to generate plasmas was a commercial disk of copper, 99.8% pure, placed on the opposite basis of EC with respect to the extraction hole. The laser beam, characterized by an energy of 12 mJ per pulse, was focused in a spot area of 0.005 cm^2, corresponding to an irradiance of 1.0×10^8 W/cm^2. In particular, the beam entered the EC through a quartz window placed on the lateral surface of EC.

To analyze the dependence of SEE yield on the angle θ, we used 4 different FCs, one made by a simple aluminium plate and three made by worked aluminium rods, cut with different tilting, corresponding to θ =30°, 45° and 55°, as shown in figure 3.

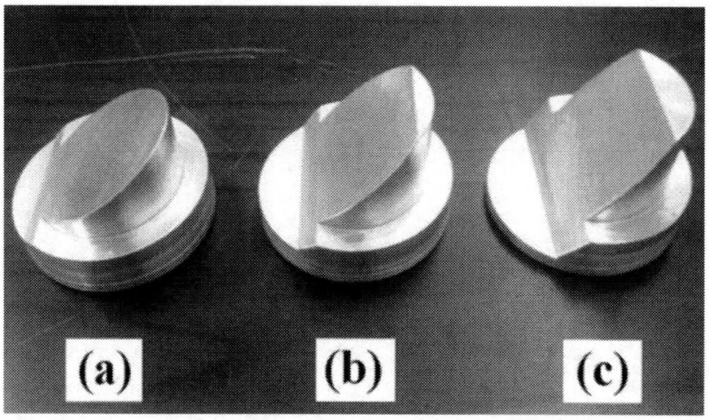

Figure 3. Picture of the tilted collectors housed within the FC used during experiments. The corresponding θ angle is: a) 30°, b) 45° and c) 55°.

For each collector, measurements have been performed both with the suppressor grid polarized in order to avoid SEE and with the grid connected to ground. This method has been used to obtain respectively Q' and Q, which have been deduced integrating the corresponding current signal obtained from the FC. Each value has been calculated 10 times, in order to obtain mean and standard deviation. Then, these values have been used to compute γ, solving equation (1) for it. Moreover, the measures have been conducted applying different accelerating voltages to EC, up to 40 kV.

4. Results and discussion

After cleaning in a ultrasound bath and then with acetone, each collector has been put inside its housing within PLATONE and used as FC with the beams obtained from the LIS. The voltage applied to EC has been raised in order to the understand how γ increases as the beam energy increases. Results of these tests are reported in figure 4.

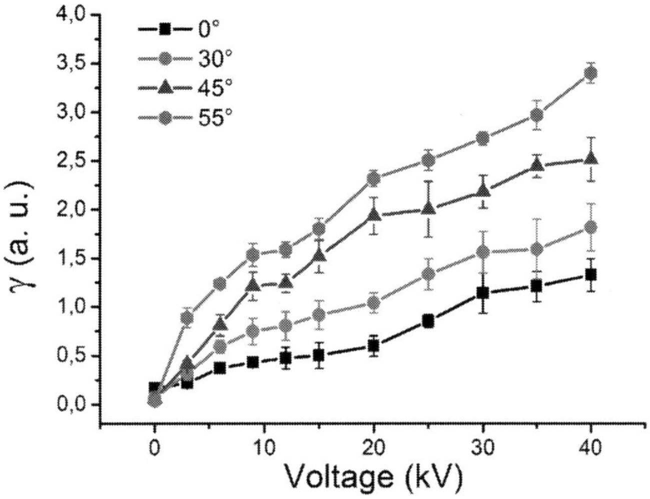

Figure 4. Behaviour of γ as a function of the accelerating voltage, for the various beam incidence angles tested.

The value of γ, in the range of voltages tested, increases as a power law, x^m with $m \approx 0.70 \pm 0.08$. Furthermore, for each voltage, we analyzed the behaviour of γ as a function of the beam incidence. For this reason, we normalized each dataset to the corresponding value obtained at normal incidence ($\theta = 0$), in this way the values should follow a $\sec^{\alpha}\theta$ law. This analysis revealed an interesting characteristic. In the range of voltages 3-20 kV, the normalized values appear to grow rapidly as a function of the angle (figure 5), while in the range 25-40 kV the growth is somewhat slower (figure 6). Instead, at 0 kV, the normalized γ decreases (figure 6).

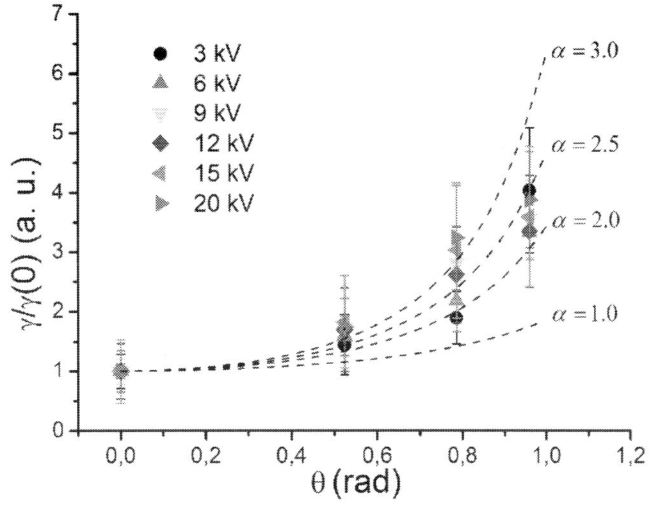

Figure 5. Incidence angle dependence of the normalized SEE yield in the range of 3-20 kV on EC. Dashed lines represent $\sec^{\alpha}\theta$ for various exponents.

Despite of the fact that the measurements have been repeated several times, this behaviour persisted. Since aluminium oxides develops spontaneously when the metal is left in air, we concluded that at low voltages SEE yield is increased by the oxide present on the surface of the collectors[8]. As the measurements proceed, the surface gets cleaned by the ion beams, reducing the SEE yield registered. Consequently, since we started measures using a lower voltage and then we progressively increased it, the growth of γ slows at the higher voltages used. Furthermore, the behaviour at 0 kV could appear unexpected. However, in this case the energy of the beam is very low (below 100 eV), consequently SEE should be ascribed only to Potential Electron Emission, which behave differently from Kinetic Electron Emission[3].

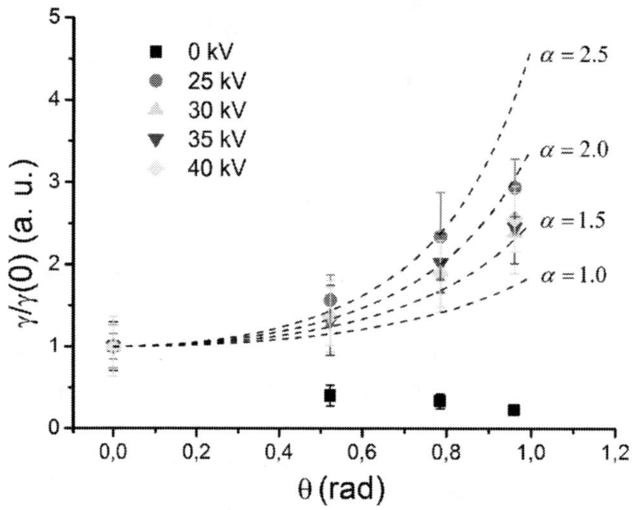

Figure 6. Incidence angle dependence of the normalized SEE yield in the range of 25-40 kV on EC. Dashed lines represent $\sec^{\alpha}\theta$ for various exponents. Moreover, it is presented also the dependence of the 0 kV case.

5. Conclusions

We showed that the SEE yield on FCs used for laser-plasma diagnosis follows a $\sec^{\alpha}\theta$ law with respect to the beam incidence angle. This suggests that whenever non-normal incidence is used, particular care should be taken in order to consider the increased SEE yield. Moreover, we showed also that a good preventive cleaning has to be performed before any measurements, due to metal oxides developed on the surface of the collectors when exposed to air.

References

[1] Luches A, Nassisi V and Perrone M R, 1985 *Rev. Sci. Instrum.* **56** 759-760
[2] Pearlman J S, 1977 *Rev. Sci. Instrum.* **48** 1064-1067
[3] Ferrón J, Alonso E V, Baragiola R A ad Oliva-Florio A, 1981 *Phys. Rev. B* **24** 4412
[4] Velardi L, Siciliano M V, Delle Side D and Nassisi V, 2012 *Rev. Sci. Instrum.* **83** 02B717
[5] Nassisi V, Delle Side D and Velardi L, 2013 *Appl. Surf. Sci.* **272** 114
[6] Dionne J F, 1975 *J. Appl. Phys.* **46** 3347

Journal of Physics: Conference Series **508** (2014) 012022 doi:10.1088/1742-6596/508/1/012022

High magnetic pulse by solenoids for intense ion beam transport

E Giuffreda[1] , L Velardi, D Delle Side, F Paladini and V Nassisi

Department of Mathematics and Physics, University of Salento, LEAS and INFN, via per Arnesano, Lecce, Italy

Abstract. A high voltage pulser up to 50 kV, with a main pulse of 2 μs time duration, was realized using a fast high voltage capacitor of 150 nF (50 kV) connected to the ground by a homemade spark-gap. The circuit, contained in a metallic box to avoid irradiation, had short electric connections in order to reduce the inductances. The diagnostic was performed by two systems: a self-integrating Rogowski coil and an integrator connected to a current transformer, both suitable for pulses longer than 1 μs. To focus ion beams with energy reaching 40 keV, we made two solenoids having an inductance of 3.6 and 16 μH which were fed by the current generated by the pulser up to 12 kA, 1 μs at 50 kV of charging voltage of capacitor. The measurements determined a magnetic field at the solenoid center of 0.61 T for the more inductive solenoid and 1.14 T for the other one.

1. Introduction

In recent years, several techniques of laser-matter interactions have been used in order to produce ion beams. They have already found useful applications in many physical environments as well as out – of – physics applications (biology, medicine etc.). The problem we try to face in this paper is how to create devices able to focus ion beams using magnetic fields. Both quadrupole[1] and solenoidal[2] beam transport schemes are feasible [3], but we will focus on the latter one, attempting to find an optimized configuration to earn the maximum magnetic field strength.

2. Theory

In classical electrodynamics theory, it's easy to show a variation along the radius of a solenoid gives rise to a force which can trap the particles in a magnetic field[4]. In a cylindrical symmetry, supposed the magnetic field **B** is axisymmetric, that means in cylindrical coordinates $B_\vartheta = 0$ and $\partial B / \partial \vartheta = 0$, it's possible to know from Maxwell's equation $\vec{\nabla} \cdot \vec{B} = 0$ written in cylindrical coordinates (r, ϑ, z) :

$$B_r = -\frac{1}{2} r \left(\frac{\partial B_z}{\partial z} \right)_{r=0}$$

assuming a negligible variation of the field with r.

Since the expression for a magnetic field for $r \to 0$ in a finite solenoid is known, it's possible to get the expression for the average force along the z axis as well as its radial component:

[1] To whom correspondence should be addressed

$$\overline{F_z} = -\frac{1}{2}\frac{mv_\perp^2}{B}\left(\frac{\partial B_z}{\partial z}\right)_{r=0}$$

$$\overline{F_r} = \frac{mv_\perp^2}{r_L B}B_z$$

where $v_\perp$ is the speed lying in the plane perpendicular to **B**, depending on the particle q charge through the relation $v_\perp = \frac{|q| r_L B}{m}$, where r_L is the Larmor radius of the particle trajectory.

The focusing of the ions is however not perfect. It depends on many factors, but our work can be simplified by the following considerations: a particle with $v_\perp = 0$ has no magnetic moment which is $\mu = \frac{1}{2}\frac{mv_\perp^2}{B}$, and won't feel any force along **B**. A particle with small rate between speed perpendicular and parallel to **B** will also escape if the maximum value of the magnetic field B is not large enough. That's why we can only focus on the maximum values reached by the magnetic field in the solenoid in order to check how good its focusing skills will be.

3. Materials and methods

We are going to test two different solenoids with different geometrical dimensions, comparing the inductance properties able to satisfy our needs. The first solenoid (denoted in the following with S1) has an inductance of $3.6\mu H$ (11 rings, length 12 cm) and the second one (S2) an inductance of $16\mu H$ (20 rings, length 20 cm). Both S1 and S2 have been realized by a copper wire with diameter 8 mm (S1) and 5 mm (S2). To generate **B**, we connected the solenoids to a pulser with a (negative) high-voltage power supply, composed by a capacitor of 150 nF (100 kV), closed on the solenoids by a homemade fast spark gap. The sketch of the pulser is shown in Fig.1.

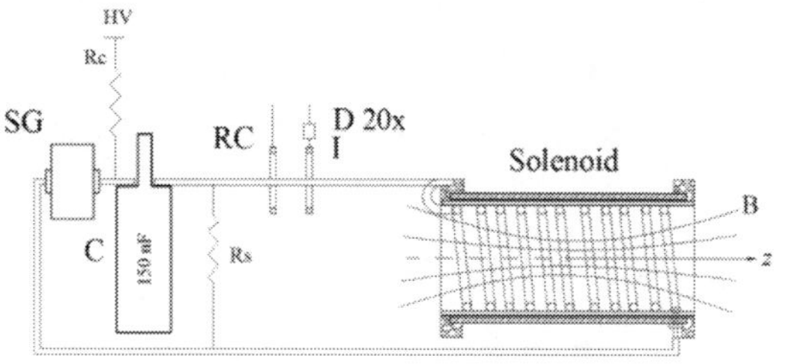

Figure 1. Experimental setup. C: capacitor; SG: spark gap; RC: Rogowski coil, I: integrator; D: divider 20x.

The power supply provided charge to the capacitor through the resistor R_C (200 kΩ) and the resistor R_S (200 kΩ). The discharge current in the capacitor is measured with two different diagnostic systems: a Rogowski coil and an integrator connected to a current transformer.

Our Rogowski coil [5] is composed of 140 rings realized by a conductive wire with diameter 0.8 mm in order to reduce the resistance. The diameter of the rings is 2 cm , while the one of the coil is 3.5 cm. Its inductance is $11\mu H$ and its resistance 0.4 Ω. In order to get the properties of a self-integrator, the coil was closed on a load resistor of 0.5 Ω (see Fig. 2a). In this conditions the integrating time $L/R = 12\mu s$. Therefore, the theory tells us the longest pulse the system can diagnose is less than 12 μs, about 2 μs.

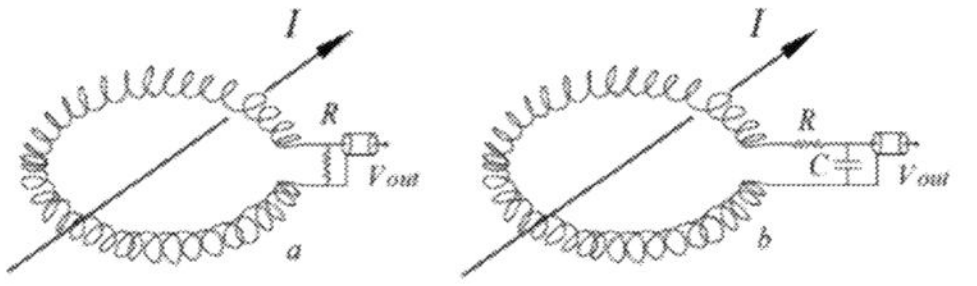

Figure 2. a) Rogowski coil; b) Transformer current with integrator.

The second diagnostic system, the integrator connected to a current transformer, consists of a coil with 38 rings connected to an integrator composed by a resistor of 331 Ω and an capacitor of 47 nF connected to ground. The relatively small number of rings allows to decrease the stray capacitance between the coil and the case, avoiding oscillations and allowing the application of the integrator circuit of high input impedance (See Fig. 2b). In this case the longest pulse the system can diagnose is also about 2 μs.

The measured signals provided by the two detectors, allow us to know the current flowing into the solenoid and consequently the magnetic field inside. As the theory predicts, the output waveform signal is a damped sinusoid.

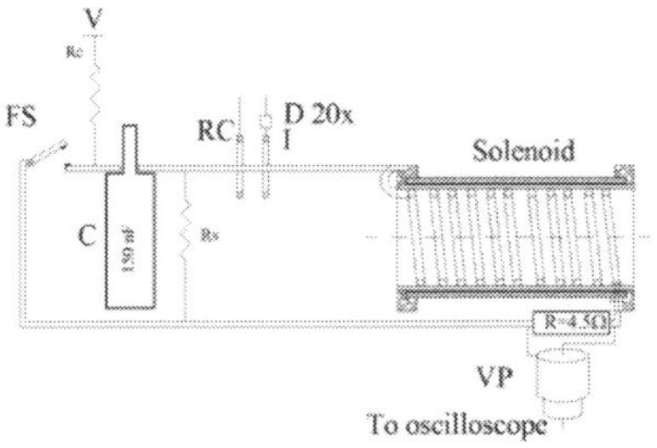

Figure 3. Calibration setup: C: capacitor; FS: fast switch; RC: Rogowski coil, I: integrator; D: divider 20x; VP: voltage probe LeCroy PPE 20kV.

The calibration of both the systems was performed by means of a sinusoidal pulse very close to the one obtained during the experiment. It's obtained by a pulser having a 150 nF capacitor connected by a fast switch to a 50 Ω coaxial structure. Additionally, during the calibration process we used an additional resistor in the circuit of 4.5 Ω to measure the current on the resistor, see Fig. 3.

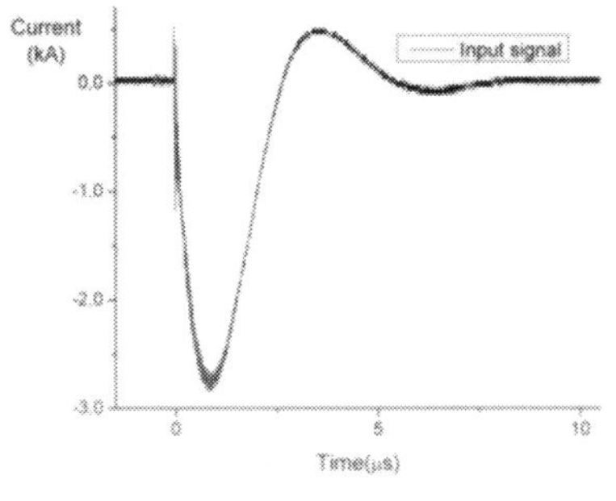

Figure 4. Input signal during calibration in S1.

From the displayed signal, we could calculate the input/output ratio. It's a time-dependent quantity, so the computation of its value was performed in coincidence of the first peak of the signal (≈2 μs). Adopting this method the extracted values are 335 A/V for the Rogowski coil and 519 A/V for the integrator connected to the current transformer at an applied voltage of 20 kV.

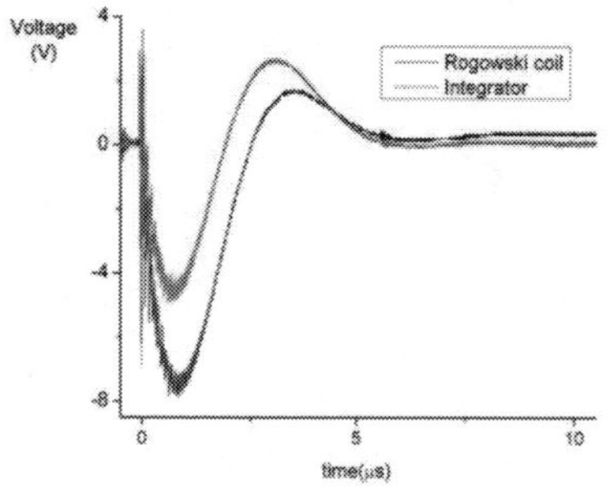

Figure 5. Diagnostic of the signal in S1.

4. Magnetic field mapping

The high current flowing into the solenoid and the presence of high voltage don't allow to measure the magnetic field. What we do to overcome this obstacle is, in a first place, to take a look at the theory: in both a finite and an infinite solenoid, the expression of the magnetic field keeps its linear dependence on the current, although its space dependence drastically changes. In addition to this, as we showed in theory section, the knowledge of the magnetic field B_z as a function of z on the solenoid's axis ($r=0$) can provide us a value for the radial magnetic field and allows us to calculate the average force exerted on the particle along z.

We show a table with the geometrical features of the examined solenoids:

Feature	S1	S2
Radius a	30 mm	45 mm
Coils number N	11	20
Length L	12 cm	20 cm

The solenoids described above have been mapped in a low – current regime (10 A) by means of a Hirst GM07 gaussmeter. The resulting plots are shown here compared with the theoretical behavior:

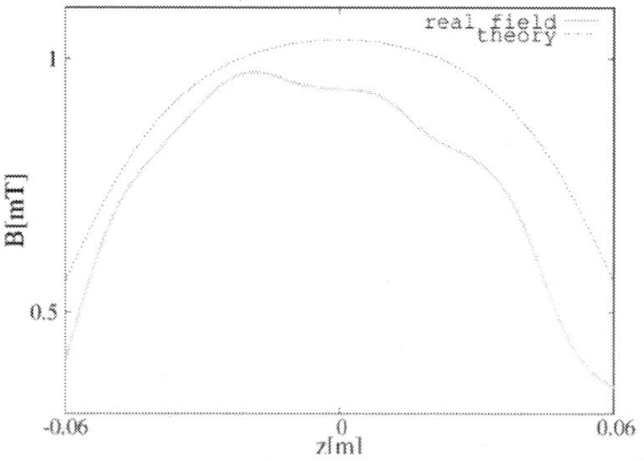

Figure 6. Magnetic field mapping in S1.

It's easy to notice the better agreement of the theory with the magnetic mapping of S2, probably due to a more uniform coil distribution per length unit.

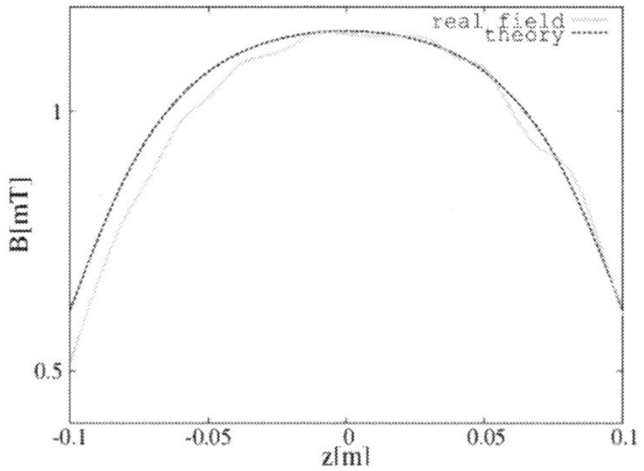

Figure 7. Magnetic field mapping in S2.

The plot takes the origin of the z axis in the middle of the solenoids, so that we can stress how close to the theoretical symmetric behavior the solenoids are. The analytical expression provided by the theory for the magnetic field is [6]:

$$B_z(z) \cong \frac{\mu_0 NI}{2L}\left[\frac{z+\frac{L}{2}}{\sqrt{\left(z+\frac{L}{2}\right)^2+a^2}}-\frac{z-\frac{L}{2}}{\sqrt{\left(z-\frac{L}{2}\right)^2+a^2}}\right] \quad \text{for } r \to 0 .$$

Now it's possible to calculate the maximum value of the magnetic field. Using the procedure described above, denoting the low - current regime with the subscript L and the high - current one with H, we can get an expression for the ratio of the magnetic fields in the same position, hence the magnetic field we're interested to:

$$\frac{B_H}{B_L}=\frac{I_H}{I_L} \Rightarrow B_H=\frac{B_L}{I_L} I_H .$$

This provides us the following values of the magnetic field strength, reached with the maximum applied voltage, 50 kV. We denote with the subscripts 1 and 2, S1 and S2 respectively:

$$B_1 = 1.14T$$
$$B_2 = 0.61T$$

while the flowing currents resulted of 11.7 kA and 5.4 kA, respectively. The magnetic field mapping helps us, using the previous formula, to calculate the magnetic field in the boundaries of the solenoid, obtaining at 50 kV, in z direction:

$$B_1 \cong 0.41T$$
$$B_2 \cong 0.28T \; .$$

Using the relation between z and r components, it's possible to get the radial component in the boundaries ($r = a, z = -L/2$) for S1 and S2:

$$B_{1_r} = 0.31T$$
$$B_{2_r} = 0.09T$$

useful to know the conditions of the particles when they enter the solenoids.

Typical waveform of the magnetic field (current) is showed in Fig. 8 recorded by the Rogowski coil for S1.

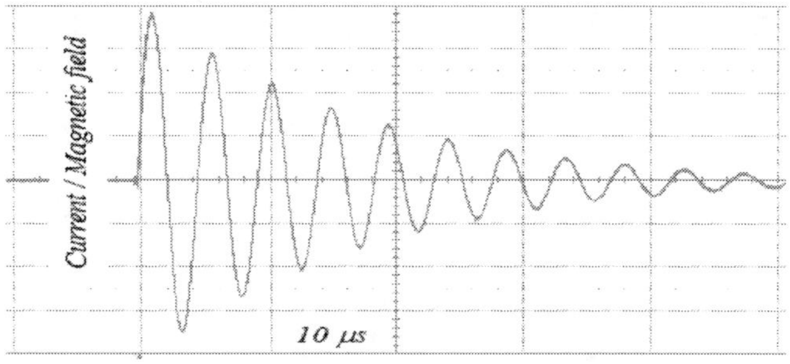

Figure 8. Typical waveform of the current/magnetic field in S1.

As predicted by the theory, the output signal is a damped sinusoid having a frequency of 215 kHz. The corresponding signal for S2 was 103 kHz.

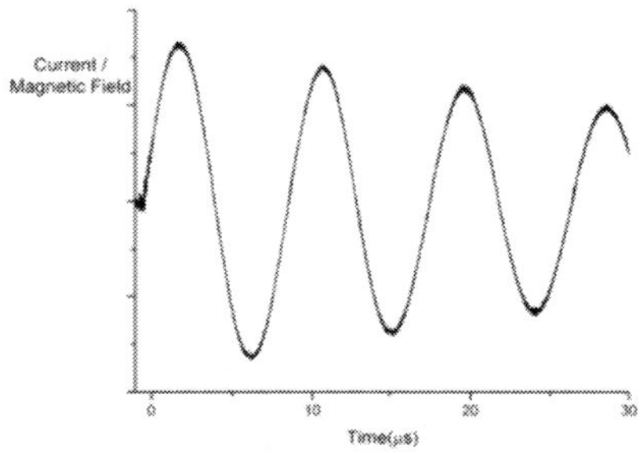

Figure 9. Typical waveform in S2.

5. Conclusions
In this paper the design of a fast high voltage circuit to feed focusing solenoids has been proposed. The experimental setup has been optimized so that the maximum intensity of the current can be achieved in the solenoid and as a consequence also high magnetic field strengths inside the solenoids able to focus ion beams[7, 8]. A rapid calculation [9] based on the solenoids dimensions and our maximum magnetic field intensities can estimate a maximum beam energy for a proton beam about 40 keV in a non-relativistic regime.

References
[1] P A Seidl, D Baca, F M Bieniosek, C M Celata, A Faltens, L R Prost, G Sabbi, W L Waldron, R Cohen, A Friedman, S M Lund, A W Molvik, *The High Current Transport Experiment for Heavy Ion Inertial Fusion,* PAC 2003.
[2] E. Lee, *Solenoid Transport for Heavy Ion Fusion*, 15th Intel Symposium on Heavy Ion Fusion 2004.
[3] D Shuman, E Henestroza, G Ritchie, W Waldron, D Vanacek, S S Yu 2005 *Proceedings of Particle Accelerator Conference, Knoxville, Tennessee,* 3798-802.
[4] F. F. Chen, *Introduction to Plasma Physics and Controlled Fusion vol.1,* Plenum Press, New York and London, 1984
[5] V Nassisi and A Luches 1979 *Rev. Sci. Instrum.* **50**, 900-902.
[6] E E Callaghan and S H Maslen 1960 *The Magnetic Field of a Finite Solenoid,* NASA Technical Note, D-465.
[7] L Velardi, D Delle Side, J Krasa, V Nassisi 2014 *Nucl. Instrum. Methods in Phy. Res. A* **735**, 564-67
[8] S. Agosteo, M.P. Anania, M. Caresana, G. A. P. Cirrone, C. De Martinis, D. Delle Side et al. 2014, to appear in *Nucl. Instrum. Methods Phy. Res. B*, doi: 10.1016/j.nimb.2013.12.035
[9] Vinit Kumar 2009 *Am. J. Phys.* **77**, 737-741.

Journal of Physics: Conference Series **508** (2014) 012023 doi:10.1088/1742-6596/508/1/012023

Neutron fluences of the D-D fusion reaction at 10^{16} W/cm^2 laser-target interactions

S Cavallaro[1,5], L Torrisi [1,2], M Cutroneo[1,2], J Krása[3] and J Ullschmied[4]
[1]INFN-LNS, Via S. Sofia 44, 95123 Catania, Italy
[2]Dip.to di Fisica, Università di Messina, V.le F.S. D'Alcontres 31, 98166 S. Agata, Messina, Italy
[3]Institute of Physics, ASCR, v.v.i., 182 21 Prague 8, Czech Republic
[4]Institute of Plasma Physics, ASCR, v.v.i., 182 00 Prague 8, Czech Republic

E-mail: cavallaro@lns.infn.it

Abstract. In last decade many studies have been carried on deuterium-deuterium nuclear reaction induced in laser-target interactions. The relationships between neutron yields and reaction mechanisms and laser-target patterns need to be further clarified. In this contribution we investigate on fusion yields by changing the target thickness and composition and the laser energy and focal position. The experiment has been performed at PALS Laboratory in Prague. Ion yields have been measured by a Thomson spectrometer and by SiC detectors placed at suitable distances in TOF configuration. Neutron fluences have been evaluated by neutron bubble dosimeters and CR39 track detectors. Results about neutron fluences and fusion process are presented and discussed.

1. Introduction

Intense pulsed lasers have been employed to generate high plasma temperature, electron density and electric field of ion acceleration. High temperatures can result in relative velocities of deuteron ions so large as to initiate nuclear fusion. Moreover, in our experiment a high intensity laser is used to irradiate in vacuum a deuterated target producing plasma from which deuterons are accelerated at MeV energies, as recently demonstrated [1,2]. These ions could induce also nuclear fusion in the same target. The reaction has two branches that occur with equal probability [3]:

$$D + D \longrightarrow T\,(1.01\,\text{MeV}) + p\,(3.03\,\text{MeV}) \tag{1}$$

$$D + D \longrightarrow 3He\,(0.82\,\text{MeV}) + n\,(2.45\,\text{MeV}) \tag{2}$$

Protons and neutrons emitted from the reaction have a characteristic energy of 3.03 MeV and 2.45 MeV, respectively.

[5] To whom correspondence should be addressed

2. Experimental setup

The measurements were carried out the Iodine PALS laser in Prague, operating at 1315 nm, 300 ps pulse duration, 70 μm laser spot diameter, and 10^{16} W/cm^2 pulse intensity [4]. Thick (3 mm thickness) and thin (10 μm thickness) targets, based on deuterated polyethylene $(CD_2)_n$ prepared at the Physics Department of Messina University, have been irradiated to generate plasma in backward and backward/forward directions, respectively. Plasmas were monitored on-line through a Thomson Parabola Spectrometer (TPS), coupled to a Multi-Channel-Plate (MCP), a phosphorous screen and a CCD camera, placed in forward direction along the normal to the target surface, and by detectors connected in time-of-flight (TOF) techniques using ion-collectors (IC) and silicon carbide (SiC) semiconductors. TOF spectra were recorded by a fast storage oscilloscope operating at 20 GS/s. Protons coming from nuclear events were recorded through the TPS, IC and SiC detectors placed in vacuum. Fusion neutrons were detected by using a plastic scintillator placed in air outside the vacuum chamber, at 2.4 m distance and angle of 30° from the target normal direction, as Fig. 1 shows. In the present work, neutron bubble dosimeter pens, with 5.2 bubbles/mrem sensitivity [5], and CR39 track detectors have been also employed to monitor and to detect neutrons. A few CR39 samples were positioned in forward direction around the laser-TPS axis, at $\theta = 4^o$, $\theta = 7.1^o$, $\theta = 7.5^o$, and in backward direction at angles of $\theta = 10.1^o$ and $\theta = 22^o$ with respect to the target normal. The efficiency of CR39 to neutron of 0.5 to 10 MeV energy has been measured with an Am-Be neutron source (34.3 10^9 Bq), 10 cm far from each CR39 frame, covered by a CH_2 film 100 μm in thickness, as proton radiator. A neutron efficiency of about 6 10^{-4} has been measured with this set-up, well taken into account by Monte Carlo simulation. Foils of CR39 have been chemically etched by a 6.25N NaOH solution at 70° C for a time of 12h. Neutron tracks were analyzed by a photo-camera coupled to an optical microscope (4X-obj-10X ocular gain).

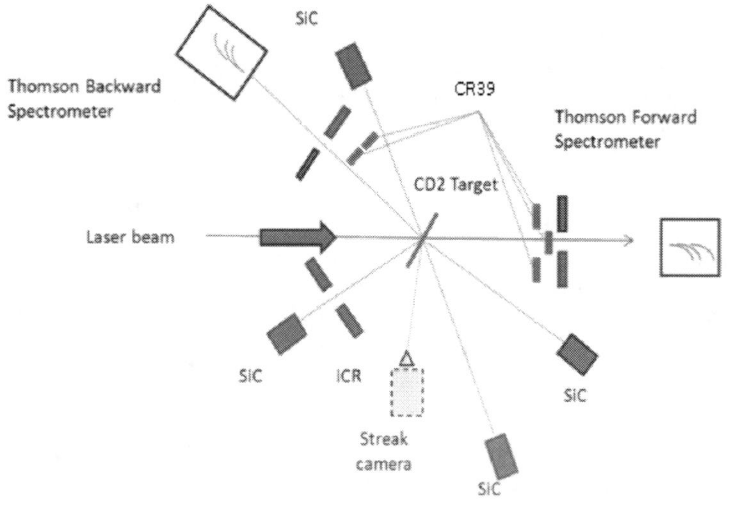

Figure 1. Scheme of the experimental set-up. The horizontal plane defined by the laser direction and the normal to target at 30° with beam axis, is seen from the top of the interaction chamber.

3. Results

The energy of the deuterons accelerated at irradiation of thick and thin CD2 targets. was obtained from TOF measurements. Fig. 2 shows a typical example of SiC-TOF spectrum, measured at 30° forward angle and 60 cm target distance, when irradiating 14-μm thin CD_2 with 606 J of laser energy at a focal position of +50 μm (inside the target surface). The spectrum shows a photopeak caused by photons coming from the plasma (start signal), and high and large slower peaks induced by ions. The front ion peak, located at 25 ns, is probably due to protons with a kinetic energy of 3.0 MeV. The second ion peak located at 40 ns may be associated with fusing deuterons, which produce the second peak of 2.4 MeV neutrons observed with a scintillation detector (Fig. 4). The presence of 3.0 MeV protons can be due to fusion events occurring during the plasma generation. The TPS ion mass analysis is shown in

Fig.3 for the forward detection. The TPS spectra acquired during thin CD2 laser irradiation show very well the deuteron and carbon parabolas being overlapped due to the same charge/mass ratio. We note that the SiC signal is affected by the Schottky contact, which decreases the energy of detected ions and thus strongly modifies the TOF signal, the latter being proportional to the ion energy deposited in the epitaxial layer [6].

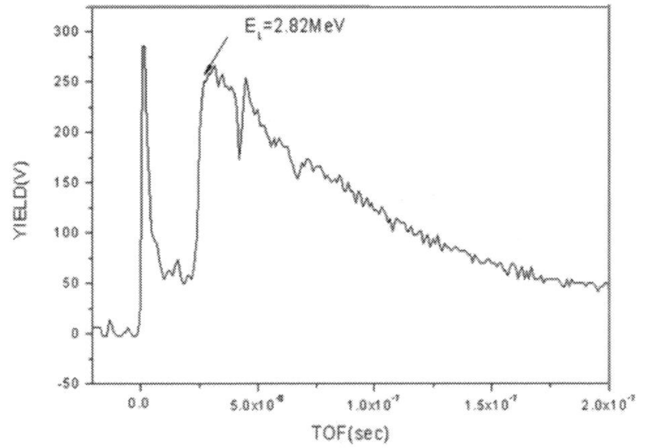

Figure 2. Typical signal of SiC detector obtained during a thin CD_2 laser irradiation.

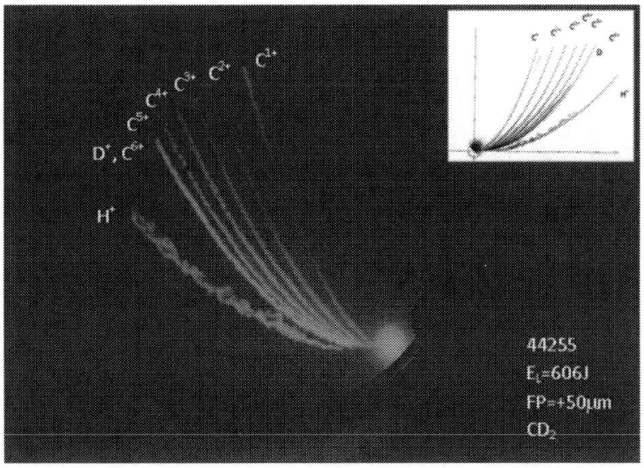

Figure 3. Typical TPS spectrum obtained by laser irradiation of thin CD_2 target.

The proton and neutron emissions from fusion events with the characteristic energy of 3.0 MeV and 2.45 MeV, respectively, were observed by SiC detectors and plastic scintillators in TOF approach when irradiating thick deuterated polyethylene targets. A typical spectrum of neutrons is seen in Fig. 4 showing the presence of a fast photopeak, due mainly to electron Bremsstrahlung in the primary target, and a broad peak centered at about 110 ns and corresponding to the 2.45 MeV neutrons thermally emitted from the targets. The (d, n) channel from the (D+D) fusion reaction with thick targets has been monitored by neutron dosimetric bubble pens (inset), and analyzed by means of CR39 detectors (inset). A direct and clear evidence of nuclear events is shown in spite of small efficiency of detection and high background noise. Table I shows the results obtained by cumulating ten laser shots. By assuming an isotropic neutron emission a value of about 4.5×10^8 neutrons per shot is obtained in good correlation with the values obtained in previous experiments [7,8].

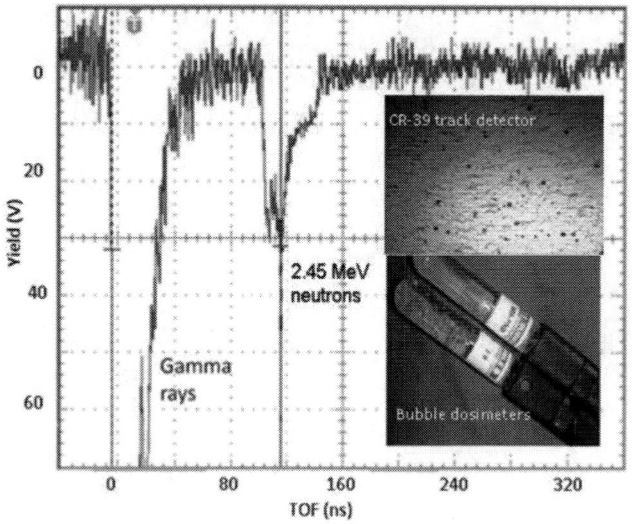

Figure 4. Typical TOF scintillator spectrum, neutron bubble dosimeters and CR-39 track detector.

Table 1. Cumulative values for 10 laser shots.

Detector	Theta (°)	Neutrons/cm^2	Solid angle (sr)	Neutrons/sr
		Forwards		
13	7.1	2.6×10^5	5.67×10^{-4}	4.7×10^8
15	7.5	2.9×10^5	5.67×10^{-4}	5.2×10^8
Stack-CR39	4.1	2.1×10^5	5.67×10^{-4}	3.6×10^8
		Backwards		
16	10.6	1.1×10^6	4.00×10^{-3}	2.6×10^8
17	22.0	4.3×10^5	2.63×10^{-3}	1.6×10^8

4. Conclusions

Deutered polyethylene targets can be irradiated by laser to study nuclear physics aspects in hot plasmas. D-D fusion can be induced in plasmas generated at sub-nanosecond regimes at laser intensities of the order of 10^{16} W/cm^2. Both the accelerated and thermal components of deuterons seem to contribute to the fusion process. An average number of about 4.5×10^8 neutrons per laser shot has been found.

References

[1] Hegelich BM, Albright BJ, Cobble J et Al., 2006 *Nature* **439**, 441-444
[2] Torrisi L, Cavallaro S, Cutroneo M et Al., 2013 *Appl. Surf. Sci.* **272**, 42-45
[3] Martin BR, 2006 Nuclear and Particle Physics (Wiley J & Sons, Ltd. ISBN:0-470-01999-9)
[4] PALS Laboratory http://www.pals.cas.cz/2012
[5] Bubble Technology Industries:
http://www.bubbletech.ca/pdfs/BTI_BUBBLE_General_May72009.pdf
[6] Margarone D, Krása J, Giuffrida L et al. 2011 *J. Appl. Phys.* **109** 103302
[7] Krasa J, Klir D, Velyhan A, et Al. 2013 *Laser Part. Beams* **31**, 395-401
[8] Torrisi L, Cavallaro S, Cutroneo M, et Al. 2012 *Rev. Sci. Instrum.* **83**, 02B11.

Potential of a laser-driven source of characteristic γ-rays for fast detection and identification of nuclear materials

Fabio Belloni

European Commission, Directorate-General for Research & Innovation, Directorate Energy, Rue du Champ de Mars 21, 1049 Brussels, Belgium

E-mail: fabio.belloni@ec.europa.eu

Abstract. Active interrogation of special nuclear materials (SNM) represents the most promising detection technique in combating their illicit trafficking and diverted use. Among the many interrogation approaches under investigation, the one based on photoinduced fission relies on large prompt and delayed radiation signatures. Pulsed induction (tens of ns) of photofission, in particular, enables to distinguish between prompt and delayed radiation emission, providing additional signatures to identify fissile materials. As the probing radiation, 6- to 7-MeV characteristic γ-rays produced by the $^{19}F(p,\alpha\gamma)^{16}O$ reaction are of particular interest; in fact, they have energy sufficient to induce fission in SNM, but insufficient to create significant radioactivity via photonuclear activation, especially in lighter elements. In this study, the feasibility of a pulsed source of characteristic γ-rays based on the interaction of laser-accelerated protons with a fluorine-rich target is assessed. For practical applications, proton energies above 2 MeV, a bunch population of the order of 10^{14} particles or higher, and a laser repetition rate of the order of 10 Hz are of interest. The development of ultrahigh-power table-top lasers could greatly contribute to the deployment of this kind of interrogation source for a variety of operations.

1. Introduction

Illicit trafficking and diverted use of SNM –notably highly enriched uranium (HEU)– represent one of the main concerns for the global safety. The techniques used for the detection and identification of nuclear materials can essentially be distinguished between passive and active interrogation. While passive interrogation lies on the detection of radiation emitted through the spontaneous decay of nuclei (e.g., γ-ray detection and imaging, neutron detection), active interrogation excites nuclei in a target – by either neutrons or γ-rays– and looks for their radiation signatures (neutrons and/or γ's) [1]. One the most interesting features of active interrogation is represented by the fact that the strength of its response is independent of the inherent activity of the target, whereas it is proportional to the intensity of the probing radiation.

Among the many active interrogation approaches under investigation, the one based on photoinduced fission relies on large prompt and delayed radiation signatures [2]. In particular, pulsed induction of photofission (on a timescale of tens of ns) enables to distinguish between prompt and delayed radiation emission, providing additional signatures to identify fissile materials [3,4]. Moreover, the prompt response leads to an improved signal-to-background ratio in a fast detector. As the probing radiation, 6- to 7-MeV characteristic γ-rays produced by the $^{19}F(p,\alpha\gamma)^{16}O$ reaction are of particular interest; in fact, they have energy sufficient to induce fission in SNM, but insufficient to create significant radioactivity via photonuclear activation, especially in lighter elements [4,5]. In stable isotopes, for example, neutron separation energies are more likely to overcome 7 MeV at low rather than at high mass number. Besides the reduced impact in terms of measurement background and radiation protection, this feature also provides the technique with additional discrimination capability between SNM and surrounding ordinary materials. Bremsstrahlung sources, on the contrary, are inherently less selective and provide higher collateral activation.

The most promising experiments on a fast, intense source of mono-energetic γ-rays via the $^{19}F(p,\alpha\gamma)^{16}O$ reaction have so far utilised pinch-reflex ion diodes to generate the proton beam [4,5]. Here, the feasibility of this kind of γ-source by using laser-driven proton beams [6] is comparatively assessed.

2. Phenomenology

The $^{19}F(p,\alpha)^{16}O$ reaction has a Q-value of 8.115 MeV and proceeds via the population of excited levels of the compound nucleus ^{20}Ne. These levels have large α-decay widths to excited states of ^{16}O [7]. The reaction cross section (calculated [8]) is plotted as a function of proton energy in figure 1. For proton energies below 4 MeV, transitions to the ground state and four excited states of ^{16}O mostly occur, through the emission of five α-groups. The second, third and fourth excited levels of ^{16}O almost exclusively decay to the ground state directly, by the emission of γ-rays with energies of 6.129, 6.917 and 7.116 MeV, respectively. For proton energies above 4 MeV, excitation of higher levels in the residual ^{16}O nucleus becomes more likely, which however leads –with high probability– to cascade decays through one of the 6.129-, 6.917-, or 7.116-MeV levels [7,9].

The number of photons, N_γ, produced over 4π sr by N_p protons impinging on a thick target with an energy distribution $f(E)$ can easily be calculated if the thick-target yield, Y, is known as a function of the incident proton energy:

$$N_\gamma = 4\pi e N_p \int_0^\infty f(E)Y(E)dE \qquad (1)$$

where e is the elementary charge and $[Y] = C^{-1}sr^{-1}$. Systematic measurements of total and fractional thick-target γ-yields for the $^{19}F(p,\alpha\gamma)^{16}O$ reaction only exist for proton energies up to 4.0 MeV [9, and references therein]. Results confirm that the total photon yield increases with increasing proton energy, and show how the relative intensity of the individual γ-lines depends on proton energy. The γ-emission is found essentially isotropic. In figure 2, total γ-yields have been plotted for CaF_2, MgF_2 and SF_6 thick targets, using the data of Micklich *et al.* [9]. These three fluorides are of particular interest as possible proton targets in an actual γ-source. Curves in figure 2 are generated by a sigmoidal function of the form $Y(E) = A\cdot\left(1-\left(1+\exp\left((E-E_0)/b_0\right)\right)^{-1}\right)$, where A, E_0 and b_0 are parameters retrieved from the experiments and tabulated in ref. 9 (we note that the expression for the sigmoid presented there is incorrect). The highest yields have been measured at $E = 4.0$ MeV, and vary from 4.7x10^7 to 6.0x10^7 μC^{-1}sr^{-1}, depending on the fluoride. Similar or even higher values are expected up to $E \approx 9$ MeV (see figure 1), for the sum of the three γ-lines of interest. Figure 2 also shows that the yield becomes considerable (e.g., within a factor of 10 from the plateau value) only for proton energies above 2 MeV. In what follows, we will retain the energy region highlighted in figure 1 –approximately ranging from 2 to 11 MeV– as the most favourable one in terms of photon yield.

Sources with an intensity of at least 10^{12} photons per pulse can offer reasonable fluences (> 10^3 cm^{-2}) on objects at large stand-off (up to 100 m) to provide a detectable signal (prompt/delayed fission signature) [5], on the basis of state-of-the-art sensitivity of the detection techniques and typical SNM target amounts (e.g., 5 kg HEU or 1 kg ^{239}Pu, with thick metal shielding). Assuming a proton spectrum bound within the 2÷11-MeV RoI, and $\langle Y\rangle_f = 5.0\times10^7$ μC^{-1}sr^{-1} as the spectrum-averaged value of Y in equation (1), $N_p = 1.0\times10^{16}$ is calculated for $N_\gamma = 1.0\times10^{12}$.

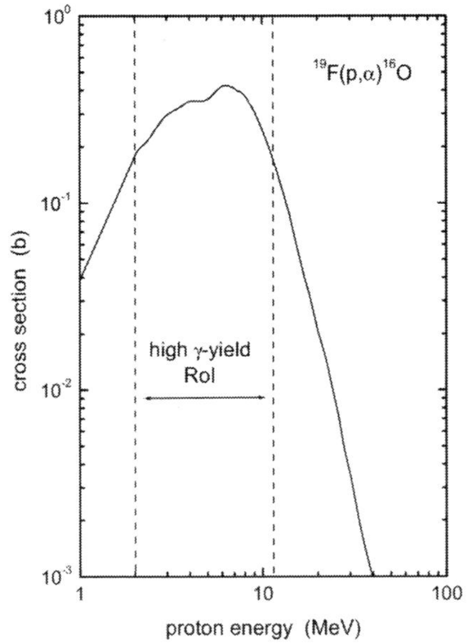

Figure 1. Cross section for the $^{19}F(p,\alpha)^{16}O$ reaction, as a function of proton energy. Calculated values from the TENDL-2009 library [8] have been plotted. A high-yield region for γ-ray emission is also indicated.

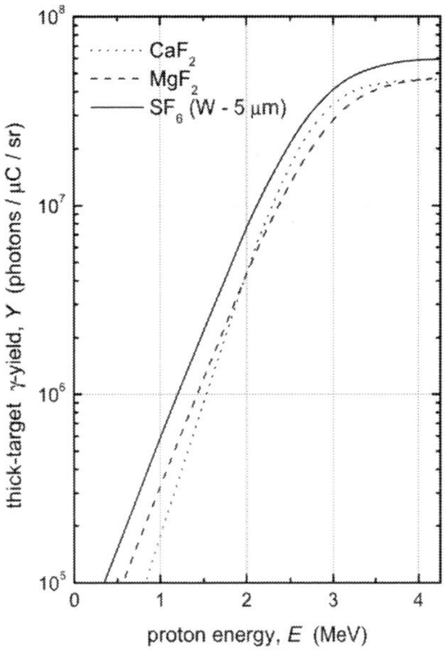

Figure 2. Total thick-target γ-ray yield (summed over the 6.129-, 6.917- and 7.116-MeV lines) for the $^{19}F(p,\alpha\gamma)^{16}O$ reaction in three fluorides of interest, as a function of the incident proton energy. For SF_6, the effective proton energy through a 5-µm W foil which windowed the gas cell is used.

3. Experimental approaches

3.1 Ion-beam diode

Ion diodes supplied by fast high-voltage generators have been used to produce intense pulsed proton beams for a variety of applications [10]. Schumer *et al.* [5] have been able to generate a 2-MeV, 270-kA proton beam of 75-ns duration, which resulted in a γ-ray pulse of 50-ns duration and intensity estimated in $3.1x10^{11}$ photons after impinging on a polytetrafluoroethylene (Teflon) target. The ion diode they used was constituted of a hollow thin-wall cylindrical stainless-steel cathode (12 cm in diameter, connected to ground) and an annular Al anode, connected to a 2-MV pulsed power generator. The anode Al ring was closed by a polyethylene (CH_2) foil, while the hollow cathode was closed by a thin polycarbonate disc. By the effect of the high-voltage pulse, electrons emitted from the cathode initially bombard the annular Al anode; as the current increases to several hundred kA, electrons self-pinch onto the polyethylene foil, producing a plasma. Plasma protons are accelerated towards the polycarbonate disc, which they pass through before entering a 1-Torr air region where the beam is charge-neutralized. In this region, the beam is ballistically transported to the Teflon target, where the γ-ray emission occurs. The Teflon target does not survive the interaction with the proton beam up to a distance of 103 cm from the anode. At this distance it however shows sever damage, characterized by surface melting within a region of 50 cm in diameter. Placing the target at shorter distances would moreover be detrimental to the purpose of active interrogation, because of the intense bremsstrahlung pulse stemming from the anode. This pulse, which can provide unwanted background, has to be well separated in time from the characteristic γ-ray pulse.

3.2 Laser-driven proton source

As has been widely demonstrated in the past few years, the interaction of ultrahigh-power laser pulses (intensity of the order of or higher than 10^{19} Wcm^{-2}, sub-ps duration) with thin foils (thickness of 1÷100 μm) is able to produce well collimated (< 10°), ultrashort (few ps) bunches of protons with energy ranging from a few MeV up to several tens of MeV (see e.g. ref. 6 for a review). In the case of metal foils, protons are mostly originated by hydrogen adsorbed onto the target surface and by hydrocarbon surface contamination, and have also been detected at lower laser intensities [11]; nevertheless, polymer foils and metal/polymer composite foils are also used. Among the possible applications of this technique, the proton beam can be directed against a secondary target, placed at a distance of a few cm from the laser target, to induce nuclear reactions. For example, (p,xn) reactions have been demonstrated in a Pb target by McKenna *et al.* [12].

For applications to a γ-interrogation-source, the results of Snavely *et al.* [13] are here considered as a benchmark. Following irradiation of a 100-μm CH polymer foil by a 1 PW laser pulse (peak intensity on the target of $3\text{x}10^{20}$ Wcm^{-2}), protons with energy up to 58 MeV were generated. The energy spectrum was found exponential in shape, with a mean energy of 4 MeV. Above 10 MeV, $2\text{x}10^{13}$ protons were measured. These figures are typical and have been replicated by other authors using similar experimental conditions [14,15]. From the data of Snavely *et al.* [13] we extrapolate $N_p(E > 2$ MeV$) = 1.5\text{x}10^{14}$; values up to $5\text{x}10^{14}$ do not seem unrealistic by making use of engineered targets.

4. Discussion

With an energy spectrum like that generated by e.g. Snavely *et al.* [13], all protons with $E > 2$ MeV would significantly contribute to the ^{19}F(p,αγ)^{16}O reaction. In fact, neglecting other kinds of nuclear reactions[1], protons with energy beyond the RoI of figure 1 will approach that region while slowing down in the target, "climbing up" the cross section curve. With $N_p = 1.5\text{x}10^{14}$ and $\langle Y \rangle = 5.0\text{x}10^7$ μC^{-1}sr^{-1}, one calculates $N_\gamma = 1.5\text{x}10^{10}$ in a single pulse, which is a value already appropriate for interrogation at close and medium distance (e.g., below 10÷20 m).

The recent development of 100-TW class Ti:sapphire lasers characterised by ultrashort pulses (around 30 fs) and operation in repetition mode (10 Hz) [16] shows the potential to bring table-top lasers to emulate the proton-acceleration performances of higher energy, longer pulse-duration, single-shot laser systems taken as a reference in this paper. In connection to active interrogation, one can notice that assuming e.g. $N_p = 5.0\text{x}10^{14}$ and $\langle Y \rangle = 5.0\text{x}10^7$ μC^{-1}sr^{-1}, $N_\gamma = 1.0\text{x}10^{12}$ could be achieved in 1 s with a 20-Hz repetition rate. The constraint on the distance over which a repetitive source could find application comes from the detection threshold of the signature signal on the single pulse; if the single-pulse response is detectable, then integration of the signal over a sequence of pulses is possible. This way, values of N_γ much higher than those achievable by single-shot sources could be built up in a few seconds. Furthermore, a repetitive source would provide a unique solution for the interrogation of (fast-)moving objects (e.g., means of transport) or series of objects (e.g., baggage on a conveyor belt, or containers on a train). Finally, table-top ultrahigh-power lasers with a repetition rate of the order of 10 Hz could match the features of compactness, operability and portability that an interrogation source should exhibit in situations of practical interest (e.g., inspection at airports, harbours and train stations, or mounting of the source on a mean of transport).

On the contrary, pinch-reflex ion diodes do not seem to show the same potential for development in terms of the abovementioned features. In particular, high-power generators with the characteristics described in section 3.1 are inherently single-shot systems. Repetitive operation of the γ-ray source developed by Schumer *et al.* [5] is further complicated by the damage or destruction of the diode foils

[1] In particular, the (p,n) reaction is possible in ^{19}F for proton energies above 4.23 MeV, with cross sections between 25 and 60 mb (calculated values [8]) in the energy region ranging from 4.5 to 10 MeV. The cross section rapidly decreases above 10 MeV. Neutron background induced by this reaction should be given careful consideration in experiments and applications.

and of the Teflon target. Moreover, the proton acceleration and transport scheme there applied results in energies close to 2 MeV, which are just at the edge of the high γ-yield region visible in figures 1 and 2. This makes that γ-source quite inefficient (one finds $N_p = 1.3\text{x}10^{17}$ vs. $N_\gamma = 3.1\text{x}10^{11}$; cp. calc. in section 2, $N_p = 1.0\text{x}10^{16}$ vs. $N_\gamma = 1.0\text{x}10^{12}$). The need to work at higher energies is recognised by the authors themselves; as a matter of fact, 4-MeV, 100-kA proton beams can be delivered by state-of-the-art pulsed power generators [5,17]. Amongst other things, a smaller bunch population and a higher-energy, broader spectrum can help mitigate the thermal and collision-damage loads on the target.

5. Conclusion

Findings of recent experiments on a fast, intense source of characteristic γ-rays based on a pinch-reflex ion diode have been reviewed, in connection to possible application to SNM active interrogation, to the phenomenology of the $^{19}F(p,\alpha\gamma)^{16}O$ reaction used for the production of the γ-radiation, and to the state of the art of laser-driven proton sources. A significant potential has been assessed for these latter in "static" and "dynamic" applications (e.g., remote inspection of containers standing in airports and harbours, or moving on means of transport), inasmuch as proton energies well above 2 MeV, a bunch population of the order of 10^{14} particles or higher, and a repetition rate of the order of 10 Hz are achievable. The development of ultrahigh-power table-top lasers could greatly contribute to reach these targets and to meet the requirements in terms of deployability –including portability– of the interrogation source.

Acknowledgements
The kind assistance of Ms. D. Ostojić is gratefully acknowledged.

References

[1] Christian J F *et al* 2008 *Proc. SPIE* **6945** 69451Q
[2] Antolak A, Doyle B, King M, McDaniel F, Morse D, Provencio P, Raber T and Rossi P 2008 A Dual Neutron/Gamma Source for the Fissmat Inspection for Nuclear Detection (FIND) System *Sandia National Laboratories Report (SAND2008-7934)*
[3] Kinlaw M T and Hunt A W 2006 *Nucl. Instr. Meth. Phys. Res. A* **562** 1081
[4] O'Malley J *et al* 2009 *Proc. SPIE* **7304**, 73041J
[5] Schumer J W *et al* 2007 *IEEE Nuclear Science Symposium Conference Record,2007. NSS '07. IEEE* vol 6 (Monterey, CA: IEEE Publishing) p. 1026
[6] Daido H, Nishiuchi M and Pirozhkov A S 2012 *Rep. Prog. Phys.* **75** 056401
[7] Fessler A, Massey T N, Micklich B J and Smith D L 2000 *Nucl. Instr. Meth. Phys. Res. A* **450** 353
[8] Koning A J and Rochman D 2009 *TENDL-2009 Library* (through the ENDF database, available at https://www-nds.iaea.org/exfor/endf.htm)
[9] Micklich B J, Smith D L, Massey T N, Fink C L and Ingram D 2003 *Nucl. Instr. Meth. Phys. Res. A* **505** 1
[10] Golden J, Kapetanakos C A, Pasour J A and Mahaffey R A 1981 *Am. Sci.* **69** 173
[11] Doria D, Lorusso A, Belloni F and Nassisi V 2006 *J. Plasma Phys.* **72** 229
[12] McKenna *et al* 2005 *Phys. Rev. Lett.* **94** 084801
[13] Snavely R A *et al* 2000 *Phys. Rev. Lett.* **85** 2945
[14] Hatchett S P *et al* 2000 *Phys. Plasmas* **7** 2076
[15] Yang J M *et al* 2004 *J. Appl. Phys.* **96** 6912
[16] Zeil K, Kraft S D, Bock S, Bussmann M, Cowan T E, Kluge T, Metzkes J, Richter T, Sauerbrey R and Schramm U 2010 *New J. Phys.* **12** 045015
[17] Commisso R J *et al* 2010 Development of an intense, pulsed source of combined characteristic-gamma-rays and neutrons to induce fission *Plasma Science, 2010 Abstracts IEEE International Conference on* (Monterey, CA: IEEE Publishing)

Laser ablation coupled to mass quadrupole spectrometry for analysis in the cultural heritage

A Torrisi[1], M Cutroneo[2], E D Castrizio[3] and L Torrisi[2]

[1]Dip.to di Fisica e Astronomia, Università di Catania, Via Santa Sofia 64, 95123, Catania, Italy

[2]Dip.to di Fisica e S.d.T., Università di Messina, Ctr. Papardo 31, 98166 S. Agata, Messina, Italy

[3]Dip.to delle Civiltà Antiche e Moderne, Univ. di Messina, Polo Annunziata, 98168, Messina, Italy

E-mail: atorrisi83@gmail.com

Abstract. A Nd:YAg laser operating at 1064 nm, 150 mJ, 3 ns pulse duration, 1-10 Hz repetition rate and 10^9 W/cm^2 intensity is employed to irradiate ancient metallic and ceramic samples in high vacuum. A mass quadrupole spectrometer (MQS), operating between 1-300 amu with sensitivity better than 0.1 ppm, analyzes elements and compounds. Repetitive laser ablation removes in controlled manner the first surface layers of the irradiated samples so that the irradiation time can be correlated to the layer depth. MQS can be fixed to peculiar masses so that during the laser irradiation the mass yields can be plotted as a function of the sample depth. The technique permits to give the depth profiles of elements, chemical compounds and isotopes characterizing the composition of the analyzed samples. The analysis of ancient coins based on bronze and silver alloys and of old vitrified colored ceramics has been investigated to identify peculiar elements of the colored layers. Particularly, the lead isotopic ratios $^{208}Pb/^{207}Pb$ and $^{206}Pb/^{207}Pb$ were measured in bronze coins. Measurements were compared with the database of lead isotopic ratios in lead minerals extracted from old mines in the Mediterranean basin. In some cases, of special interest for Archeologists, the comparison has indicated that the lead employed for the coin production could have been extracted from mines of particular geographic sites.

1. Introduction

Ancient bronze coins coming from Mediterranean basin are of special interest in numismatic and archeology in dependence of the historic period of their coinage and use. The archeological bronze generally is based on an alloy containing: copper, lead, iron, tin and zinc. These coins are the best indicator for tracing the ore sources and the trading of the huge amount of copper used to realize the bronze alloy, thus their investigations permit to improve the archeological knowledge. Recently the use of the physical technique LAMQS (Laser Ablation coupled to Mass Spectroscopy) can be employed to analyze the bronze elemental composition both in the bulk and in the superficial patina, to detect special chemical compounds vs. depth and to measure the isotopic ratios of some peculiar elements [1]. Particularly interesting is referred to the Pb content of the bronze alloy, that can be different depending on the geographic area and on the historical period of coinage and to the lead isotopic ratios that permits to characterize the specific mineral used and sometimes to go back to the geographic area of the origin mine [2, 3]. The analysis is applied to some coins from IV – X century

Journal of Physics: Conference Series **508** (2014) 012025 doi:10.1088/1742-6596/508/1/012025

AD coming from different Mediterranean sites: Egypt, Greece and Italy. LAMQS is applied also to ancient silver coins and vitrified ceramics of the Mediterranean basin.

2. Materials and methods

In this section we discuss how to format the title, authors and affiliations. Please follow these instructions as carefully as possible so all articles within a conference have the same style to the title page. This paragraph follows a section title so it should not be indented.

A set of analyzed bronze coins are the reported in the photos of Fig. 1. Characteristic-X-ray fluorescence (XRF) technique, induced by 20 keV electrons of a SEM microscope (Zeiss EVO), was employed to identify the elemental composition of the sample surface.

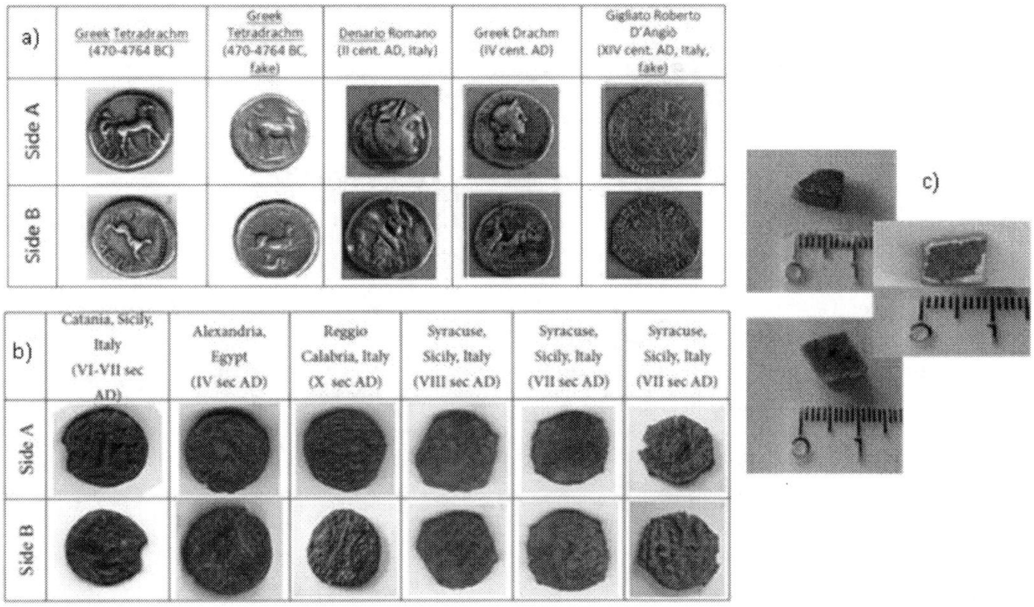

Figure 1. Photos of the analyzed coins base on Silver (a) and Bronze (b) and of the vetrified ancient pottery (c).

A Nd:Yag laser, operating at 1064 nm, with 3 ns pulse width, 1-150 mJ settable pulse energy, single pulse or 1-10 Hz repetition rate, was employed. The laser was focussed on the target (sample) up to a spot size of 0.5 mm^2. A preventive calibration on a bronze alloy permitted to evaluate the ablation yield in terms of removed thickness per laser shot. The calibration uses a surface profile meter to measure the ablation depth with a resolution of 10 nm [4]. Generally the analyses are performed at the pulse energy of 30 mJ at which the ablation yield is of about 10 nm per laser shot.

A mass quadrupole spectrometer (MQS), Pfeiffer Vacuum-Prisma 300, with 1-300 amu mass range, mass resolution less than 1 amu and sensitivity less than 0.1 ppm was employed to measure the mass of the particles emitted from the laser ablation process. Depth profiles can be measured operating with a laser ablation at 10 Hz repetition rate and analyzing, contemporary, the mass yields of the element of interest. A special attention is given to the detection of the masses of the four Pb isotopes: ^{204}Pb, ^{206}Pb, ^{207}Pb and ^{208}Pb. Fig. 2 shows a photo of the experimental set-up (a) and the Cu ablation yield vs. laser fluence (b). Measurements were performed at 10 J/cm^2 at which the removal depth is of 10 nm/mm^2 shot.

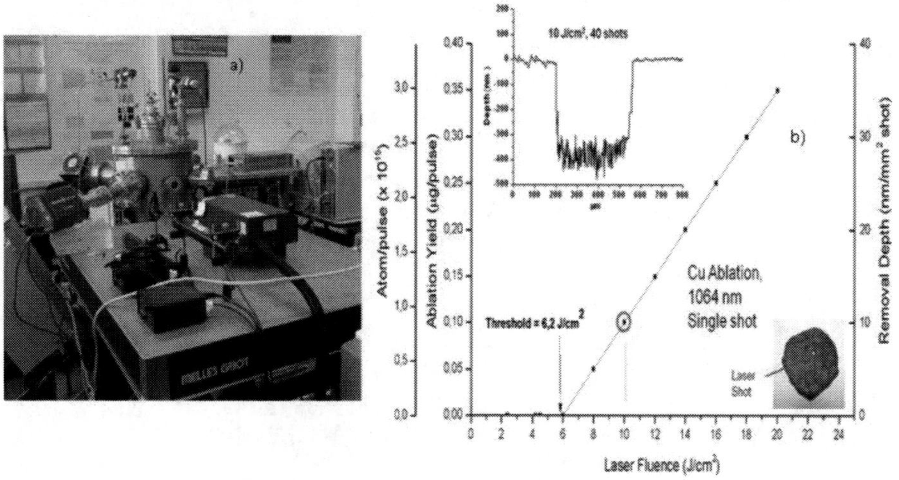

Figure 2. Photos of the experimental set-up (a) and Cu ablation yield vs. laser fluence (b).

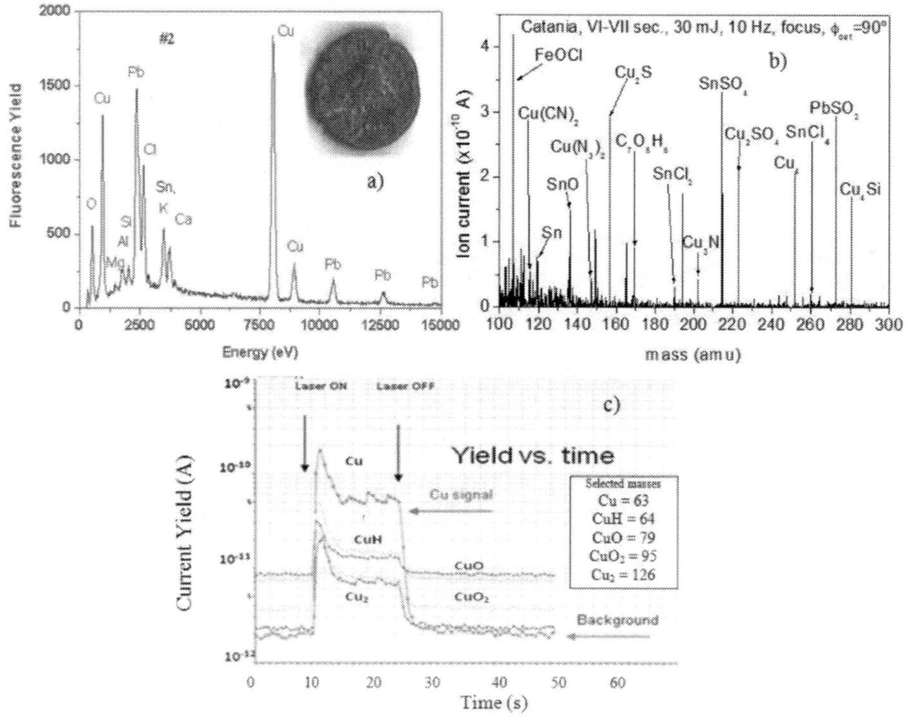

Figure 3. XRF spectrum typical of a bronze coin (a), LAMQ spectrum in bronze coin (b) and relative to some elements and compound of interest (c).

3. Results

Fig. 3a shows a typical X-ray spectrum relative to the sample #2 in bronze (a) which composition is based on Cu (~70-80%), Pb (~10-20%) and Sn (~6-10%) and that is rich in oxygen, carbon and trace elements in the superficial patina.

The higher lead concentration is found in the bronze samples of Catania #1 and Syracusae #4. Fig. 3b shows a typical LAMQS spectrum, in the mass range 100-300 amu, of the analysis relative to the bronze sample #1. The spectrum shows many chemical compounds based on Cu, Fe, Sn, Pb, O, S, Cl and N [4]. Spectrum can be acquired during laser ablation to have information about special masses as a function of the laser irradiation time (3c).

Fig. 4a shows an example of mass spectrum in the range 205-210 amu obtained reporting the yield as a function of the mass in order to measure the Pb isotopic ratios. Fig. 4b shows an example of mass spectra obtained by fixing the masses of interest, relative to the Pb isotopes 206, 207 and 208 amu, and reporting the yields as a function of the laser ablating time. Low ablation times, below 60 s, are used to reduce the laser invasively. Measurements indicate that the main Pb compounds in the surface of bronze coins are due to PbO, $PbSO_2$, and $PbCl_2$. The range of variability for the isotope ratios $^{208}Pb/^{206}Pb$ and $^{207}Pb/^{206}Pb$ is in the range 1.86 - 2.70 and 0.64-1.13, respectively. The lead isotope masses may interfere with different compounds, such as the Pb-H and Cu_3NH, however an accurate analysis using the maximum MQS resolution reduces the uncertainty. Fig. 4c shows the comparison of experimental data (points) with literature (ellipses) [5, 6, 7]. Measurements indicate that the lead used to realize the coins may comes from different mines giving minerals, such as galena, with similar isotopic ratios but their geographic place may be different with respect to the literature data.

Further investigations are in progress in order to investigate about the literature-experimental data differences that can be attributed or to the lead employed for the coinage of the analyzed samples or as colorant of the vitrified ceramics. A better investigation will be performed increasing the number of collected data and comparing isotopic ratio measurements with literature mineralogical data.

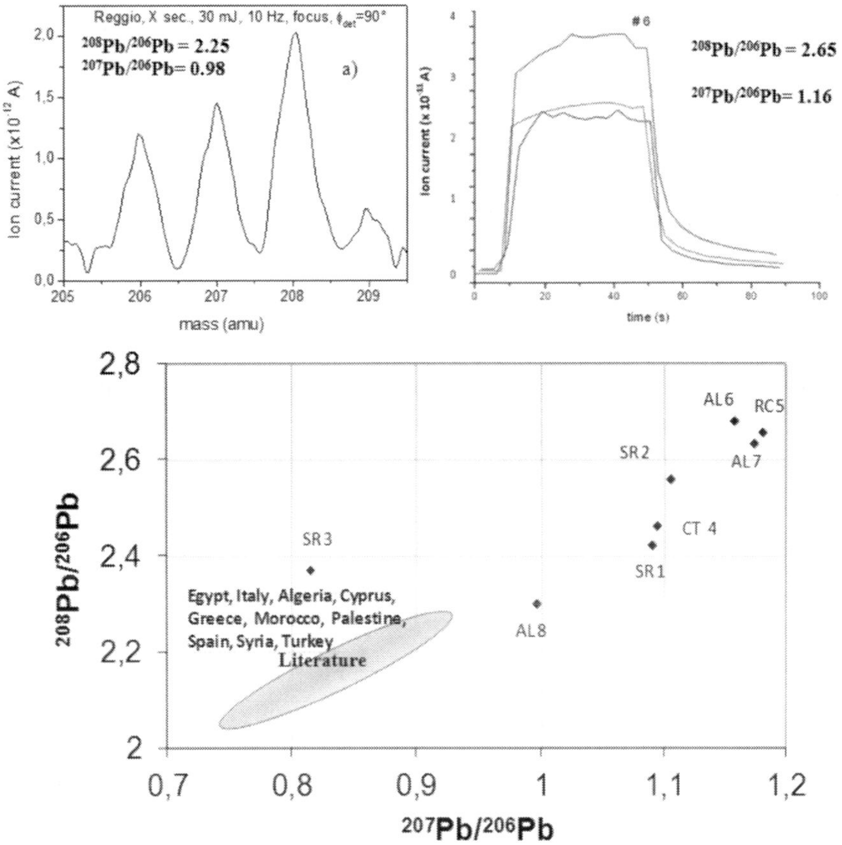

Figure 4. Mass spectra (a), time (depth) profile for the ^{206}Pb, ^{207}Pb, ^{208}Pb isotopes (b) and measure comparison with literature data (c).

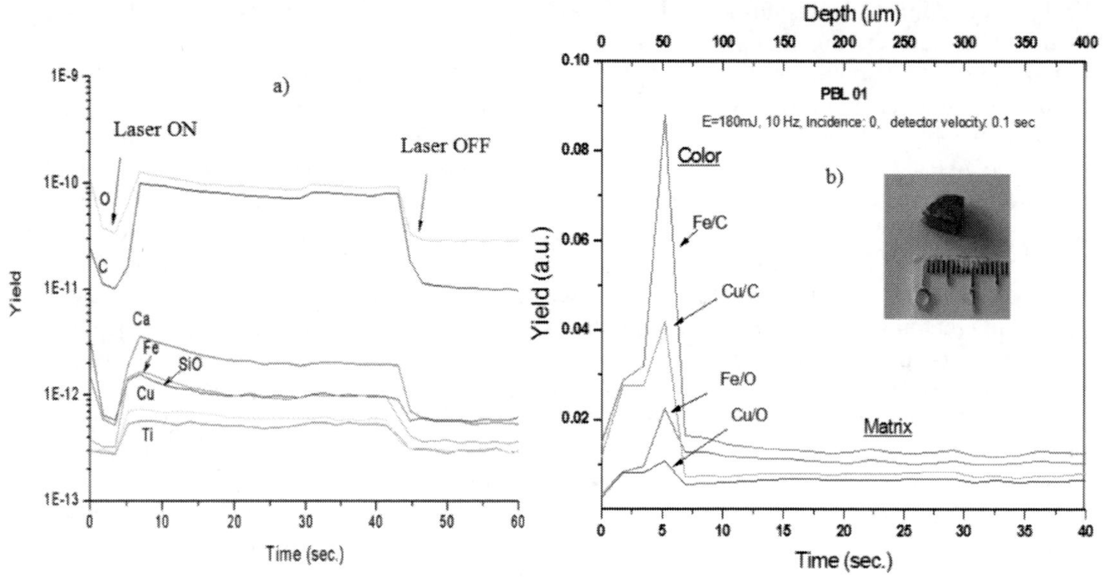

Figure 5. Typical LAMQS spectra of elements in vitrified ceramics (a) and measurement of thee colour layer thickness (b).

4. Discussion and conclusions

The LAMQS analysis can be employed to analyze the elemental composition, chemical compounds and isotopic ratios of ancient coins and ceramics. Obtained results demonstrated that coins composition is different between patina and bulk, that the Cu, Pb and Sn concentration in bronze is different for different set of coins. Major differences in the elemental composition is found for the sample #5 from Syracusae (VII sec A.D.) that have higher Cu content, for #1 from Catania (VI-VII sec A.D.) that has the higher Pb content, for the sample #2 from Alexandria (IV sec A.D.), that has higher concentrations of $PbSO_2$ and $PbCl_2$. The XRF analyses only to the first micron of the patina coin surface (corresponding to the range of 20 keV electrons) while the LAMQS analyses about 10 ns per laser shot, thus, operating at 10 Hz repetition rate, the yield is depending on the laser irradiation times. After 60 s irradiation the composition is relative to a depth of about 6 microns. The major difference are found for the coin #3 from Reggio Calabria (X sec A.D.), having higher $^{208}Pb/^{206}Pb$ and $^{207}Pb/^{206}Pb$ ratios, and for the coin #6 from Syracusae (VII sec A.D.), having the lower $^{207}Pb/^{206}Pb$ ratio. Other analyses with LAMQS technique has permitted to study the elemental composition of some ancient Sicilian colored pottery and measured their thickness. Further measurements are in progress in order to improve the statistical data.

References

[1] L. Torrisi, G. Mondio, A.M. Mezzasalma et Al., Eur. Phys. J. D 54, 225–232 (2009)
[2] L. Torrisi, F. Caridi, L. Giuffrida et Al., Nucl. Instr. And methods B, in press, 2010.
[3] F. Caridi, L. Torrisi, A. Borrielli et Al., Rad. Effects & Defects in Solids 165, 6, 1–12, 2010
[4] Brettscaife.net, 2010, Lead isotope analysis in Archeology: http://www.brettscaife.net/lead/data/index.html
[5] S.Klein, Y. Lahaye, G. P. Brey, H. M. Von Kanhel; Archeometry 46, 3(2004) 469-480;
[6] M. Resano, E. Garcia-Ruiz, F. Vanhaecke; Mass Spectrometry Reviews, 2010, 29, 55-78.
[7] C. G. Eiseman, "Greek Lead", PhD Thesis-Univ. of Pennsylvania, 1980; www.penn.museum/documents/publications/expedition/PDFs/222/Eiseman.pdf

Root canals decontamination by coherent photons initiated photoacustic streaming (PIPS) of irrigants: an ex-vivo study

E Pedullà[1], C Genovese[2], C Scolaro[3], M Cutroneo[3], G Tempera[2], E Rapisarda[1], L Torrisi[3]

[1] Dipartimento di Chirurgia, Università di Catania, via Plebiscito, 628 – 95100 (CT)
[2] Dip.to di Scienze Biomediche, Università di Catania, via Androne, 81 - 95100 (CT)
[3] Dip.to di Fisica e S.d.T., Università di Messina, S.F. d'Alcontres 31, 98166 S. Agata, Messina Italy

E-mail: eugeniopedulla@gmail.com

Abstract. The aim of this *ex vivo* study was to assess the antibacterial effectiveness of coherent photon initiated photoacoustic streaming (PIPS) of irrigants using an Er:YAG laser equipped with a newly designed, stripped and tapered, tip in extracted teeth with infected root canals. One hundred-forty-eight single-rooted extracted teeth were prepared using a rotary abrasive instrument providing a root channel with a suitable size. The samples were sterilized and all teeth except ten (negative control group) were inoculated with *Enterococcus faecalis* and incubated in a CO_2 chamber at 37°C for 15 days in Eppendorff tubes filled with trypticase soy broth medium changed every 2 days. Infected teeth were then randomly divided into 4 test groups (n=32 for each): pulsed erbium:YAG laser at non-ablative settings for 30 seconds with sterile bi-distilled water (Group A) or 5% sodium hypochlorite (NaOCl) (Group B); without laser activated sterile bi-distilled water irrigation for 30 seconds (Group C) or 5% NaOCl irrigation for 30 seconds (Group D); the positive control group received no treatment in infected teeth (n=10). Colony-forming units (CFUs) were counted from bacteriologic samples taken before (S1) and after treatment (S2). Data were analyzed by Kruskal-Wallis and post hoc Dunn's multiple comparison tests. CFU counts were significantly lower in groups B and D than in group C ($P<0.001$). Moreover, there was a significant difference between Group A and C ($P<0.001$). Group B showed the highest CFU reduction, which was significantly greater than that evident in groups A or C ($P<0.001$). There were no statistically significant differences between group B and D ($P>0.05$). None of the four groups predictably generated negative samples. Under the conditions of this *ex vivo* study, statistically significant difference wasn't found in planktonic bacteria reduction between the laser and NaOCl or NaOCl alone groups.

1. Introduction

The first laser specifically designed for dentistry was marketed in 1972. Less than ten percent of dentists world wide own laser, but there are many indications for use for oral procedures [1]. All dental lasers emit either a visible light wavelength or an invisible infrared light wavelength in the portion of that no ionizing spectrum called thermal radiation.

Since human dental tissues are composed of a combination of compounds, the clinician must choose the best laser for each treatment.For soft tissue treatments, the practitioner can use any available wavelength, because all dental lasers absorb one or more of the soft tissue components. For

Journal of Physics: Conference Series **508** (2014) 012026 doi:10.1088/1742-6596/508/1/012026

hard tissue, however, the Erbium lasers with very short pulse durations easily ablate layers of calcified tissue with minimal thermal effects [2].

The Erbium lasers are absorbed only in a few to several microns of the target tissue's surface, whereas diode lasers can reach a few millimetres deeper into the soft tissues [2].

Dental tissue interaction with laser energy is affected by several factors: wavelength, beam diameter, focused or defocused distance, pulse energy, and providing tissue cooling. The correct combination of all of those parameters should ensure an efficient and beneficial outcome [3]. The currently used dental lasers and their tissue interaction are depicted on the absorption coefficient vs. wavelength of the electromagnetic wave in biological tissues, as reported in Figure 1.

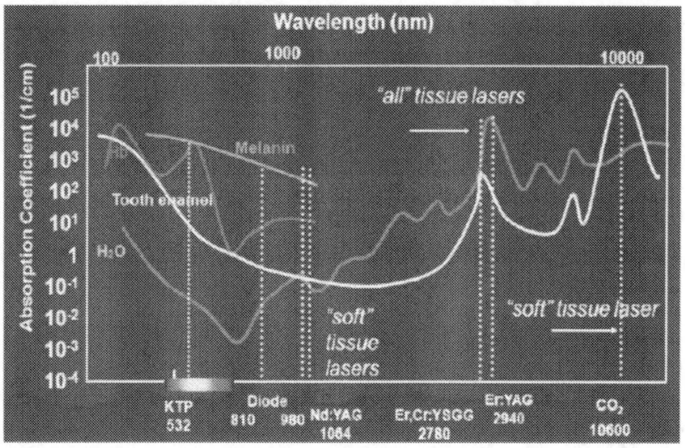

Figure 1. Absorption coefficients of the principal dental tissue components as a function of the laser wavelength.

Recently, the use of different lasers (diode or Er:YAG laser) was proposed in endodontics that is the branch of dentistry for root canal therapy. Microorganisms are the major causative factor associated with endodontic treatment failure. Numerous endodontic protocols use irrigant solutions and mechanical preparation to clean and decontaminate the root canal system [4, 5]. Sodium hypochlorite (NaOCl) has been the most commonly used irrigation solution in endodontics since the early 1900s. It is used alone or in combination with a chelating solution to reduce bacteria and eliminate smear layer. However, studies have shown that the irrigants have a limited ability to effectively reach all internal aspects of seemingly complicated root canal architecture. Different agitation techniques have been proposed to improve the efficacy of irrigation solutions, including agitation with hand files, plastic instruments, sonic, ultrasonic and, more recently, laser devices. Lasers have the ability to clean and effectively disinfect root canals, including elimination of highly resistant species such as *Enterococcus faecalis* [6, 7]. Several authors have investigated different lasers for use during root canal therapy. Currently, visible (532nm), invisible infrared (from 803 to 1064nm) and invisible medium infrared (2780 and 2940nm) wavelengths are commonly used in endodontics. The Er:YAG laser wavelength (2940nm) has the highest absorption in water and high affinity to hydroxyapatite, which makes it suitable for use in root canal therapy. George *et al.* [3] reported the first *in vitro* study to examine the capacity of lasers to activate irrigants inside root canal systems to increase its action on smear layers.

Sahar-Helft *et al.* reported that a combined regimen of a low concentration of chlorhexidine with laser irradiation could be a potential means of inhibiting *Enterococcus faecalis* growth.

Other studies by De Moor *et al.* compared the laser activated irrigation (LAI) to passive ultrasonic irrigation (PUI) on dentine debris. A recent study [8-10] reports the debriding and cleaning efficacy of irrigation enhanced by a new Erbium laser technique that minimizes the morphological thermal effects of infrared radiation. This laser technique uses a photon induced photoacustic streaming (PIPS) of irrigants produced by a newly designed tapered and stripped tip with specific minimally ablative laser setting: low energy (20 mJ); pulse repetition rate (15 Hz) and short pulse duration (50 μs) [11]. The aim of this study was to compare the efficacy of root canal disinfection of *E. faecalis* in extracted teeth using PIPS [12] (generated by a new Er:YAG laser technique with low energy laser settings plus a new tip for endodontic therapy) or the conventional syringe irrigation technique.

2. Materials and Method

One hundred-forty-eight single-rooted human teeth extracted for periodontal reasons were used in this study. The study protocol was approved by the Ethics Committee of the Catania University. After a conventional access cavities all teeth were instrumented to the anatomical apex by mechanical preparation with Mtwo nickel titanium (NiTi) rotary abrasive instruments (Sweden & Martina, DueCarrare-Pd, Italy) up to the instrument with a tip diameter of 0.25 mm and a taper (final size of the cone) of 60 μm. A coronal reservoir for irrigant placement was created with a Gates Glidden drill #5 placed 5 mm into the canal. Conventional endodontic irrigation with 5% sodium hypochlorite (NaOCl; Niclor 5, OGNA Laboratory, Muggiò, Milan, Italy) and 17% EDTA (OGNA Laboratory, Muggiò, Milan, Italy) was used throughout instrumentation; then all canals were dried using air and sterile paper points. At this point the teeth were placed individually in sterilizer pouches, autoclaved at 134°C for 17 minutes, and stored until use. Ten teeth were not infected and formed the negative control group. All the other 138 teeth were inserted apart in an upright position in the holes of individual sterile Eppendorf vials (Eppendorf Italia, Milan, Italy) filled with trypticase soy broth so that only the apical 2 to 3 mm of the teeth were immersed in the solution. The root canal of each tooth, except for the negative control teeth, were inoculated with 10 μL of a pure culture suspension of *Enterococcus faecalis* (American Type Culture Collection [ATCC] 29212, Oxoid Limited, Basingstoke, Hampshire, United Kingdom - equivalent to Mac Farland 0.5 - 1.5×10^8 CFU /mL) by sterile syringe and a 30-G irrigation needle (Max-i-Probe; Dentsply Rinn, Elgin, IL, USA) placed as close to anatomical apex as possible. After inoculation, all teeth were incubated at 37°C in a CO_2 chamber to allow growth of *Enterococcus faecalis* for 15 days and each culture medium was replenished every 2 days. Ten out of 138 infected teeth were not treated and formed the positive control group. Then 100 μL of sterile ringer solution were deposited into the root canals and a sterile paper point, with a tip diameter of 0.25 mm and a taper of 20 μm, was placed to working length of each tooth, allowed to saturate and then each placed in a sterile vial containing 2 mL of Trypticase Soy Broth medium (TSB) which was immediately vortexed to collect the initial microbiological samplings (Sampling 1). The 128 experimental teeth were randomly divided into 4 groups (n=32 each). The teeth in group A were treated with bi-distilled sterile water irrigation and laser irradiation for 30 seconds; those in group B were treated with 5% NaOCl irrigation and laser irradiation for 30 seconds; the teeth in groups C were irrigated with only sterile bi-distilled water and those in group D only with 5% NaOCl for 30 seconds. The standardized irrigation protocol involved the use of a 30-G irrigation needle tip introduced as close to anatomical apex as possible without binding and deposition of 3mL of irrigant solution. A final flush was performed with 5mL of sterile 2M sodium thiosulphate solution to inactivate the NaOCl. The laser irradiation protocol was performed by an Er:YAG laser with a wavelength of 2940 nm (Fidelis AT, Fotona, Ljubljana, Slovenia) equipped with a newly designed 12 mm long, 400 μm quartz tip. The tip was tapered and had 4 mm of the polyamide sheath stripped back from its end. Laser operating parameters (using the free-running emission mode) of 20 mJ per pulse, 15 Hz, 50 μs pulse duration, were employed for all of the treatment groups where lasers were used. The co-axial water spray feature of the handpiece was turned off. The tip was placed into the coronal reservoir only and activated for 30s. After canal irrigation had been completed, the remaining intracanal NaOCl was neutralized with 5 mL 2M sodium thiosulphate for 30 s. Then, 10μL of a sterile ringer solution was placed in all root canals and sterile abrasive paper points, with a tip diameter of 0.25 mm and a taper of 20 μm, were inserted to the anatomical apex and left for 60 seconds to soak up the contents of the canals (Sampling 2). The wet paper points were then dropped into sterile Eppendorf tubes (Eppendorf Italia, Milan, Italy) with 1 mL of sterile ringer solution and sonicated for 30 seconds to free the bacteria in solution. Serial dilutions were performed up to a concentration of 10^6. Dilutions from 10^4 to 10^6 were plated on blood agar and incubated at 37°C in a CO_2 chamber for 48 hours. The plates containing more than 30 colonies were counted. The mean number of CFUs for the 10^5 dilution plates was then calculated, and the results were analyzed by statistical software (MedCalc Software, Mariakerke, Belgium) using Kruskal-Wallis and analysis of variance with Dunn's multiple comparison *post hoc* tests and a P value of <.05 was considered statistically significant.

Fig. 2 shows a photo of the used Er:YAG laser (a), a typical NiTi rotary abrasive instruments to enlarge the root channel (b), a scheme of the procedure of root channel preparation (c), the fibre tip transporting the laser light (d) and, finally, a scheme of the chemically cleaning and debriding of the root canal system using the Er:YAG laser energy at sub-ablative power levels (e).

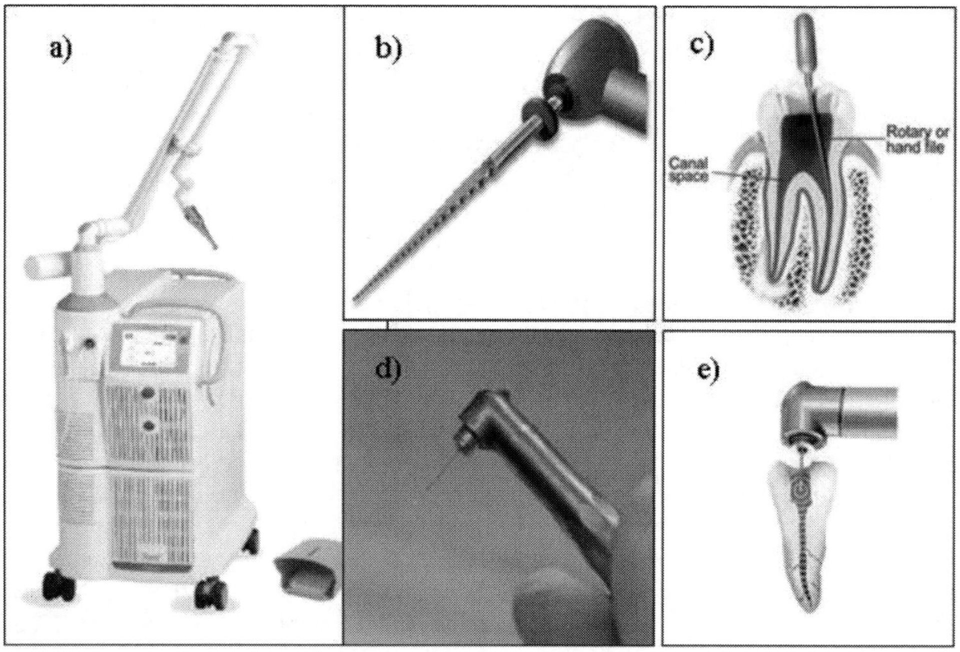

Figure 2. The Er:YAG laser application.

3. Results

Table 1 shows a summary of the results. All of the positive control teeth exhibited high bacterial growth, whereas none of the negative controls showed growth.

After treatment, the number and proportion of negative samples was as follows: 15 for group A (46,87%); 30 for group B (93,75%); 0 for group C (0%) and 25 for group D (78,12%). CFU counts were significantly lower in groups B and D compared to group C (P<0.001). Moreover, there was a significant difference between Group A and C (P<0.001). Group B (NaOCl irrigation with laser irradiation) showed the greatest percentage bacterial reduction among the groups tested. Moreover, there was a significant reduction of bacteria CFU counts in Group B compared with Groups A or C (P<0.001), although no statistically significant differences were found in comparison with group D (P>0.05). However, laser-activated irrigation with NaOCl (Group B) was significantly superior to conventional irrigation alone with NaOCl (Group D) in generating negative samples (P<01).

After treatment, the number and proportion of negative samples was as follows: 15 for group A (46,87%); 30 for group B (93,75%); 0 for group C (0%) and 25 for group D (78,12%). CFU counts were significantly lower in groups B and D compared to group C (P<0.001). Moreover, there was a significant difference between Group A and C (P<0.001). Group B (NaOCl irrigation with laser irradiation) showed the greatest percentage bacterial reduction among the groups tested. Moreover, there was a significant reduction of bacteria CFU counts in Group B compared with Groups A or C (P<0.001), although no statistically significant differences were found in comparison with group D (P>0.05).

Table 1. Microbiologic mean counts before and after root canal irrigation for the experimental (n=32 each) and control groups (n=10 each).

	Counts before ($x10^6$)		(S1) ($x10^6$)	Counts after ($x10^4$)		(S2) ($x10^4$)	Reduction (%)	Negative culture
	mean	median	range	mean	median	range	median	
Group A	2.84	1.58	7.68 – 0.125	45.9	7.66	0 – 64.5	73.01	15/32
Group B	2.83	1.22	9.15 – 0.247	0.346	0	0 – 5.25	99.80	30/32
Group C	2.58	1.41	13.2– 0.388	139	92.3	18.7– 489	13.04	0/32
Group D	2.92	1.34	12.1– 0.0263	2.54	0	0 – 35.2	97.13	25/32
Ctrl+	2.91	2.72	9.28– 0.797	-	-	-	-	0/10
Ctrl-	0	0	0 - 0	-	-	-	-	10/10

4. Discussion

The present study compared the reduction of *E. faecalis* in straight canals with conventional irrigation to that obtained using photon induced photoacustic streaming (PIPS) of irrigants by an Er:YAG laser with low energy parameters and equipped with newly designed stripped and tapered tips. Four different irrigation protocols were tested: sterile bi-distilled water with or without activation by Er:YAG laser irradiation, or 5% NaOCl irrigation with or without activation by Er:YAG laser irradiation. Many recent publications have focused on the use of the Er:YAG laser. Some have shown positive results, whereas others have reported questionable findings on debris removal or bactericidal effect [7, 10-13]. The results of the present study are suggestive of a positive bactericidal effect of photon - induced photoacustic streaming (PIPS) of irrigants, generated by a pulsed Er:YAG laser equipped with a newly designed quartz tip (400micron, tapered and with 4mm of the polyamide sheath stripped back from its end) but located in the coronal reservoir only. When activated in a limited volume of fluid, the high absorption of the Er:YAG wavelength in water, combined with the high peak power derived from the short pulse duration that was used (50 μs), resulted in a photomechanical phenomenon [11]. Moreover the use of subablative laser settings prevents damage of root canal dentine by a vaporized water mediated explosion [2]. It is suggested that this allows easy access for the photomechanical effects to occur within the root canal, which may assist in cleaning canals of various shapes. Using the tapered and stripped design tips with PIPS, the apex can be reached without the need to negotiate the tip close to the apex. The bactericidal effect of PIPS is non-thermal and does not depend on the canal size as the laser tip is maintained without wall contact in the coronal reservoir [11]. One possible explanation for the effect of laser activation in the present study may be increased NaOCl reaction kinetics [14]. The significant difference between CFU counts of groups A and C ($P<0.001$) seems to show that the PIPS technique is capable of quite well effectively disinfecting the root canals even with sterile water washing by a photomechanical action. However, the significant difference between groups A and B confirms the importance of sodium hypochlorite to obtain a root canals high disinfection (Table I). None of the four groups showed negative samples predictably. Statistically significant differences were not found comparing groups B (NaOCl irrigation + PIPS) and D (NaOCl irrigation only). However, the fact that laser activation generated more negative bacterial samples than conventional irrigation warrants further investigation. In fact, changes in the irrigation protocol could increase the number of bacteria-free cases. Moreover, a longer activation time and possibly changes in the chemical composition of the irrigant, such as adding surface-active agents, might be helpful to enhance deep penetration into dentinal tubules and increase the disinfection of root canals.

5. Conclusions

Under the conditions of this ex vivo study, none of the test groups have removed bacteria from infected root canal systems completely. Water irrigation alone was statistically less effective compared to the other groups in its bacterial load reduction. There were statistically significant differences between bacteria load reduction with or without laser activated bi-distilled water, while there were no statistically significant differences between bacteria load reduction with or without laser activated NaOCl.

References

[1] Coluzzi D. (2008) Fundamental of lasers in dentistry: Basic science, tissue interaction and instrumentation. Journal of Laser Dentistry 16(Spec Issue), 4-10

[2] Coluzzi D. (2009) Lasers in Dentistry: From Fundamentals to Clinical Procedures. American Dental Association Seminar seriers, 2-16

[3] Coluzzi DJ. (2004) Fundamentals of dental lasers: science and instruments. *Dental clinics of North America* 48, 751–70.

[4] Siqueira Jr JF. (2002). *Oral surgery, oral medicine, oral pathology, oral radiology, and endodontics* 94, 281-93.

[5] Siqueira Jr JF, Rocas IN. (2004). *Oral surgery, oral medicine, oral pathology, oral radiology, and endodontics* 97, 85-94.

[6] Schoop U, Moritz A, Kluger W, Patruta S, Goharkhay K, Sperr W, Wernisch J,Gattringer R, Mrass P, Georgopoulos A. (2002). *Lasers in surgery and medicine* 30, 360–4.

[7] Gordon W, Atabakhsh VA, Meza F, Doms A, Nissan R, Rizoiu I, Stevens RH. (2007). *Journal of American Dental Association* **138**, 992-1002.

[8] George R, Meyers IA, Walsh LJ. (2008). *Journal of Endodontics* **34**, 1524-7.

[9] Sahar-Helft S, Slutzky-Gopldberg I, Moshonov J, Stabholtz A, Jacobovitz M, Tam A, Steinberg D. (2011). *Photomedicine and laser surgery* **29**, 753–8.

[10] De Moor RJ, Blanken J, Meire M, Verdaasdonk R. (2009). *Lasers in Surgery and Medicine* **41**, 520-3.

[11] DiVito E, Peters OA, Olivi G. (2010). *Lasers in medical science* Dec. 1 [Epub ahead of Print].

[12] Schoop U, Goharkhay K, Klimscha J, Zagler M, Wernisch J, Georgopoulos A, Sperr W, Moritz A. (2007). *Journal of American Dental Association* **138**, 949-55.

[13] Peters OA, Bardsley S, Fong J, Pandher G, Divito E. (2011). *Journal of Endodontics* **37**, 1008-12.

[14] Macedo RG, Wesselink PR, Zaccheo F, Fanali D, van der Sluis LW. (2010). *International Endodontic Journal* **43**, 1108-15.

Nd:Yag laser irradiation of single lap joints made by polyethylene and polyethylene doped by carbon nanomaterials

A M Visco[1], V Brancato[1], M Cutroneo[2] and L Torrisi[2]

[1]Dept. of Electronic Engineering, Chemistry and Industrial Engineering, University of Messina, C.da Di Dio, 98166 Messina
[2]Dept. of Physics And Earth Science, University of Messina, C.da Papardo 31, 98166 Messina

E-mail: avisco@unime.it

Abstract. Thermoplastic polyethylene can be welded by the transmission laser welding technique (TTLW) that exhibits some process related benefits with respect other conventional joining methods. This justifies its large use in wide fields, from the automotive to medical or domestic appliances. In this research, we studied single lap joints made by polyethylene pure and filled with carbon nanomaterials (0.2% in weight) to make the polymer laser absorbent. The joints were irradiated by a Nd:YAG laser operating at 1064 nm (first harmonic) with an intensity of 10^7 W/cm^2 and 1÷ 30Hz, a maximum pulse energy of 300mJ and a laser spot of ≈1 cm^2 (no focusing lens were employed). The joints were characterized by morphological analysis, mechanical shear tests and calorimetric analysis. The results suggested that the laser exposition time must be opportunely balanced in order to avoid a poor adhesion between the polymer sheets and to realized efficient joints. In particular the mechanical test showed that the laser exposition time of 40 seconds is the best conditions to obtain the highest shear strength of the joints of 140 N. After too prolonged laser exposure times, degrading phenomena starts.

1. Introduction

The laser welding of polymers exhibits some process-related benefits compared to conventional joining methods such as bonding, ultrasonic, vibration or hot plate welding . For example, gluing of plastics mostly requires surface pre-treatment and needs organic solvents. Hot plate or hot air welding is cost-efficient, but slow and subject to wear. Ultrasonic or vibration welding expose the work pieces to high mechanical load [1].

Laser polymer welding has great advantages since it means low thermal and mechanical load to the component. It is clean, accurate and it has a high automation degree, but it does not uses any solvents. Moreover, laser welding does not generate any micro particles and this is a significant advantage in particular for fluid reservoirs and medical components [2]. Also, there is no melt ejection and therefore no distortion with laser welding. Moreover, the components do not have to be preprocessed before welding; this fact also contributes to a constant welding quality.

The main applications of laser welding of polymeric materials are in automotive engineering, electronics, communications, medical device technology, human care and domestic appliances (such as food and beverage packaging) [3,4].

During the laser welding process, the laser beam passes through the upper transparent polymeric layer until it hits the welding zone, where it is absorbed by the lower welding partner. The pressure of the tooling clamps enhances thermal conduction into the transparent joining partner, which also causes a local plasticization of the polymer to create a secure adhesive bond [5]. So this technique is suitable for applications where very high weld seam quality is specified [6].

In particular, in this work we applied the laser welding technique to a wide use thermoplastic polymer, ultra-high molecular weight polyethylene (UHMWPE). The special interest for this specific kind of polyethylene derives from its peculiar chemical and physical properties and the mechanical resistance. The polymer is a white, semitransparent and semicrystalline and it has a low absorption coefficient for visible light. In order to increase its absorption the UHMWPE was filled with pigments, like carbon nanostructures. We have employed a Nd:Yag laser at the wave length of 1064 nm and checked the mechanical strength of the joints welded at different times of laser exposure. The results showed that a good sealing action can be obtained at low exposure time. The highest shear load was of ≈140 N, and it was exhibited in the joint realized with a laser exposition for 40seconds, due to a deep and intimate interpenetration of the melted polymer at the sheet interfaces. We highlighted that higher exposition times induce severe damages of the polymeric structure.

2. Materials and methods

Ultra High Molecular Weight Poly Ethylene (UHMWPE-GUR 1020 ρ= 0.930 g/cm^3, $M_w \approx 3\times10^6$ g/mol) was supplied by Ticona and referred as "UH". Powder of carbon nano materials (supplied by Good Fellow) with particle size of 50-100 nm order, was employed as filler and referred as "NC".
Nanocomposites were made by mixing the powder of UHMWPE and 0.2% weight percentages of NC with pure ethanol (Fluka), as dispersing medium; it was referred as "UH-NC.

The mixing was kept in ultrasound bath at 25°C/ 2h. Than the solvent was separated and both the pure UH and the UH-NC nanocomposites were moulded in a hot press at 200°C /20 minutes (P= 20MPa), obtaining sheets 60mm×60mm and 1mm of thickness by using releasing thin teflon films (123 micron) supplied by P.A.T.I. s.p.a.. The UH sheets had an appearance semitransparent while the UH-NC ones, black.

The welded joint was obtained by coupling two rectangular polymer sheets (30x20 mm,1 mm thick), partially overlapped for a length of 15 mm and pressed between each other with pressure of about 70 kPa. The UH/UH-NC joints were then laser exposed for different times, within the range 30-90 seconds at 25°C/2 minutes by a 9 ns Nd:Yag laser operating at 1064 nm (first harmonic) in single pulse or at 1÷30 Hz repetition rate, with an intensity of about 10^7 W/cm^2, a maximum pulse energy of about 300 mJ and a laser spot of ~ 1 cm^2 (no focusing lens were employed). The laser pulse energy had a Gaussian shape with a diametric FWHM of about 5 mm. The incident angle of the laser beam was 0°. They were referred as "UH-NC30, UH-NC40, UH-NC60, UH-NC75, UH-NC90" where the last number indicates the laser radiation time.

The Absorption Coefficient (μ) of the laser light transmission in polyethylene was measured through the transmitted energy in thin films (90 μm thickness) with 1064 nm wavelength and 10 mJ pulse energy.

The shear test was carried out on the joints at 25°C by a LLOYD LR 10K universal testing machine with a crosshead speed of 5 mm/min. The specimens had a rectangular geometry, 20mm x30 mm and 1 mm of thickness. For each irradiation dose 10 specimens were tested in order to give the average value.

Differential Scanning Calorimeter, DSC mod. SDT Q600 (TA Instruments) was performed from 30°C up to +180°C (heating rate of 5°C/min) under nitrogen flow (100 ml/min). The measurements gave an endothermic peak (melting peak) whose minimum value gave the melting temperature (T_m). Lamellar thickness (L_c) was calculated according to the Thomson-Gibbs equation [5].

The morphological observation of the laser spot was performed at 40x magnification by means of an optical microscope Mod. ZEISS Stemi 2000 C.

3. Results and discussion

The welding mechanism occurs at the interface where the laser highly light pass through the laser-transparent UH sheet placed on the top (irradiated face) and is absorbed by the dark UH-NC sheet laser-absorbent placed on the bottom (un-irradiated face; see figure 1a). The presence of NC powder in the polymeric sheet is necessary in order to make it laser-adsorbent [7]. The absorption coefficient values (μ) of the pure UH and the nanocomposites film containing the NC are 25 cm^{-1} and 81.1 cm^{-1}, respectively. This suggests that the carbon nano materials at λ =1064 nm gives an absorption power to the UHMWPE more than double its starting value.

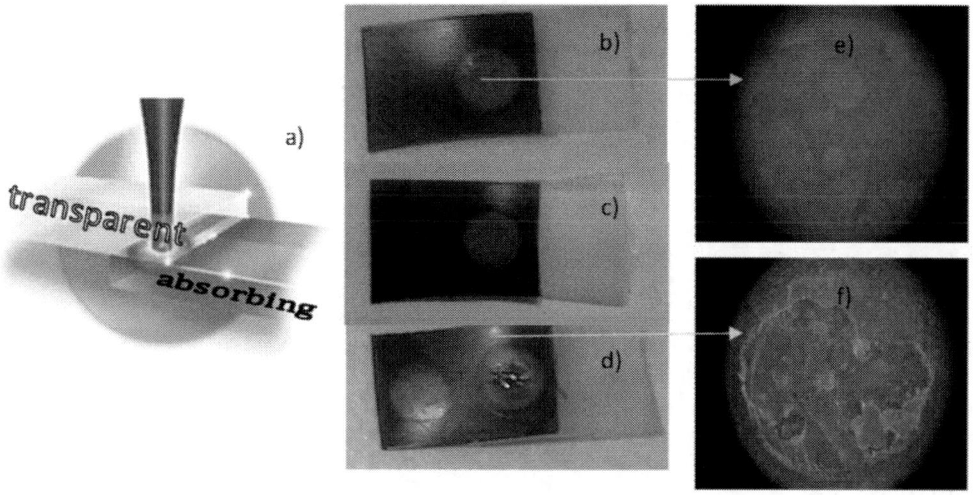

Figure 1. Scheme of a single lap joint irradiation by a laser source (a). Image of the UH/UH-NC30 (b), UH /UH-NC75 (c), UH /UH-NC90 joint (c). Optical microscope image of the of the laser spot in the UH-NC30 (e) and UH-NC90 joint (f).

In order to verify if the absorption power of the doped polyethylene sheets can be enough to produce a good weld, several joints were prepared changing the laser radiation exposure time and then they were checked by means of shear tests. Three examples of joints with the dimensions of the sheets geometry and of the overlapped area are shown in figure 1 b-d. In particular the image of figure 1b,c shows the UH/UH-CN30 and UH/UH-CN75 joints where are present the circular spots in the black sheets. Their observation suggests that the laser light pass through both the two overlapped polymeric sheets. The laser spot has a regular circular shape, yellow-orange colored; its diameter is of about 11 mm. Its difference in color suggest that the NC structures are ejected from the polymer matrix by the laser energy. Besides, its morphology is homogeneous suggesting that the change of the material induced by the laser energy absorption is perfectly distributed in all the area. The heat can be regularly distributed in all the laser spot area so that no any visible degrading phenomena occurs. In figure 1d is presented the image of the UH/UH-CN90 joint; here the spot appears irregular and damaged, suggesting the starting of degrading phenomena after protracted exposition times such as 90 seconds. The laser spot diameter is of 10.1 mm, shorter compared to that of the UH/UH-CN30 joint above described. The optical images of the inner laser spot area of the UH-NC30 joint (fig.1e) and of UH-NC90 one (fig.1f) detached after the shear stress are shown in figure 1e,f . The images at 40x magnification highlights the presence of some little swollen areas in the spot of the UH-NC30 joint that became highly rough in the UH-NC90 joint with well visible damages. All the above described images suggest that the laser exposure time of 30-75 seconds is enough to produce a well sealed area among the two polymeric sheets while the time of 90 could be excessive.

The average load/displacement curves of the joints: UH /UH-NC30-60-75-90 are shown in figure 2. The curves indicates that the shear resistance of the joint improves with the laser exposure time in the order : UH /UH-NC30< UH /UH-NC60< UH /UH-NC75, since the maximum load grows from 61 N to 153 N, enhancing of about 167%. Then it decreases again in the UH /UH-NC90 joint, since it results of 138N.

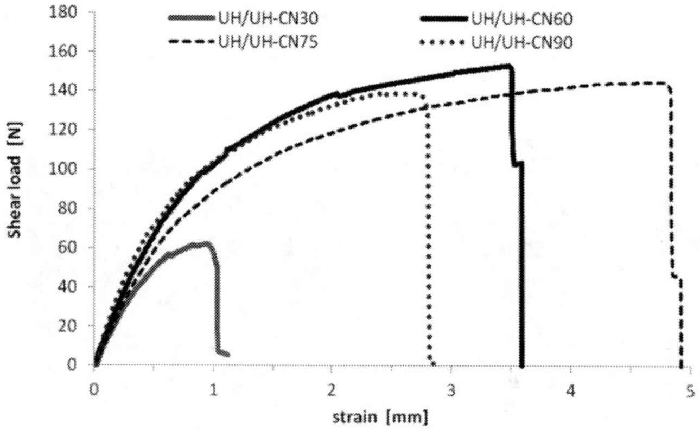

Figure 2. Shear load-strain curves of the UH/UH-NC30-60-75-90 joints.

The laser spot area of joints laser exposed for different times, within the range 30-90 seconds, progressively decreases from 105.3 mm^2 to 80.5 mm^2 and the shear stress values in the same range of radiation time increases from 1.58 MPa to 1.78 MPa.

In particular, the shear load is initially very low suggesting that the adsorbed energy is enough to give an appreciable weld. So, in these conditions, the welded joints exhibits a "poor" mechanical strength. After longer irradiation times, higher than 40 seconds, the shear stress value reaches a maximum that retains high up to 60 seconds of exposure. So, the sealing action induced by low irradiation exposure times (within the range 10-20 seconds) is not strong; the shear stresses are low, around 0.6 MPa. But when the exposure time overcome a threshold value, such as 40 seconds, the adsorption power of the sheet become very high so that the heat developed at the interface is enough to engage a more intimate interconnection between the two polymeric sheets and the joint effectiveness grows reaching the maximum value. At very high exposure times, such as 90 second the shear load little decreases, due to the starting of degrading phenomena. So in the exposure range 40-75 seconds the sealing action is maximum. Anyway, since the shear strength values are very close, the time of 40 seconds can be considered enough to achieve a good joint. Therefore, it is considered as the optimal exposure time.

The difference in the welding area among the samples suggests that a low exposition time involves the more external layers of the polymeric sheet where the heat developed can be better distributed on a wider area. Instead a higher exposure time, such as 90 seconds, produces a great amount of heat during the irradiation; this generates a smaller welding area but deeper compared to the UH-NC30 one where the resulting shear strength are higher. More investigation tests will be necessary in order to demonstrate the correlation among the strength and the contact deepness of the layers involved in the lacerated joints. We have other experimental evidences already observed in iron oxide doped polyethylene sheets, red in appearance, studied in a previous paper [8].

Finally, the calorimetric DSC analysis was performed inside the laser spot of the colored sheets with the aim to observe the laser effect on the polymer with increasing the laser exposure time (figure 3). The measured heat flux versus the temperature gave the endothermic peaks whose minimum value indicates the melting temperature (T_m). The calorimetric parameters (melting temperature and lamellar thickness) were measured as a function of the radiation time of 0-90 seconds. The T_m decreased from 134.6°C to 132.5°C (figure 3a) while the lamellar thickness improved from 23.9 nm to about 20.9 nm with increasing the laser exposition time (figure 3b). These changes highlighted a modify in the structural organization of the polymer after the laser irradiation due to the high temperatures reached during the irradiation. In particular the polymer organize itself in a different way since the crystals seems to be bigger in wideness but few in amount so that the melting temperature generally decreases.

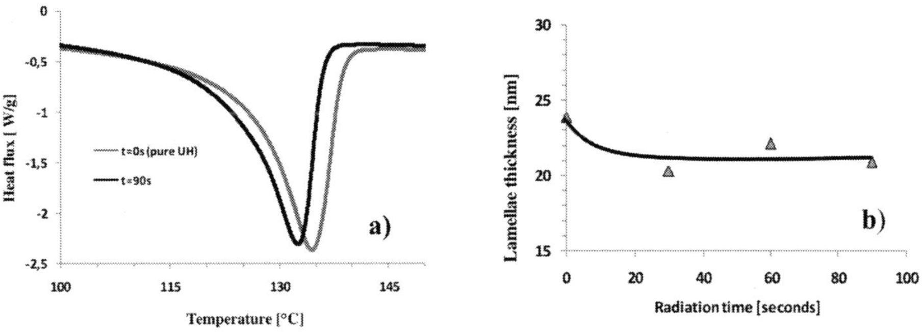

Figure 3. DSC curves (a) and lamellar thickness (b) vs the radiation time.

4. Conclusions

In this paper were prepared joints based on UHMWPE and carbon nanostructures (0,2 weight%), welded by a Nd:Yag laser operating at λ=1064 nm at different laser exposition times.Carbon nanomaterials can be well employed as colored pigment to produce single lap polymeric joints. A good sealing action in polymeric joint can be obtained after a laser exposition of 40 seconds. The resulting joint showed a 150 N shear strength and a regular welded area. Structural polymeric modifications are induced during the laser irradiation in the polymeric material that lowers its melting temperature and enhance the lamellar wideness. A higher laser exposure times worsen the mechanical strength of the joint since damages are produced in the laser spot. Works are in progress in order to deep investigate about the carbon nanostructure amount affect in the laser welding process of the polyethylene.

References

[1] Yarlagadda P K and Chai T C 1998 *Journal of Materials Processing Technology* **74** 199 – 212
[2] Amanat N, James N L and McKenzie D R 2010 *Medical Engineering & Physics* **32** 690-699
[3] Boglea A, Olowinsky A and Gillner A 2007 *Applied Surf. Sci.* **254** 1174-1178
[4] Brown N, Kerr D, Jackson M R and Parkin M R 2000 *Optics & laser technology* **32** 139-146
[5] Torrisi L, Caridi F, Visco A M and Campo N 2011 *Applied Surface Science* **257** 2567-2575
[6] Visco A M, Campo N, Torrisi L and Caridi F.2011 *Applied physic A- Materials Science and processing* : **103(2)** 439-445.
[7] http://www.lpkfusa.com/lq/processprincipals.htm
[8] Visco AM , Brancato V, Campo N , Torrisi L , Caridi F and Cutroneo M 2013 *Appl. surface Science* **272** 99-10

Journal of Physics: Conference Series **508** (2014) 012028 doi:10.1088/1742-6596/508/1/012028

UV-VIS-NIR spectral optical properties of silver iodide borate glasses

C Crupi[2], G Di Marco[2], L Torrisi[1], C Branca[1], G Carini[1], U Wanderlingh[1] and G D'Angelo[1]

[1]Dipartimento di Fisica, Università di Messina, Viale Ferdinando Stagno d'Alcontres 31, I-98166 Messina, Italy
[2]IPCF del C.N.R., Sezione di Messina, Viale Ferdinando Stagno d'Alcontres 37, I-98166 Messina, Italy

E-mail: gdangelo@unime.it

Abstract. We present a study of optical properties of a series of silver iodide borate glasses$(AgI)_x(Ag_2O{\cdot}B_2O_3)_{1-x}$by UV–VIS–NIR spectroscopy. The results show an increased absorbance in the whole analysed spectral range when the AgI concentration is augmented. In particular, the enhanced intensity of the wavelength band at 400-500 nm with silver iodine content suggests that this band arises from plasmon-related absorption, describing the formation of silver nanoparticles. With respect to this study, our results could motivate novel target designs consisting of ternary silver boron based bulk glasses for generating resonant absorption of laser light by plasma.

1. Introduction

The laser-plasma interaction that occurs when a high power laser pulse is focussed on a solid target is used for different applications and in particular for the ion acceleration. Over the last few years, significant attention has been paid to the improvement of the efficiency of the coupling between the laser pulse and the plasma obtained by the ionization of the target, aiming to enhance the ion acceleration. Enhanced energy absorption and ion acceleration to higher energies has been shown to occur in a number of configurations. In particular, if a target with a periodic surface modulation on the irradiated side is considered, a resonant coupling of the laser pulse with surface waves [1] can be achieved at a particular angle of incidence. Different strategies have been proved to be useful in producing increment of the laser absorption by resonance absorption at the critical surface. The insertion in dielectric target of absorbent nano-and micrometric particles with size comparable with the laser wavelength [2], the production of nano- and microscopic topological defects on the surface of the target by ion sputtering , rolling burnishing or chemical etching [3], the employment of nano-size foam target [4] or of targets consisting of aligned nano-tube/wires [5] are relevant examples.
These "nano-structured" targets enable a concentration of the electromagnetic field at the target edges and excite plasmons around the structure, resulting in an efficient energy transfer from the laser to the material.

Special attention has also been devoted to the study of optical properties of nanometer-sized particles dispersed in glass matrices, specially when noble metal particle are considered[6,7]. These optically active glasses offer a broad range of applications, also because their functionality can be modified by appropriate doping and post synthesis thermal processing [8]. In particular, borate glasses are good candidates as optically active glasses. They in fact are robust, non flexible and inexpensive. More importantly, doping the borate glasses with silver oxide and performing subsequent heat treatment causes the reduction of the doped silver ions leading to the formation of metallic silver nanoparticles. Ultraviolet-visible absorption spectroscopy shows a plasmon-related absorption band between 400 and 500 nm indicating the formation of silver nanoparticles [9,10].

Owing to their structural characteristics and optical properties, doped borate glasses appear to be promising systems to be used as solid target to increase effects of resonant absorption in laser-generated plasmas. Furthermore they can be obtained as thin film (thickness lower than 50μm) to be employed at high laser intensity to generate forward directed plasmas.

Here we present a preliminar spectroscopy study on a serie of silver borate glasses doped with silver iodide. In these systems the insertion of iodide ions expands the network, resulting in opened up porous structureand formation of AgI aggregates [10,11]. Both characteristics make these systems promising solid targets, able to increase the effets of laser resonant absorption following the interaction with laser ligth.

2. Experimental details

Three ternary borate glasses belonging to the family $(AgI)_x(Ag_2O{\cdot}B_2O_3)_{1-x}$ with an AgI molar fraction, x, of 0.2, 0.4 and 0.6 and pure boron oxide, B_2O_3, were investigated. These glasses were prepared by the melt quenching technique, heating at about 1100-1200 K weighted amounts of silver iodide, silver oxide and boron oxide and pouring the melts in stainless steel molds kept at room temperature [12]. No evidence of crystalline structures was found by X-ray diffraction analysis.

The absorption spectra of borate glasses were recorded by using a Perkin Elmer Lambda25 UV-Vis-NIR spectrophotometer (scan speed 120 nm/sec, step 2 nm, range 450 nm - 1100 nm) and applying the diffuse reflectance technique and using an integrating sphere. Diffuse reflectance spectroscopy has proven to be a fast and sensitive technique for qualitative information. However, according to different theories, quantitative data can be derived. The best known Kubelka–Munk theory [13] states that for thick layers, a correlation is obtained between the absolute reflectance R_∞ and the absorption coefficient k of the layer at a given wavelength, according to the following Kubelka–Munk function $F(R_\infty)$:

$$F(R_\infty) = \frac{(1-R_\infty)^2}{2R_\infty} = \frac{k}{S} = \frac{\varepsilon c}{S} \tag{1}$$

where S stands for the scattering coefficient (depending on the size and form of the particles), ε for the molar extinction coefficient of the analyte and c for its molar concentration. The Kubelka–Munk transformation thus converts a reflectance spectrum into a spectrum similar to a conventional absorbance spectrum for solution samples. In particular, the data reported in this paper have been obtained converting the $F(R_\infty)$ function into transmission spectra by means of the UV KinLab software.

3. Results and discussion

In figure 1 the transmittance curves as a function of wavelength in the visible and near-infrared regions for pure B_2O_3 and for silver borate doped with three different silver iodine content glasses are reported. The investigated samples show transmittance values between 15% and 50% up to 1000 nm whereas at higher wavelengths a strong slope towards enhanced values is observed. This trend clearly indicates a strong absorption ability of these borate glasses in nearly the whole analyzed λ range. It's worth to note that, a marked transmittance decrease for the doped samples is revealed at wavelength below 650 nm where the pure borate glass shows a rather constant T value. This occurrence suggests the existence of absorption mechanisms which are activated only in the ternary silver borate glasses but not in the pure glassy boron oxide. Otherwise, at 913 nm a transmittance dip is observed for the investigated systems as shown in the inset of figure 1. This finding indicates the presence of an absorption band whose origin should be inquired in the electronic states of the B-O network of the borate matrix since present in all these glasses.

With the aim to exalt such experimental results, focusing the attention on doped silver borate glasses, the electronic absorption spectra of these samples are shown in figure 2 (a).

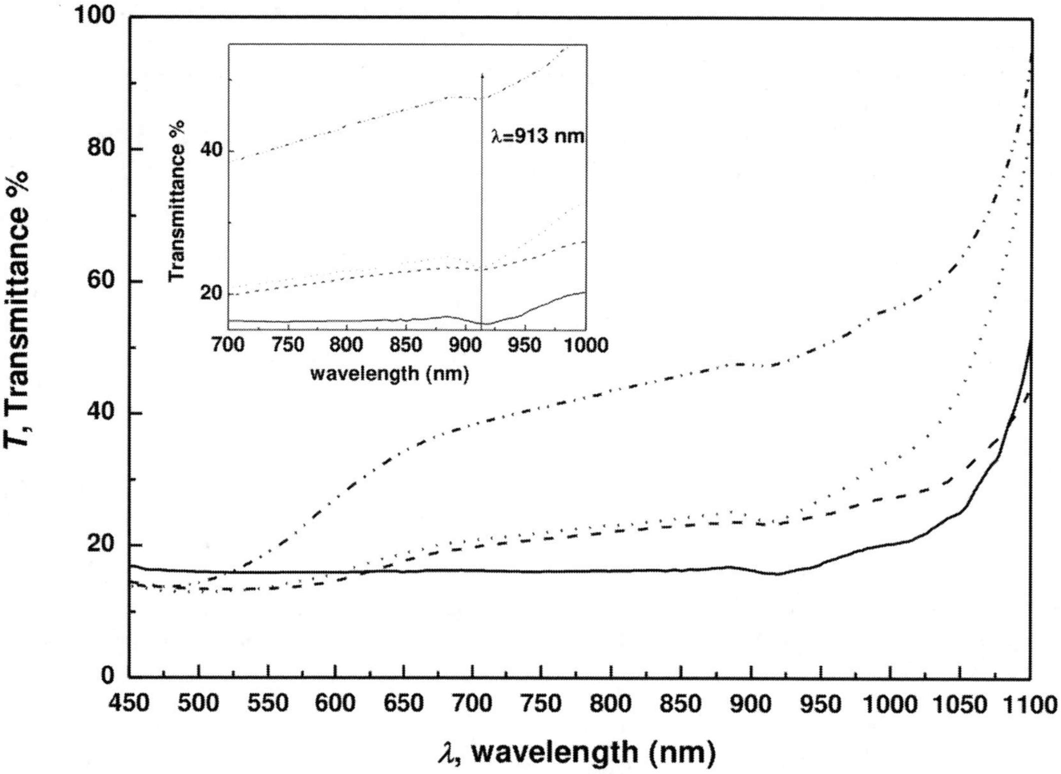

Figure 1. Transmittance curves of pure boron oxide (solid line) and of silver borate glasses doped with different AgI molar concentrations x: $x = 0.2$ (dashed line); $x=0.4$ (dotted line) and $x=0.5$ (dash-dotted line). In the inset a magnification of the spectral range at wavelength above 700 nm is shown.

As expected from the observed T behaviour, the absorbance of these three glasses decreases with increasing λ value in nearly all the investigated wavelength range showing two bumps in the visible and near infrared regions. According to literature data [8,9], the marked band at lower wavelength can be ascribed to the surface plasmon resonance from silver nanoparticles present in the studied systems.

In figure 2 (b) the three acquired absorption spectra normalized to the absorbance value at the highest investigated wavelength, where all samples show the lowest absorption, are plotted. This graphic representation allows the comparison of the relative absorbance intensity among all the investigated systems. An increased absorbance with augmented AgI concentration is observed in the whole analyzed spectral range. Most importantly the enhanced intensity of the lower wavelength band with silver iodine content suggests that just the Ag particles belonging to this binary compound give rise to an increased plasmon surface resonance with a consequent higher absorption.

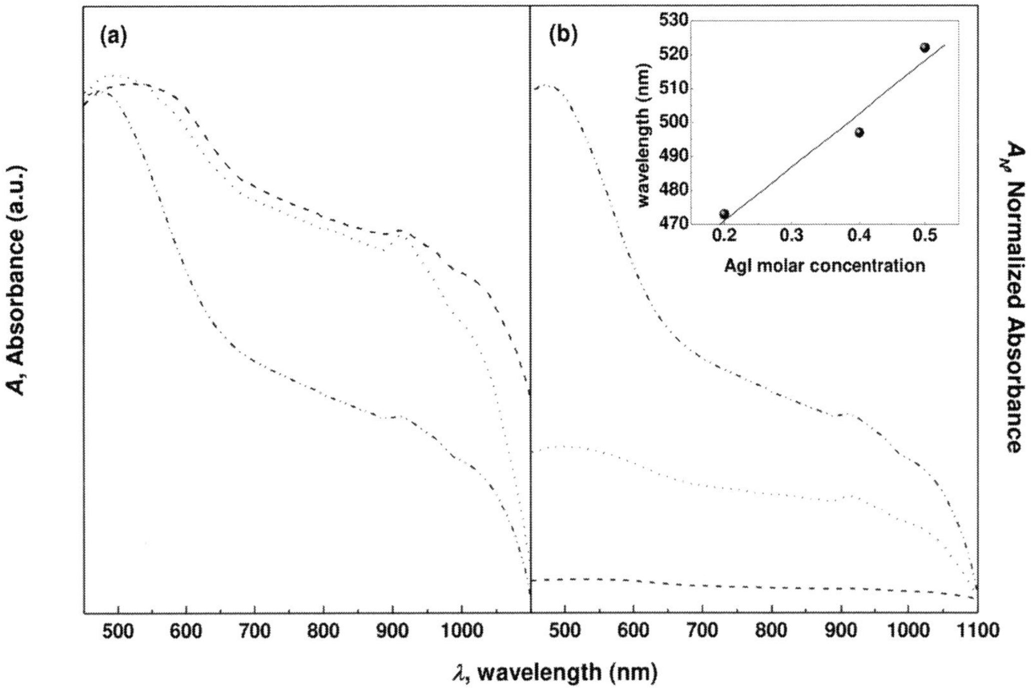

Figure 2. Absorbance spectra (a) and normalized absorbance spectra (b) of silver borate glasses doped with different AgI molar concentrations x: x= 0.2 (dashed line); x= 0.4 (dotted line) and x =0.5 (dash-dotted line). In the inset the λ position of the visible absorption peak as a function of AgI molar content is shown.

It is widely accepted that for mean particle radius less than 10 nm, no significant shift in the surface plasmon resonance peak is observed [14,15] while for larger nanoparticle size this band is significantly red-shifted [8,16,17]. A previous study has evidenced that for silver particles having a radius of 10 nm the low wavelength band is located at 409 nm [17]. As a consequence, the marked red-shift of the samples investigated in this paper would suggest a silver particle size greater than 10 nm. In this regard, a future scanning electron microscopy analysis on these system has been planned which would give more information on this issue.

Currently no decisive attribution can be put forward for the absorption band revealed at 913 nm since literature data are defective. For this reason, further detailed analysis will be performed on ternary borate glasses.

The next steps will be to investigate the size distribution of the silver nanoparticles as well as the distribution of the nanoparticles in the glass itself as a function of annealing time and temperature and to test the employment of these glasses as solid target for generating resonant absorption of laser light by plasmas.

References

[1] Maier S A and Atwater H A 2005 *J. Appl. Phys.* **98**, 011101
[2] Cutroneo M, Torrisi L, Margarone D and Picciotto A 2013 *Applied Surface Science* **272** 50
[3] Torrisi L, Cutroneo M and Caridi F 2012 *Laser and Particle Beams* **31** 37
[4] Lei A L et al. 2006 *Phys. Rev. Lett.* **96**, 255006
[5] Nishikawa T et al. 2001 *Appl. Phys. B* **73** 185
[6] Yamane M and Asahara Y in: *Glasses for Photonics*, Cambridge University Press, UK, 2000.
[7] Kellermann G and Craievich AF 2004 *Phys. Rev. B* **70** 054106

[8] Jimenez J A, Lysenko S, Zhang G and Liu H 2007 *J. Mater. Sci.* **42** 1856
[9] Dyrba M, Miclea P T and Schweizer S 2010 *Radiation Measurements* **45** 314
[10] Swenson J, McGreevy R L, Borjesson L and Wicks J D 1998 *Solid State Ionics* **105** 55
[11] Swenson J, Borjesson L and Howells W S 1995 *Physica Scripta* **Vol. T57** 117
[12] Chiodelli G, Magistris A, Villa M and Bjorkstam J 1983 *J. Non-Cryst. Solids* **51** 142
[13] Kortüm G and Lohr J E in: Reflectance Spectroscopy: principles, methods and application 1969, New York (Springer-Verlag)
[14] Hayakawa T, Tamil Selvan S and Nogami M 1999 *J. Non-Cryst. Solids* **259** 16
[15] Kreibig U and Vollmer M In "Optical Properties of Metal Clusters", Springer, Berlin 1995
[16] Sönnichsen C, Franzl T, Wilk T, Von Plessen G and Feldman J 2002 *New J. Phys.* **4** 93
[17] Arnold G W and Borders J A 1977 *J. Appl. Phys.* **48** 1488

An antibacterial coating obtained through implantation of titanium ions

D Delle Side[1], V Nassisi[1], E Giuffreda[1], L Velardi[1], P Alifano[2], A Talà[2] and S M Tredici[2]

[1] LEAS, Dipartimento di Matematica e Fisica *"Ennio De Giorgi"*, Università del Salento and INFN Section of Lecce, Lecce, Italy
[2] Dipartimento di Scienze e Tecnologie Biologiche ed Ambientali, Università del Salento, Lecce, Italy

E-mail: `domenico.delleside@le.infn.it`

Abstract. Everyday life is exposed to the risks of contracting severe diseases due to the diffusion of severe pathogens. For this reason, efficient antimicrobial surfaces becomes a need of primary importance. In this work we report the first evidences of a new technique to synthesize an antibacterial coating on Ultra High Molecular Weight Polyethylene (UHMWPE)samples, based on a non-stoichiometric, visible light responsive, titanium oxide. The coating was obtained through laser ablation of a titanium target, then the resulting ions were accelerated and implanted on the samples. The samples where tested against a *Staphylococcus aureus* strain, in order to assay their antimicrobial efficacy. Results show that this treatment strongly discourages bacterial adhesion to the treated surfaces.

1. Introduction

During the last years, the occurrence of severe threats to public health raised the attention on the conditions under which pathogens spread across the population. For example, in the cases of Avian Influenza[1] or of the Severe Acute Respiratory Syndrome[2], it has been shown that infections are likely in healthcare workers, although correct antiseptic strategies would ensure good safety levels. In general, several reports pointed out that infectious agents are nested in common use objects such as shopping carts handlebar[3], computer keyboards[4], clinical surfaces[5], faucet handles[4], door handles[6]. Although a proper treatment with antiseptic detergents would greatly reduce risks, there are situations in which such a strategy is inapplicable. Consequently, clearly emerges the necessity of permanent antimicrobial coatings.

Since 1985, the year in which Matsunaga et al.[7] reported its antimicrobial activity, there has been a great interest in the application of titanium dioxide for sanitization purposes. The antimicrobial effect of TiO_2 arises from its photocatalytic activity. In this material, an electron could be excited by ultraviolet (UV) radiation; the corresponding excess energy promotes it to the conduction band, creating a couple *electron-hole*. Both the electrons and the holes generated in such processes have a strong reducing and oxidizing activity and consequently react with H_2O and/or O_2 to give reactive oxygen species (ROS). These ROS are extremely reactive when in contact whit organic compounds and are supposed to perform a key role in TiO_2 antimicrobial activity. In any case, it has been found that titanium dioxide performs an efficient, long-lasting and costs-effective antimicrobial effect[8, 9].

Despite of the fascinating properties of TiO_2, the requirement of the biohazardous UV light severely limits its application. Moreover, it has to be considered that UV radiations represent only at most 4% of the solar energy spectrum reaching daily the Earth surface[10]. For these reasons, a great scientific effort has been devoted to the development of methods that could activate TiO_2 photocatalysis in the visible range. Currently, impurity doping is the most common technique for expanding the spectral response of pure titania[11]. Moreover, several reports indicate that non-stoichiometric titanium oxides develop excellent photocatalytic properties under visible light exposure[10, 12]. These findings suggest the possibility of a new technique for an easy synthetization of non-stoichiometric titanium oxides coatings based on ion implantation over impurities adsorbing materials.

In general, materials contain on their surfaces oxygen-rich compounds (CO or H_2O are clear examples). Using energetic Ti ions that impinge on the surface of such materials, it is possible to break the bonds that tie oxygen with the other elements. In such a way we consequently obtain the chance to stimulate the formation of non-stoichiometric titanium oxides on that substrate, as it is customary in ion implantation[13].

In this work we present the first proof-of-concept of this technique, using UHMWPE as substrate. It is indeed known that in this polymer at least 11% of surface contaminants elements is represented by oxygen[14]. Moreover, in order to reach our goal, it is essential that the energetic ion beams used for the bombardment process have a broad energy spread. In this way, the ions with higher energy will break the bonds, while the lower energy ones could create new bonds with the free oxygen. For this reason we choose a setup in which large energy spread ion beams obtained by laser ablation are accelerated to bombard the samples surface.

2. Materials and Methods

The main apparatus used to reach our scope is the "PLATONE" accelerator[15], which is a laser ion source (LIS) coupled to a double stage electrostatic accelerator. The LIS uses a KrF excimer laser (Lambda Physik, Mod. COMPEX 205) and a stainless steel vacuum chamber as accelerating chamber (AC), Fig. 1. The accelerating chamber has inside an expansion one (EC) that enables the hydrodynamic expansion of the plasma before the ion extraction. The EC forms a hermetic contact with the target support T. A base of the EC, together with T, is fixed to the AC by an insulating flange (IF) that allows the application of a positive high voltage to the EC. At the opposite EC side there is a $1.5\,cm$ diameter hole in order to extract ions. The distance T-EC is fixed at $18\,cm$. A ground electrode (GE) is placed at a distance of $3\,cm$ from EC, having the center drilled by a hole of the same diameter of the EC one. After the GE, at a distance of $2\,cm$, it is placed a further third electrode. This electrode was utilized either as Faraday cup (FC) or sample support. The vacuum was operated through the use of two turbo-molecular pumps, reaching a pressure of the order of $10^{-6}\,mbar$. During the experiments

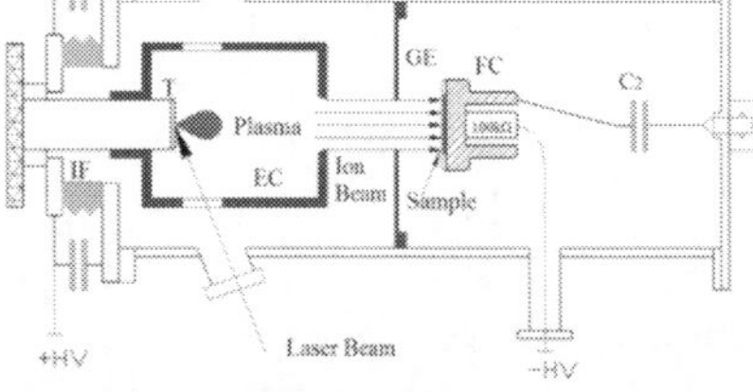

Figure 1. Cross section of the *PLATONE* accelerator. IF: insulating flange; AC: acceleration chamber; GE: ground electrode; EC: expansion chamber; FC: Faraday cup.

the accelerating voltages of the first (T+EC-GE) and second (GE-FC) stage were fixed at +40 and $-20\,kV$ respectively, providing a maximum ion beam energy of $60\,keV$ per charge state. We used such an apparatus because the resulting ion beams are known to have the broad energetic spectrum that we require[16].

The target used was a commercial thick Ti disk with a diameter of $2\,cm$, 99.99% pure. The UHMWPE samples used as substrates, instead, have a density of $0.93\,g/cm^3$, while their dimensions are $20 \times 20\,mm^2$ for a thickness of $1\,mm$. These samples have been fixed on the FC and implanted with 22000 laser shots, for a total dose of about 8.8×10^{15} ions$/cm^2$. After the treatment, the samples were analyzed through scanning electron microscopy (SEM), energy dispersive spectroscopy (EDS), atomic force microscopy (AFM) and UV-Visible (UV-Vis) spectroscopy.

Moreover, in order to test the efficacy of the treatment, we challenged the samples with 10^4 CFU$/ml$ of *Staphylococcus aureus* SA-1 in micro-wells filled with $4\,ml$ of Nutrient Broth at $37°C$, under moderate shaking. The *S. aureus* strain was isolated from a catheter-related bloodstream infection[17]. Before biological testing, samples were UV sterilized for $1\,h$. Treated and untreated UHMWPE samples were placed on direct day/artificial light exposure in the same tank, sealed to prevent any contaminations. During incubation, the samples have been washed 3 times with saline solution (NaCl 0.9%) in order to remove from samples the bacteria that werent firmly adherent to the surfaces. After a week of incubation, the biofilm maturated on the samples surface was stained using green-fluorescent nucleic acid stain (SYTO9; Molecular Probes, USA). After 15 minutes of dark incubation, the biofilm development was viewed with a Nikon Optiphot-2 microscope with an episcopic-fluorescence attachment (EFD-3, Nikon). This technique enabled us to have a direct evidence of the bacteria effectively adhered to the surfaces. In particular, the survival shares have been computed by means of the formula

$$\text{survival } \% = \frac{\text{microbe count on treated sample}}{\text{microbe count on blank sample}} \times 100. \tag{1}$$

3. Results

In order to obtain insights on the effect of the technique under development, we performed some measurements to understand the status of the UHMWPE surfaces before and after treatment. In Fig. 2, the SEM images of a blank sample and of a treated one are shown. These images show qualitatively that surface morphology undergoes drastic changes. In order to estimate the morphology changes illustrated before, we performed also an AFM analysis. Such measurements (Fig. 3) showed that the root mean square surface roughness increased from 23.2 to $94.6nm$, the average height from 172 to $637nm$ and the average roughness from 18.4 to $73.7nm$. Moreover,

Figure 2. SEM images of the blank (left) and treated samples (right).

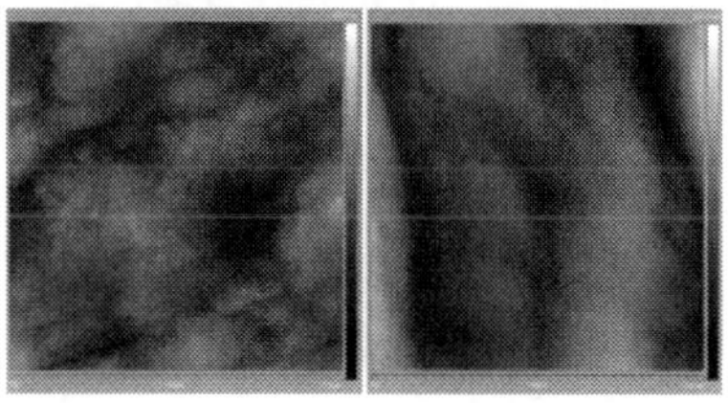

Figure 3. 3D AFM images of the blank (left) and treated samples (right).

the elemental composition maps obtained through an EDS analysis has shown that in many cases titanium and oxygen are found together in the same sites of the treated surfaces (Fig. 4).

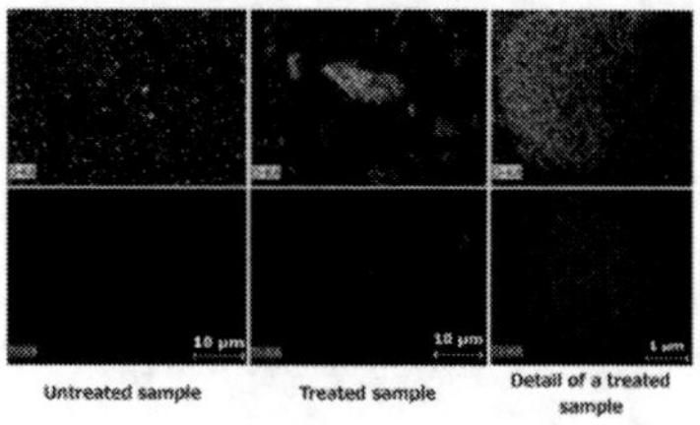

Figure 4. EDS elemental maps of the blank (left) and treated (middle) samples. On the right a detail of a site in which Ti and O are present together. O (cyan) is on top and Ti (blue) on the bottom.

Together with the measurements focused at the surface changes, we performed also a qualitative investigation of the visible-light response of the treated material with respect to the blank one. For this reason we carried out an UV-Vis spectra, measuring the absorbance of the samples, as shown in Fig. 5. Finally, we performed a bacterial surface adhesion test on a blank and 3 different treated samples, which differed for the exposure time on the normal air atmosphere after treatment (respectively: 2, 4 and 7 days). The resulting fluorescence microscope images are shown in Fig. 6. Quantitative analysis, performed by counting the bacterial cells observed in 50 microscopic fields randomly selected, revealed that the mean values of percentages of adherence to the substrate with respect to the blank were $50 \pm 14\%$ (2 days), $32 \pm 11\%$ (4 days) and $10 \pm 5\%$ (7 days). Fig. 7 shows the histogram of the results with the relative standard deviation.

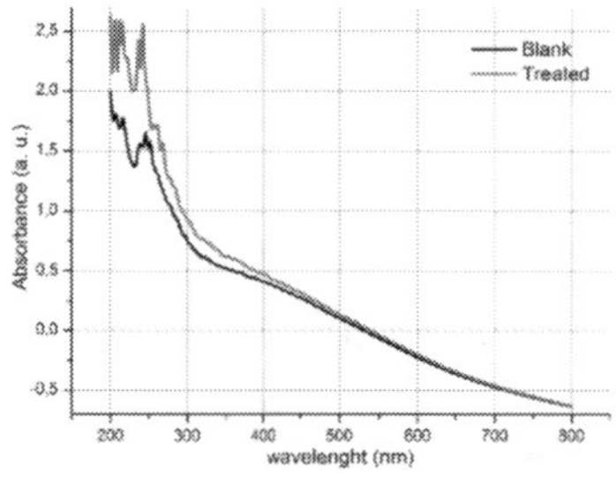

Figure 5. UV-Vis spectra of the surfaces of a blank and treated sample.

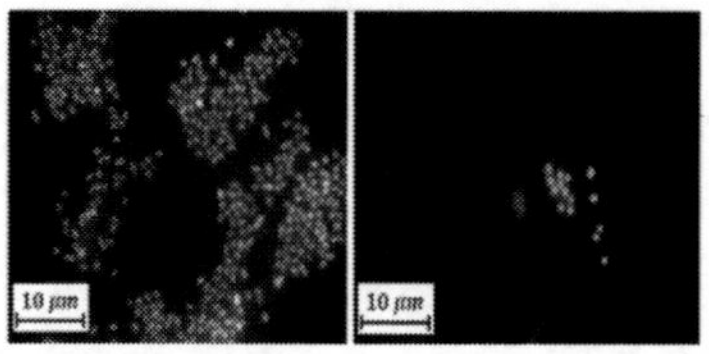

Figure 6. Representative fluorescence microscopy images of a blank (left) and treated samples (right). The treated sample was the one exposed for 7 days in the atmosphere.

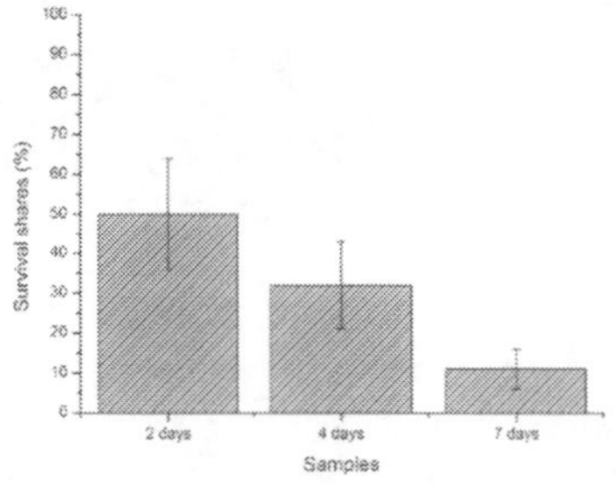

Figure 7. Bacteria survival shares obtained from samples analisys using equation (1).

4. Discussion and Conclusions

As shown by surface analysis, the treated polymer underwent significant morphology changes. Important is the fact that EDS revealed that Ti and O are often together on the treated samples, giving an indication of the oxide formation. The UV-Vis spectra showed that the modified polymer has a slight, although sensible, increase of the absorbance in the visible range, indicating that the treatment went into the right direction.Antimicrobial tests are really interesting, since gave good results on the efficacy of the resulting coatings. Moreover, it has been clearly shown that the effectiveness increases with the time of the exposure to air. This is probably due to titanium natural tendency to develop an oxide layer on its surface. This circumstance is important for applications, since it ensures that the efficacy of the treatment is not compromised while the surfaces are in the atmosphere.

Concluding, we showed that bombarding an oxygen rich surfaces with a broad energy spread dose of Ti ions the formation of a (non-stoichiometric) titanium oxide film with high antimicrobial activity in the visible range is induced. This technique seems to be promising and could simplify the commonly used treatment modality. Despite of this, these results are just preliminary and deserve more attention. In particular more insights on the nature of the titanium oxide and on the experimental parameters affecting the results are needed.

Acknowledgments
Authors are grateful to Dr. Barbara Cortese for her useful suggestions.

References

[1] The Writing Committee of World Health Organization 2005 *N. Eng. J. Med.* **353** 1374–1385 pMID: 16192482 URL `http://www.nejm.org/doi/full/10.1056/NEJMra052211`
[2] Peiris J S, Yuen K Y, Osterhaus A D and Stöhr K 2003 *N. Eng. J. Med.* **349** 2431–2441 pMID: 14681510 URL `http://www.nejm.org/doi/full/10.1056/NEJMra032498`
[3] Gerba C P and Maxwell S 2012 *Food Protection Trends* **32** 747–749 URL `http://www.foodprotection.org/publications/food-protection-trends/article-archive/2012-12bacterial-contamination-of-shopping-carts-and-approaches-to-control/`
[4] Bures S, Fishbain J T, Uyehara C F, Parker J M and Berg B W 2000 *Am. J. Infect. Control* **28** 465–471 URL `http://www.ajicjournal.org/article/S0196-6553%2800%2990655-2/abstract`
[5] Chopra I and Hacker K 1992 *J. Antimicrob. Chemother.* **29** 19–25 URL `http://jac.oxfordjournals.org/content/29/1/19.abstract`
[6] Wojgani H, Kehsa C, Cloutman-Green E, Gray C, Gant V and Klein N 2012 *PLoS ONE* **7** e40171 URL `http://www.plosone.org/article/info%3Adoi%2F10.1371%2Fjournal.pone.0040171`
[7] Matsunaga T, Tomoda R, Nakajima T and Wake H 1985 *FEMS Microbiol. Lett.* **29** 211–214 ISSN 1574-6968 URL `http://dx.doi.org/10.1111/j.1574-6968.1985.tb00864.x`
[8] Liou J W and Chang H H 2012 *Arch. Immunol. Ther. Exp.* **60** 267–275 ISSN 0004-069X URL `http://dx.doi.org/10.1007/s00005-012-0178-x`
[9] Visai L, De Nardo L, Punta C, Melone L, Cigada A, Imbriani M and Arciola C R 2011 *Int. J. Artif. Organs* **34** 929–946 URL `http://www.artificial-organs.com/article/titanium-oxide-antibacterial-surfaces-in-biomedical-devices-ijao-d-11-00132`
[10] Dholam R, Patel N, Adami M and Miotello A 2008 *Int. J. Hydrogen Energ.* **33** 6896 – 6903 ISSN 0360-3199 URL `http://www.sciencedirect.com/science/article/pii/S0360319908011208`
[11] Pelaez M, Nolan N T, Pillai S C, Seery M K, Falaras P, Kontos A G, Dunlop P S M, Hamilton J W, Byrne J, O'Shea K, Entezari M H and Dionysiou D D 2012 *Appl. Catal. B: Environ.* **125** 331 – 349 URL `http://www.sciencedirect.com/science/article/pii/S0926337312002391`
[12] Kitano M, Matsuoka M, Ueshima M and Anpo M 2007 *Appl. Catal. A: Gen.* **325** 1 – 14 ISSN 0926-860X URL `http://www.sciencedirect.com/science/article/pii/S0926860X07001937`
[13] Chen J, Wan G, Leng Y, Yang P, Sun H, Wang J and Huang N 2004 *Surf. Coat. Tech.* **186** 270 – 276 URL `http://www.sciencedirect.com/science/article/pii/S0257897204002828`
[14] Mittal K (ed) 2004 *Polymer Surface Modification: Relevance to Adhesion* vol 3 (Taylor & Francis Group)
[15] V. Nassisi, D. Delle Side and L. Velardi 2013 *App. Surf. Sci.* **272** 114 – 118 URL `http://dx.doi.org/10.1016/j.apsusc.2012.03.128`
[16] L. Velardi, M. V. Siciliano, D. Delle Side and V. Nassisi 2012 *Rev. Sci. Instrum.* **83** 02B717 URL `http://dx.doi.org/10.1063/1.3672476`
[17] Paladini F, Pollini M, Talà A, Alifano P and Sannino A 2012 *J. Mater. Sci. - Mater. Med.* **23** 1983–1990 ISSN 0957-4530 URL `http://dx.doi.org/10.1007/s10856-012-4674-7`

Journal of Physics: Conference Series **508** (2014) 012030 doi:10.1088/1742-6596/508/1/012030

Wetting ability modifications in biocompatible polymers induced by pulsed lasers

C Scolaro[1], L Torrisi[1], M Cutroneo[1] and L Velardi[2]

[1] Department of Physics and Earth Sciences, University of Messina, V.le F. S. D'Alcontres 31, 98166 S. Agata, Messina, Italy
[2] Department of Mathematics and Physics, University of Salento, Via per Arnesano, 73100 Lecce, Italy

E-mail: cscolaro@unime.it

Abstract. Wetting ability was measured in the surface of different biocompatible polymers, such as mylar, polyethylene, poly-methyl-methacrylate and teflon. Nanosecond pulse lasers at intensities of the order of 10^8 W/cm^2 were employed at different doses to irradiate the polymeric surfaces and to induce wetting ability modifications due to the chemical and physical surface changes vs. irradiation time and laser wavelength. In particular, the contact angle as a function of the surface roughness was investigated, as will be presented and discussed.

1. Introduction

In the field of Biomaterials it is important to know the wet ability and the roughness of the material to be implanted in the human environment [1, 2]. Polymeric materials have hydrophilic and hydrophobic behaviour, however their original properties can be strongly modified using surface treatments such as the high intensity laser irradiation. The laser treatments may change the surface temperature, inducing scissions, cross-linking and chemistry activity, modifying roughness and hardness and developing new surface properties depending on the irradiation conditions. In this work we study the changes and the dependence of the wet ability of polymeric materials subjected to irradiation of pulsed lasers as a function of the laser properties, i.e. as a function of the amount of energy released to the material.

2. Materials and Methods

The polymers used in this work are: Mylar $(C_{10}H_8O_4)_n$, polyethylene $(-CH_2-)$, polytetrafluoroethylene $(-CF_2-)$ and polymethylmethacrylate $((C_5O_2H_8)n)$. A Q-switched Nd: YAG laser, operating at the fundamental wavelength of 1064 nm (IR), with pulse duration of 3-9 ns, energy of 180 mJ-300 mJ, spot diameter Ø 3mm in diameter has been used at the Department of Physics of Messina. The ArF excimer laser operating at the fundamental wavelength of 193 nm (UV), also beam energy of 160mJ, pulse duration of 18 ns, spot diameter Ø 2mm was employed at the Department of Surgical Specialties, UOC of Ophthalmology of Messina. The excimer laser KrF, operating at the fundamental wavelength of 248 nm, beam energy of 165mJ, pulse duration of 25 ns, spot diameter Ø 8mm has been used at the Department of Physics of Lecce. A surface profiler Tencor P-10 using whit a 1 mg tip force, scansion length of 500 mm and scan speed of 100 μm/s was employed at the at the INFN-LNS of Catania to perform roughness measurements [3]. The technique of the sessile drop, consisting in the measurement of "'contact angle" between the tangent to the profile of a drop, deposited on the sample surface, was employed.

The experimental setup which allows direct measurement of the contact angle consists of four elements: a mobile platform that houses the sample; an optical microscope; a webcam interfaced to a computer; a calibrated syringe that allows to deposit on the surface of the drops of liquid samples. Distilled water (1 μl) and physiological solution (1 μl) were used as test liquid. The contact angle θ (degree) was calculated from the height h (mm) and the base diameter d (mm) of the droplet itself [4] as in equation:

$$\theta = 2arctg(\frac{2h}{d}) \qquad (1)$$

Liquids wet surfaces when their contact angle is less than 90°.

From surface profiles it is possible to determine the surface vertical and horizontal roughness (Fig. 1).

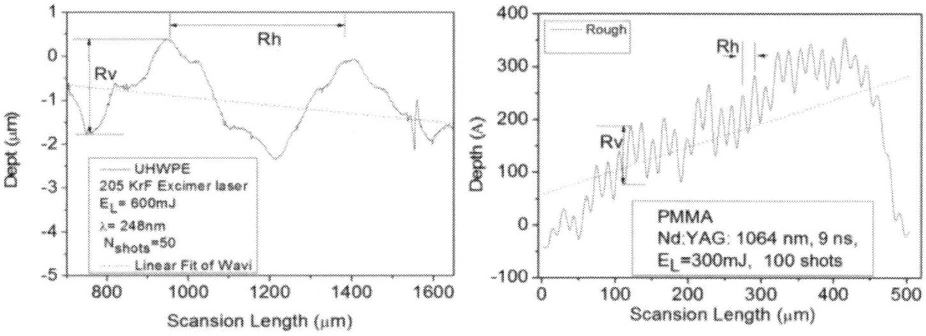

Figure 1. Vertical and horizontal Roughness.

Assuming as axis the average line of the profile, the average vertical roughness R_v depends on the vertical dimension y_i of the profile with respect to the average line as equation (2):

$$R_v = \frac{1}{n}\sum_{i=1}^{n}|y_i| \qquad (2)$$

3. Result and Discussion

PMMA was irradiated in air with Nd:YAG and excimer laser. The laser energy was changed from 60 mJ to 180 mJ for IR irradiation and from 60mJ to 160 mJ for the UV ablation. The number of laser shots was 60 (Fig. 2.a,c). PMMA was also irradiated maintaining the maximum energy constant and varying the number of laser shots (Fig. 2.b,d). The contact angle of distilled water on the pure PMMA is approximately 77°. The action of UV laser decreases the angle of wet-ability making the PPMA hydrophilic, while the lasers in the IR increases the wetting angle making the PMMA hydrophobic [5-6]. The variations of the angle due to the surface wettability may be due to changes in surface morphology. PMMA and UHMWPE were irradiated in air with the Nd:Yag laser maintaining constant the energy (E_L=300 mJ) and by varying the number of laser shots. By the graphs of Fig. 3 it is clear that both polymers show an increment of the wettability by increasing the vertical and horizontal roughness R_v and R_h, making the materials more hydrophobic. Mylar and PTFE were irradiated in air with the laser Nd:Yag at the maximum energy (E_L=300 mJ) and by varying the number of laser shots. The graphs of Fig. 4 show that for both samples, initially hydrophobic, the contact angle first decreases and then increases with the roughnesses R_v and R_h and making them hydrophilic. PMMA and UHMWPE were irradiated in air with the two different excimer lasers (λ=193nm and λ=248nm) at energy E_L=165 mJ and E_L=300 mJ, respectively, and varying the number laser shots (Fig.5). The graphs show a linear reduction of the wettability as a function of roughness R_v and R_h. Experimental results demonstrate that the action of UV laser change the surfaces from hydrophobic to hydrophilic.

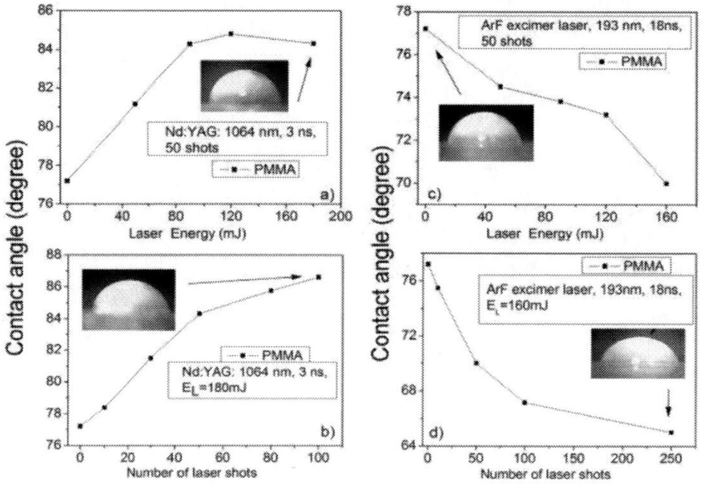

Figure 2. Contact angle of PMMA vs. IR laser energy (a) and IR number of laser shots (b) and UV laser energy (c) and UV number of laser shots (d).

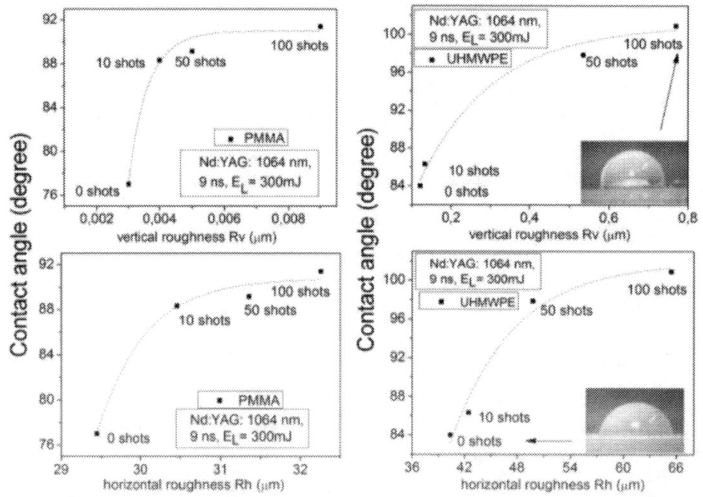

Figure 3. Roughness vertical and horizontal of the materials PMMA and UHMWPE before and after the irradiation with the laser of Nd:YAG.

3. Conclusions

The results obtained reveal that the surface modifications change the values of wet ability by increasing or decreasing the contact angle both of the distilled water and physiological solutions. The roughness created by laser irradiation increases or decreases the angle of wettability making the investigated material hydrophobic or hydrophilic i.e. controlling an important biocompatibility factor in the preparation of biomaterials and prosthesis. By varying the photons energy and the laser properties, the surface polymer modification can be controlled. This result is extremely important in the medical field because it means to improve the biocompatibility and the functionality of biomaterials and functional prostheses. For example, the surfaces of the materials used for the mechanical valve prostheses and therefore in contact with the blood, as it coagulates on rough surfaces (hydrophobic surfaces: low wettability), must be smooth and polished to facilitate removal of microscopic thrombotic formations before they reach dangerous proportions. Instead, in the artificial valve prosthesis (made with fabrics of polymeric fibers) is important to promote clotting within the porous wall of the prosthesis in order to avoid or limit the loss of blood.

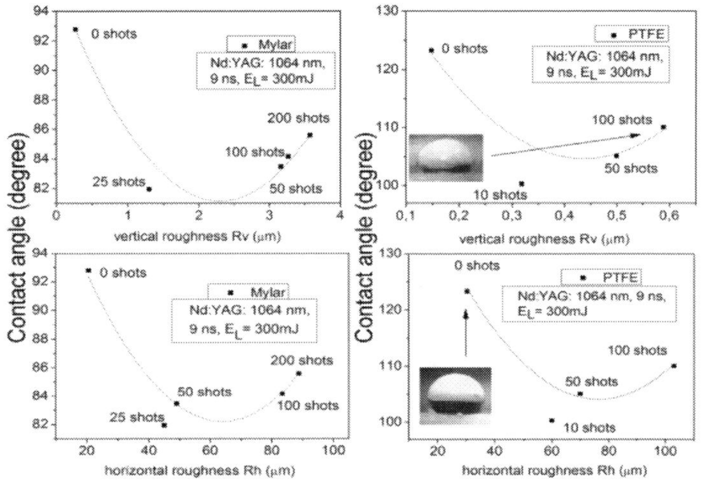

Figure 4. Roughness vertical and horizontal of the materials Mylar and PTFE before and after the irradiation with the laser of Nd:YAG.

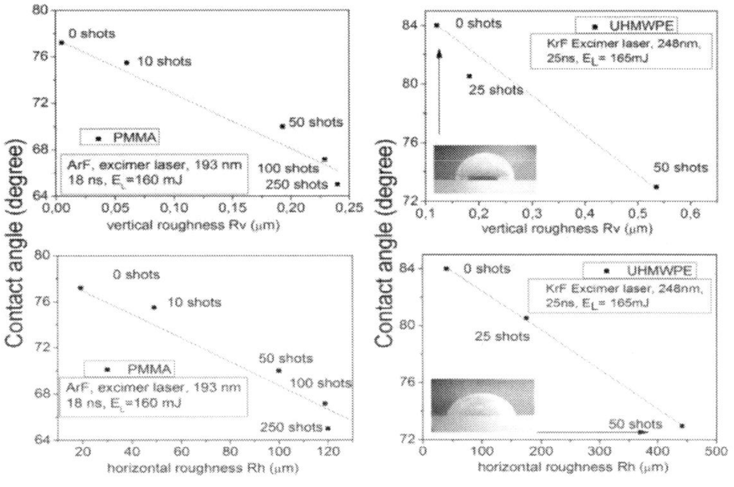

Figure 5. Roughness Rv and Rh of PMMA and UHMWPE before and after the irradiation with the different excimer lasers (λ=193nm & 248nm).

References

[1] J. Black, Biological performance of materials: fundamentals of biocompatibility, New York: Marcel Dekker, 1999

[2] T.R. Thomas (ed.), Rough Surfaces, Longman, London, 1982.
[3] P-10 Surface Profiler, Tencor (1996): http://www.zfm.ethz.ch/e/mems/eq_tencor.htm
[4] K. L. Mittal (Ed), Contact Angle, Wettability and Adhesion, Vol.6. Brill, Leiden (2009)
[5] C. Scolaro et Al., Proc. C.N."III° Workshop Plasmi, Sorgenti, Biofisica ed Applicazioni", Lecce, 19/10/2012. ISBN: 978-88-8305-101-2.
[6] C. Scolaro et Al.,Proc. XI C.N. Bioingegneria "Aspetti Clinico - Fisici ed Ingegneristici Applicati alle Scienze della Vita", Messina, 05/07/2013, in press

A pulsed magnetic stress applied to *Drosophila melanogaster* flies

D Delle Side[1,3], M P Bozzetti[2], A Friscini[2], E Giuffreda[1], V Nassisi[1], V Specchia[2] and L Velardi[1]

[1] LEAS, Dipartimento di Matematica e Fisica "*Ennio De Giorgi*", Università del Salento, SS Monteroni-Lecce – 73100 Lecce, Italy
[2] Laboratorio di Genetica, DiSTeBA, Università del Salento, SS Monteroni-Lecce – 73100 Lecce, Italy

E-mail: domenico.delleside@le.infn.it

Abstract. We report the development of a system to feed pulsed magnetic stress to biological samples. The device is based on a RLC circuit that transforms the energy stored in a high voltage capacitor into a magnetic field inside a coil. The field has been characterized and we found that charging the capacitor with 24 kV results in a peak field of 0.4 T. In order to test its effect, we applied such a stress to the *Drosophila melanogaster* model and we examined its bio-effects. We analysed, in the germ cells, the effects on the control of specific DNA repetitive sequences that are activated after different environmental stresses. The deregulation of these sequences causes genomic instability and chromosomes breaks leading to sterility. The magnetic field treatment did not produce effects on repetitive sequences in the germ cells of Drosophila. Hence, this field doesn't produce deleterious effects linked to repetitive sequences derepression.

1. Introduction

Magnetic field effects on biological matter are fairly common among scientific literature. A wide range of reports dealt with the effects of static magnetic fields of various intensities. In this context, particular emphasis has been devoted to understand the mechanisms of action of such fields on biochemical reactions[1] and on cells[2,3]. In particular, biochemical reactions could change their kinetics if they involve a stage in which radical pairs are formed[4]. A magnetic field, interacting with radicals' spin, could indeed remove the energy degeneration of the pairs, offering a preferred pathway for a specific energy level. This in turn could accelerate the corresponding reaction rate. On the other hand, cells could be affected by magnetic field due to *diamagnetic anisotropy*[5,6] of their membrane (lipid bilayers). When the anisotropy is big enough and the lipid bilayer molecules are sufficiently free in movements (for example in an external temperature greater than 20°C), the reorientation of the membrane become energetically favorable and this could influence the operation of embedded proteins, such as ion channels.

In the case of pulsed magnetic fields (PMFs) the situation is somewhat less explored. Despite of this, PMF are widely used, for example, in medical practice[7]. Currently, no mechanism of action

[3] To whom correspondence should be addressed

could be considered as definitively accepted in order to explain the action of PMFs, although some models have been suggested[8].

In this work we examined the biological effects of an intense PMF on the Drosophila model. To test if the applied magnetic field mimics stress conditions, we analyzed the effects on the control of specific DNA sequences that are "repetitive sequences" and "transposons". In physiological conditions, these sequences are maintained blocked in their chromosomal positions. Many environmental stress conditions cause their deregulation [9,10] and the "transposons" can move in the genome causing gene mutations [11,12]. We used a simple genetic system to test the deregulation of these sequences. Particularly, we searched for the presence of crystalline aggregates composed by Stellate protein in testes, since their presence is a symptom of transposon movement.

2. Materials and methods

2.1. The experimental apparatus

PMF was generated through an ad-hoc device consisting of a pulsed current generator and a solenoid (figure 1). The pulsed current generator, together with the solenoid, constitutes an RLC circuit. A high voltage generator charges the capacitor C (150 nF, 100 kV). Through a fast switch (a spark-gap), the circuit could be closed, in order to transform the electric energy stored in the capacitor into magnetic energy inside the solenoid. The solenoid is made by a 0.8 cm round copper wire. It has a radius a of 3.0 cm and an height l of 12.0 cm for a total of 11 loops (N). By means of the formula

$$L = \mu_0 \frac{\pi a^2 N^2}{l}, \tag{1}$$

its inductance has been estimated as 3.6 μH. In order to obtain an high current, no explicit resistor has been inserted within the circuit. Instead, we used as resistor the internal impedance of the whole circuit. From measurements through an high voltage probe, we estimated experimentally both L as 3.8 μH (slightly increased by parasitic inductance) and R as 0.5 Ω. This resulted in a damping time $\tau = 2L/R$ of 15.2 μs.

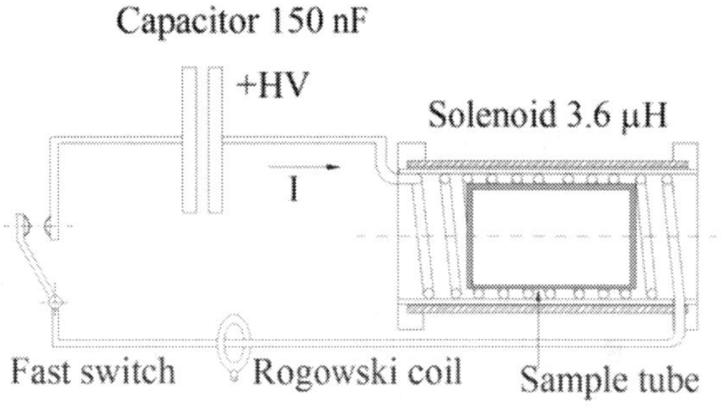

Figure 1. Sketch of the experimental apparatus. Within the solenoid is represented also the container in which the Drosophila flies were kept during the treatment.

The circuit of figure 1, for which $R < 2\sqrt{L/C}$, generates a current $I(t)$ that is described by the equation

$$I(t) = \frac{V}{\omega L} e^{-\frac{Rt}{2L}} \sin \omega t, \tag{2}$$

where V is the voltage at which the capacitor is charged, while $\omega = \sqrt{\frac{1}{LC} - \frac{R^2}{4L^2}} \approx \frac{1}{\sqrt{LC}}$ is the angular frequency at which the circuit oscillates (figure 2). In our setup, it corresponds to 210 kHz. This value is the result of both geometrical and duration constraints. In effect, we had to realize a coil that could fit the samples (a constraint on the inductance). Moreover, we decided to obtain a main pulse that lasts for about 1 μs at least (constraint on the capacitor).

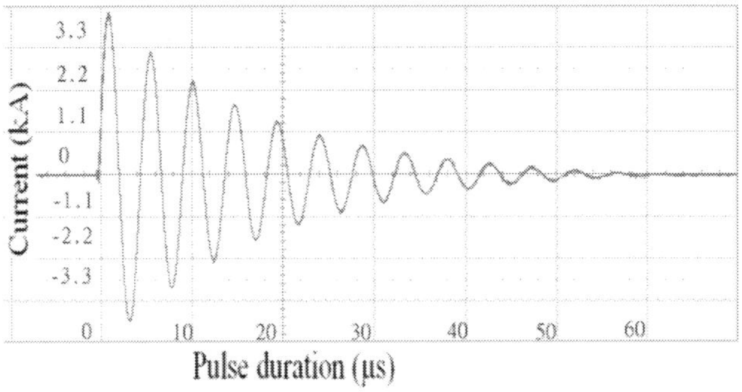

Figure 2. Screenshot of a current pulse diagnosed with the Rogowski coil on a digital oscilloscope. The signal has been taken using a 15x voltage attenuator. The charging voltage used was 24 kV.

The measurement of the current flowing in the circuit has been performed by means of a self-integrating Rogowski coil[13]. This device could be used to obtain a reliable measure of voltage that is roughly proportional to the current flowing within the circuit[14]. Since the magnetic field inside the solenoid depends linearly with respect to the applied current[14], it has been mapped with a Hall probe at low constant current (10 A). Consequently, fixing at 24 kV the charging voltage, we determined a peak current of 4.3 kA, that in turn corresponds to a peak field of about 0.4 T. The map of the axial magnetic field inside the solenoid at the peak current is presented in figure 3.

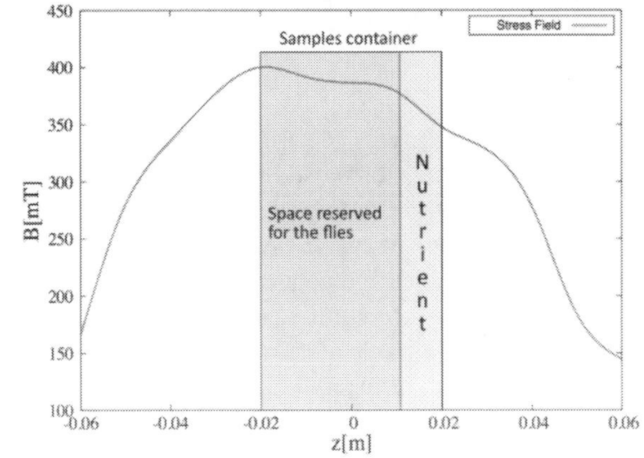

Figure 3. Axial magnetic field at the peak current within the solenoid as a function of the position; z = 0 coincides with the symmetry center of the solenoid. The figure shows also the height of the container in which the samples were maintained during experiments

2.2. Biological samples and analytical procedures

The experiments has been conducted using a Drosophila *Oregon-R* wild type strain. The flies were housed within a cylindrical plastic container of height 4.0 cm and diameter 1.5 cm. One of the extremity of the cylinder was filled with nutrient (1 cm height) for sustaining the flies.

2.2.1. Immunofluorescence on spermatocites

Testes were dissected in Ringer's solution (182 mM KCl, 46 mM NaCl, 3 mM $CaCl_2$, 10 mM Tris-HCl, pH 7.5) and immediately visualized under phase contrast optics in a Zeiss photomicroscope. Testes were fixed with methanol, washed with Phosphate buffered saline (PBS) 1X and 0,5% Triton for 15min, washed with PBS1x for 5min and incubated in primary antibody (polyclonal mouse anti-Stellate 1:50) over night at 4°C. Testes were washed with PBS 1X and 0,5% Triton and incubated with Fluorescein isothiocyanate (FITC)-conjugated anti-mouse (1:100, Jackson Immunoresearch). After washes testes were stained with DAPI or 4,6-diamidino-2-phenilindole and mounted.

2.3. Experimental procedure

In order to test if the magnetic stress can have a reliable effect on the repetitive sequences control, we enclosed within the plastic cylinder 70 samples of Drosophila adult males. The container was placed within the solenoid such that the symmetry center of the space reserved for the flies coincides with that of the solenoid (figure 3). The flies underwent a magnetic stress (0.4 T peak) at a repetition rate of 1 Hz. We left the flies under magnetic field for 2 hours, for a total of 7200 pulses.

3. Results and discussion

We analyzed the spermatocytes of flies subjected to magnetic field looking for the presence of crystalline aggregates. Indeed, the presence of such aggregates is a symptom of the deregulation of repetitive sequences. The spermatocytes did not exhibit the presence of crystalline aggregates (figure 4). We can hypothesize that the treatment does not produce effects on the deregulation of repetitive sequences in the germ cells of Drosophila.

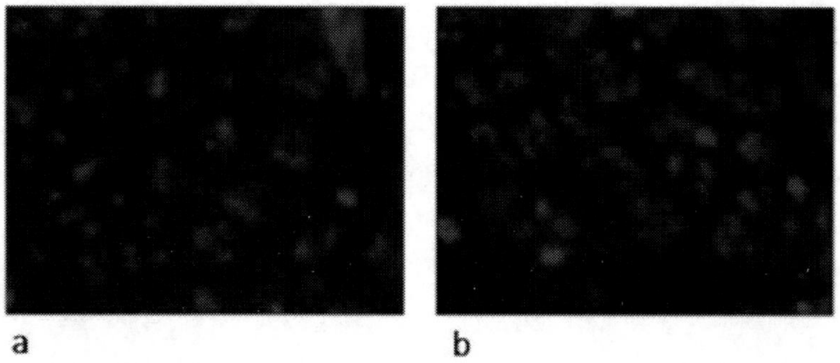

Figure 4. In blue the nuclei of spermatocytes from testes: a) control; b) stressed males. The spermatocytes of stressed males do not present crystalline aggregates.

4. Conclusions

We presented the characteristic of an intense pulsed magnetic fields used to induce stress in biological samples. Magnetic field pulses of 0.4 mT (peak) and repetition rate of 1 Hz were applied to a set of Drosophila melanogaster adult males. To test if the applied magnetic field mimics stress conditions, we analyzed the effects on the control of specific DNA repetitive sequences that are activated after different environmental stresses. We also looked at the structural state of the chromosomes in individuals subjected to stress. The magnetic field treatment did not produce effects on the deregulation of repetitive sequences in the germ cells of Drosophila. This field doesn't produce deleterious effects normally linked to repetitive sequences derepression, such as gene mutations, DNA damage and genomic instability that lead to sterility.

Acknowledgements

This work has been supported by FIRB RBFR10V8K6 grant from the Italian *Ministero dell'Istruzione, dell'Università e della Ricerca* appointed to VS.

References

[1] Blankenship R E, Schaafsma T J and Parson W W 1977 *BBA-Bionergetics* **461** 297-305
[2] Binhi V N, Alipov Y D and Belyaev I Y 2001 *Bioelectromagnetics* **22** 79-86
[3] Sullivan K, Balin A K and Allen R G 2011 *Bioelectromagnetics* **32** 140-147
[4] Grissom C B 1995 *Chem. Rev.* **95** 3-24
[5] Rosen A D 2003 *Cell Biochem. Biophys.* **39** 163-173

[6] Scholz F, Boroske E and Helfrich W 1996 *Biophys. J.* **45** 589-592
[7] Shupak N M 2003 *The Radio Science Bulletin* **307** 9-32
[8] Binhi V N 1998 *Bioelectroch. Bioen.* **45** 73-81
[9] Casacuberta E and González J 2013 *Mol. Ecology* **22**, 1503-1517
[10] Hunter R G, McEwen B S and Pfaff D W 2013 *Mob. Gen. Elements* **3** e24555
[11] Specchia V, Piacentini L, Tritto P, Fanti L, D'Alessandro R, Palumbo G, Pimpinelli S and Bozzetti M P 2010 *Nature* **463** 662-665
[12] Perrat P N, DasGupta S, Wang J, Theurkauf W, Weng Z, Rosbash M and Waddell S 2013 *Science* **340** 91-95
[13] Nassisi V and Luches A 1979 *Rev. Sci. Instrum.* **50** 900-903
[14] Giuffreda E, Velardi L, Delle Side D, Paladini F, Nassisi V 2014 *J. Phys.: Conf. Ser.* in this volume.

Analysis of selective laser cleaning of *patina* on bronze coins

G Buccolieri[1,4], V Nassisi[1], L Torrisi[3], A Buccolieri[2], A Castellano[1], M Di Giulio[1], E Giuffreda[1], D Delle Side[1] and L Velardi[1]

[1] Università del Salento, Dipartimento di Matematica e Fisica, via Arnesano, 73100, Lecce, Italy

[2] Università del Salento, Dipartimento di Scienze e Tecnologie Biologiche e Ambientali, via Monteroni, 73100, Lecce, Italy

[3] Università di Messina, Dipartimento di Fisica e Scienze della Terra, V. S. d'Alcontres 31, 98166 S. Agata (ME), Italy

E-mail: giovanni.buccolieri@unisalento.it

Abstract. The *patina*, is the result of a large number of chemical, electrochemical and physical processes which occur spontaneously during interaction of metal surfaces with the environment. In this work we want to analyze and remove the patina in artefacts, exposed to atmosphere for various decades. Here, experimental results about the laser cleaning of bronze coins by KrF (248 nm) and Nd:YAG (532 nm) lasers are reported. Both laser wavelengths were efficient to reduce the chlorine concentration on the surface of the coins more than 80 %, as demonstrated by Energy Dispersive X-Ray Fluorescence analyses.

1. Introduction

In the past the conservation of cultural heritage and in particular of metallic artefacts was considered a secondary problem either for the poor economic resources of countries or for the low technologic level to get restoration and conservation of art works without damage. Nowadays, the new technologic devices available in many laboratories have also increased the interest for the scientific applications in archaeological field. Particularly, since the seventies, due to technological progresses achieved in the field of power lasers and their great marketing and diffusion, new frontiers in the study of pulsed laser ablation have been opened [1-3]. Laser is applied in cultural heritage field, but its application is not so common. In 1997 a pulsed CO_2 laser was used to investigate samples of various metals showing that laser cleaning is efficient in removing organic paints and stains without damaging the bulk metal [4].

Recently, other pulsed lasers such as excimer lasers, Nd-Yag lasers, Er-Yag lasers, double frequency Nd-Yag lasers and triple frequency Nd-Yag lasers [5, 6], have been employed in the cleaning field too. It is well known in fact that pulsed laser beams, by choosing appropriate fluences, can induce a fast and controlled vaporisation of a few surface layers of the target. So this straightforward technique can reach meaningful results in the cleaning field reducing the risk of damage on artefact surface. As a whole, the laser cleaning is the result of many processes which depend on the irradiation conditions and artefact components.

In this work we focus our activity to study the laser cleaning of ancient bronze coins, undergone to corrosion. It is well known that a great variety of coins of metal or metallic alloys, under specific

[4] To whom correspondence should be addressed

conditions, produce thin layers of corrosion products, called patina. Specific bronze patina present on outdoor exposed artefacts is the result of a large number of chemical, electrochemical and physical processes which occur spontaneously during interactions of metals with the environment.

In particular, bronze and copper alloys, exposed to urban atmospheres (with presence of CO, NO_2, SO, pollution, acid rains, ...) for various decades, exhibit a greenish patina containing several compounds. Moreover, the artefacts, located in proximity of the sea, are in contact with marine aerosol which is constituted by small drops of seawater (mainly H_2O + NaCl). The chlorine reacts in the atmosphere with nitrogen oxide and forms hydrochloric acid as described by equation (1):

$$2\,NaCl + 3\,NO_2 + H_2O \rightarrow 2\,NaNO_3 + NO + 2\,HCl \tag{1}$$

The hydrochloric acid formation is the main responsible for a particular corrosion of bronzes called "*bronze's cancer*" [7]. This process induces the formation of copper chloride, which is soluble in water and forms copper (I) oxide (red cuprite) and hydrochloric acid as described by the equation (2) and (3), respectively:

$$4\,HCl + 4\,Cu + O_2 \rightarrow 4\,CuCl + 2\,H_2O \tag{2}$$

$$4\,CuCl + 2H_2O \rightarrow 2\,Cu_2O + 4\,HCl \tag{3}$$

Then, hydrochloric acid reacts again with the copper forming cyclic reactions which propagates the corrosion inside the coin. Besides, copper chloride hydroxides are formed in different, rarely solvable, allotropic forms delineated in the equation (4) and (5) and characterised by a blue-green pigmentation on the bronze surfaces.

$$2\,Cu + HCl + H_2O + O_2 \rightarrow Cu_2Cl(OH)_3 \text{ (atacamite)} \tag{4}$$

$$2\,Cu + HCl + H_2O + O_2 \rightarrow Cu_2Cl(OH)_3 \text{ (paratacamite)} \tag{5}$$

From the above chemical reactions it is possible to understand how the chlorine concentration is one of the main responsible of the deterioration of ancient bronze coins.

2. Experimental section

2.1. Theoretical basis

Lasers could be applied in the cultural heritage field, but it is indispensable to study carefully the process in order to determine the optimal conditions of irradiation: ablation of the patina and protection of the bulk. Depending on the the laser parameters, the photoablation can produce three kinds of photoinduced phenomena: photochemical processes (direct bond breaking and photodissociation), photothermal processes (heating by vaporation and heat conduction) and photomechanical processes (photoionization, plasma formation, fast expansion and shock wave propagation). Depending on the properties of the sample and on the irradiation conditions, such as wavelength and pulse duration, one of them may become dominant.

The heating region, produced by photothermal effects, depends on the duration of the laser pulse and on the thermal diffusion length, z_{th}, which is linked to the heat propagation by the following equation:

$$z_{th} \approx \sqrt{2D\tau} \tag{6}$$

where D is the heat diffusivity of the sample and τ is the laser pulse duration. Therefore, using short duration laser pulse, the heating propagation becomes short as well as the depth of the layer which could be ablated. The photomechanical effect is represented by the mechanical coupling coefficient, C_m, which is described by the equation (7) [8]:

$$C_m = \frac{P_a}{I} \quad (7)$$

where P_a is the ablation pressure (the explosion pressure produced by the shock wave provoked by the laser interaction) and I is the laser intensity. Moreover, photochemical effects are induced by UV laser photons. In effect, they have an energy near 5.0 eV, i.e. sufficient to produce chemical etching due to the break of the chemical bond Cu-Cl (that needs 3.76 eV at least), cleaning the bronze surface from Cl. Therefore, by the above considerations, it is difficult to determine *a priori* the optimal irradiation conditions in laser cleaning of coins exhibiting *patina* layers. So it is noteworthy to say that the region of laser cleaning can be controlled mainly by the laser wavelength, laser energy, pulse duration and target material.

In this work, preliminary results about the laser cleaning of bronze *patina* are reported by using two different nanosecond laser beams at UV and visible region. Both lasers operated at the same laser fluence just to observe how the choice of the wavelength can influence the cleaning process.

2.2. Instruments

We used two different pulsed lasers: a KrF excimer laser operating at 248 nm, 25 ns pulse duration and a Nd-Yag laser operating at 532 nm (second harmonic), 10 ns pulse duration. A 50 cm focal length lens was utilised to concentrate the laser energy on irradiating spot of 0.6 cm^2. A portable apparatus for Energy Dispersive X-Ray Fluorescence (EDXRF) analysis was used in order to evaluate the variation of the chlorine concentration before and after the treatment of laser cleaning. The characteristics of the EDXRF instrumentation is described in the references [9,10]

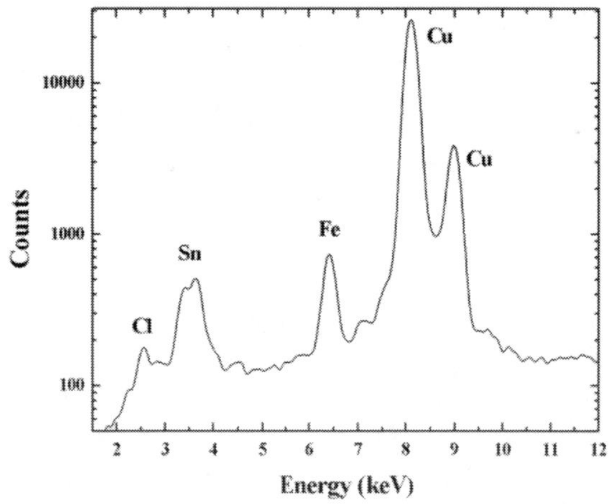

Figure 1. EDXRF spectrum before laser cleaning.

3. Results and discussion

Laser cleaning tests were carried on bronze coins made in Italy and aged at the first years of twentieth century. Sample surfaces presented a thin greenish alteration layer together with soil deposits. EDXRF analysis revealed that the corrosion layer was mainly due to chlorine (Fig. 1). Moreover, the characteristic X-ray L-lines of Sn, at 3.4 keV (Lα) and 3.7 keV (Lβ), and K-lines of Cu, at 8.04 keV (Kα) and 8.9 keV (Kβ), proved that the alloy constituting the coins was based on a Cu alloy containing Sn and Fe, i.e. it is a typical bronze.

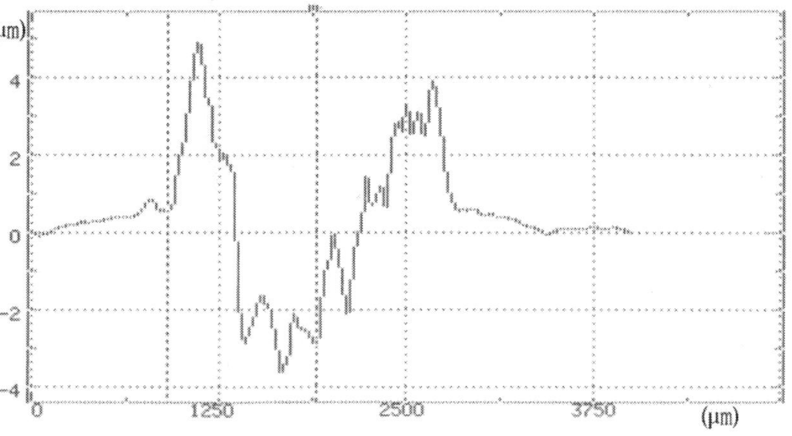

Figure 2. Typical result of the profiler.

In order to remove only the *patina* film without damages the coin surface, it was necessary to operate below the ablation threshold conditions related to bulk. They were determined experimentally as follows: utilizing a bronze sheet, with the same compositional characteristics of the coins (controlled by EDXRF analysis), having a flat area, we applied in different position at different laser pulse energies 50 laser shots. The laser spot was fixed at about 0.6 cm^2. "*Off line*" measurements of the ablated material were performed with a stylus surface profiler (Tencor Instruments ALPHA-STEP 200), in order to measure the crater profiles and to determine the ablation yield with different irradiance conditions.

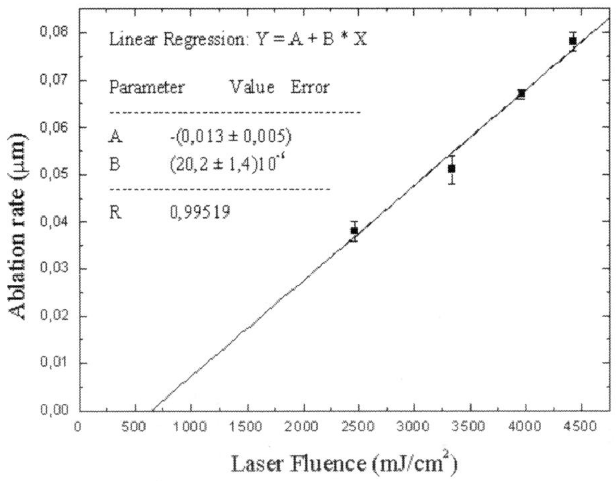

Figure 3. Etching rate of bronze sample as a function of the laser fluence.

Figure 2 shows a typical result of the profiler obtained by KrF laser for a total fluence of 3900 mJ/cm^2. At high laser fluences the spot exhibits enhanced edges. Fig. 3 shows the ablation rate as a function of the laser fluence for the KrF laser. Assuming that (the ablation rate is linearly dependent to the laser fluence) [12], the bronze threshold fluence was estimated of 655 mJ/cm^2 by the fit of the experimental data. This procedure was carried out also for visible laser obtaining a threshold fluence of 1000 mJ/cm^2. Therefore, the values of the fluence to operate laser cleaning were fixed at 300 mJ/cm^2 and 500 mJ/cm^2 per single pulse for KrF and visible laser respectively. The laser spot was about 0.6 cm^2. In order to ablate the entire surface, the coin has been shifted horizontally and vertically with respect to the laser beam.

Figure 4 shows the EDXRF spectrum, in the low X-ray energy region, before and after the cleaning treatment by KrF laser. The characteristic X-ray K-lines of Cl at 2.6 keV (Kα) and 2.8 keV (Kβ) and L-lines of Sn at 3.4 keV (Lα) and 3.7 keV (Lβ), useful for this investigation, have intensities proportional to the elements concentration in the analyzed surface. The proportionality takes in consideration the X-ray production cross section and the detection efficiency of the detector.

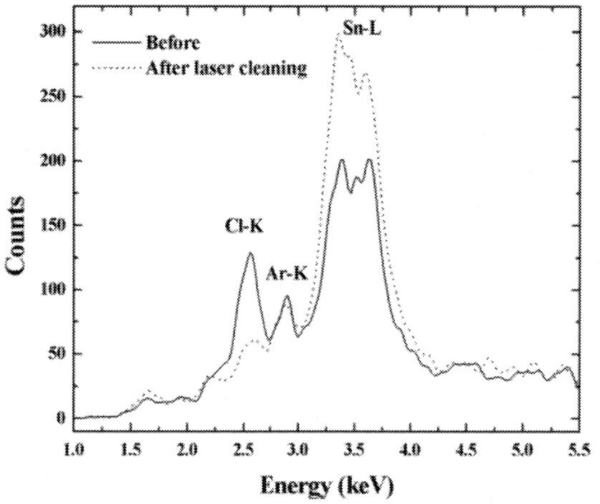

Figure 4. EDXRF spectrum before and after UV laser cleaning.

After 30 laser shots, with a repetition rate of 1Hz, the chlorine concentration decreases from 2.6 % w/w to a value below 0.5 % w/w (detection limit) against the increasing signal of tin due to its surface emersion. The analysis was repeated on several samples, and it has shown similar results. This confirms that the technique is reproducible.

These results confirm that the laser cleaning is able to remove safely the *patina* film, i.e. corrosion products of the coin surface without damaging the coin matrix.

In Fig.5 the area of a bronze coin irradiated by UV laser is showed. it is evident how the green colour of the patina disappears. The figure shows the bronze coin during the ablation process which subsequently affected the whole coin.

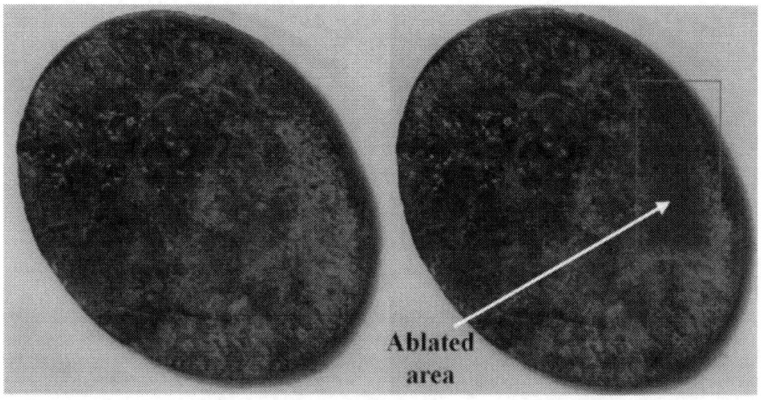

Figure 5. Photo of a virgin (left) and a processed coin (right).

Studies performed by using visible laser gave also good results on the removing of chlorine concentration from the surface of bronze coins (Fig. 6). In fact, after 30 laser shots, the chlorine concentration decreases from 2.8 % w/w to a value below 0.5 % w/w (detection limit). The decreasing of tin signal reveals that visible laser cleaning is not selective because the removal of *patina* is accompanied by a reduction of tin.

Moreover an interesting observation was derived by comparing the irradiated areas at the two wavelengths. It was possible to observe a colour changing of the coin surface. The coin cleaned by KrF laser exhibited a remarkable darkening. Same change was not observed for the visible laser treatment: in this case the ablated areas appeared very bright and polished [11].

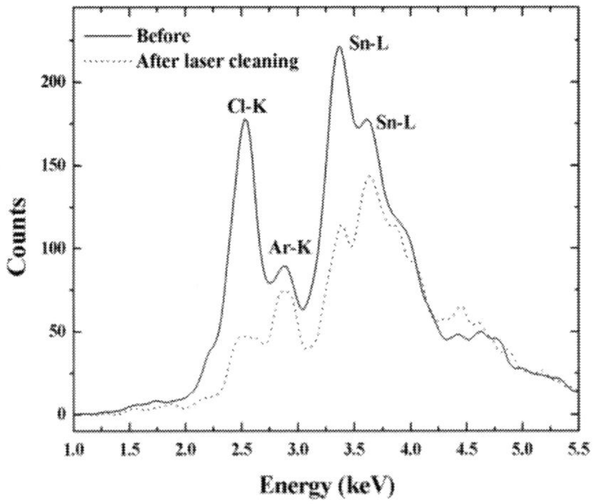

Figure 6. EDXRF spectrum before and after visible laser cleaning.

4. Conclusions

Experimental results about the laser cleaning of bronze coins show that the pulsed lasers can be used safety as a powerful tool to remove the *patina* of bronze coins. Operating below the threshold laser fluence, the UV and visible laser actions are able to remove the greenish patina from the surface of the coins more than 80 %. Besides, the obtained results prove that UV laser cleaning is able to preserve the bulk of the coin while that is not true for visible laser cleaning.

References

[1] Salimbeni R, Pini R and Siano S 2003 *J. Cult. Herit.* **4** 72–76.
[2] Siano S, Salimbeni R, Pini R, Giusti A and Matteini M 2003 *J. Cult. Herit.* **4** 140–146.
[3] Buccolieri G, Nassisi V, Buccolieri A, Vona F and Castellano A 2013 *Applied Surface Science* **272** 55–58.
[4] Cottam CA, Emmony DC, Larson J and Newman S 1997 *Laser in Conservation of Arteworks*, (LACONA II), Restauratorenblätter (Special issue), Mayer & Comp., Vienna.
[5] Pini R, Siano S, Salimbeni R, Pasquinucci M and Miccio M 2000, *J. Cult. Herit.* **1** S129–S137
[6] Drakaki E, Karydas AG, Klinkenberg B, Kokkoris M, Serafetinides AA, Stavrou E, Vlastou R and Zarkadas C 2004, *Appl. Phys. A* **79** 1111–1115.
[7] Scott DA 2002 *Copper and Bronze in Art*, Getty Publications, Los Angeles.
[8] Lee JM and Watkins KG 2001, *J. Appl. Phys.* **89** 6496–6500.
[9] Cesareo R, Castellano A, Buccolieri G, Quarta S and Marabelli M, *X-Ray Spectrometry* **33** 289–293.
[10] Castellano A, Buccolieri G, Quarta S, Cesareo R, Marabelli M, Santopadre P, Ioele M, Gigante GE and Ridolfi S 2005 *Cultural Heritage Conservation and Environmental Impact Assessment by non-destructive testing and micro analysis*, A.A. Balkema Publishers, London.
[11] Crina Anca Sandu I, Helena de Sà M, Costa Pereira M, 2011 *Surface and Interface Analysis* **43** 1134–1151.
[12] Torrisi L, Gammino S, Andò L, Nassisi V, Doria D, Pedone A, 2003, *Appl. Surf. Sci.* **210** 262–273.

Research activity in the laboratory for inertial confinement fusion in ENEA - Frascati

R De Angelis[1], P Andreoli, F Consoli, G Cristofari and G Di Giorgio
Associazione ENEA-Euratom sulla Fusione, Frascati, Italy

E-mail: riccardo.deangelis@enea.it

Abstract. Research on the use of nuclear fusion for energy production is carried out at ENEA mainly in the Frascati Research Center with laboratories dedicated to the study of magnetic confinement (Tokamak FTU) and inertial confinement (ABC). This paper summarizes the on going experimental programs and diagnostic development related to the studies of laser produced plasmas.

1. Introduction

The laboratory for inertial confinement fusion is dedicated to the study of laser produced plasmas in the 2 beams laser facility ABC, which can deliver up to 200 J in pulses a few nanoseconds duration. The system is suitable for irradiating planar targets at I >1014 W/cm2 allowing to access intensity regimes close to the requirements of fusion targets. Present ABC experimental programs are dedicated to the investigation of transmission properties of porous materials and to the study of a-neutronic fusion reactions and to the analysis of electromagnetic pulses accompanying the laser-target interaction.

Porous materials, such as polystyrene foams, have been proposed as random absorbers of radiation in laser plasma experiments; their capability of homogenizing the energy transferred from the laser to the target is particularly interesting for inertial confinement experiments, in order to limit the onset of instabilities, which can downgrade the implosion gain. The average densities of the foam absorbers used were larger than the critical density at the laser wavelength; in these conditions absorption is mainly distributed on the absorber's volume.

A-neutronic fusion reactions in solid targets irradiated by laser are also studied. Using plastic targets doped with Boron we got an appreciable number of counts of α particles at the laser powers available in ABC, indicating the occurrence of p + 11B = 3α reactions. Similar experiments were also conducted in collaboration with the picosecond laser facility LULI (Laboratoire pour l' Utilisation de Laser intense – Ecole Polytechnique, Palaiseau). In order to better understand these effects it is necessary to study the behaviour of fast particles escaping the plasma, for this purpose diagnostics such as Thomson parabolas and time of flight detectors have been employed.

Electromagnetic pulses generated in the interaction of fast laser pulses with matter can be relevant since they can produce high current noise capable of disturbing signals and damage instrumentation. An analysis of these effects has recently been started in ABC.

[1] To whom correspondence should be addressed

2. The ABC installation

The main beam parameters of the ABC laser (Fig. 1) are reported in Table I.

Figure 1. The ABC laser facility.

Table 1. The main parameters of the ABC laser

Parameter	Value
Active medium	Glass Neodimium Phospate
Wavelength	1054 nm
Beam energy	≈ 2*100 J
Pulse duration	≈ 2-5 ns
Max power/beam	$\approx 50\times10^{9}$ W
Rising time	≈ 1 ns
Rod diameters	⌀ 10/20/50/75 mm
Output beam diameter	≈ ⌀ 75 mm
Intensity	$\approx 10^{13} \div 10^{15}$ W/cm^{2}
Minimum focal spot	≈⌀ 40 μm
Beam integrator Induced Space Incoherence	500×500 μm^{2}

The focusing optics is characterized by large aperture, short focal length lenses (f/1, f=10cm) (fig. 2) and the possibility of using Induced Space Incoherence integrators (Fig.3) to smooth the beam profile.

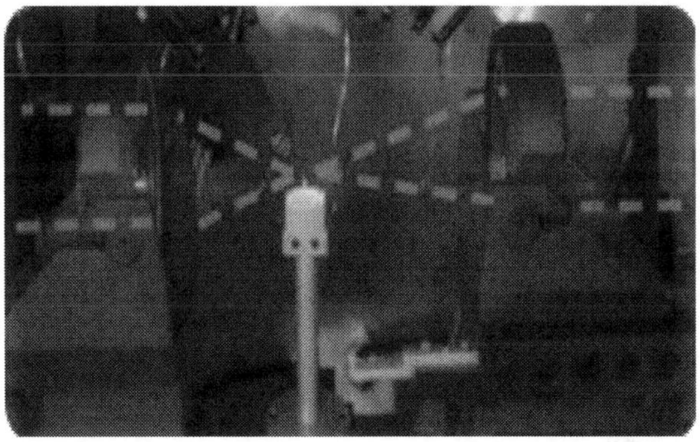

Figure 2. The focusing lenses on the target.

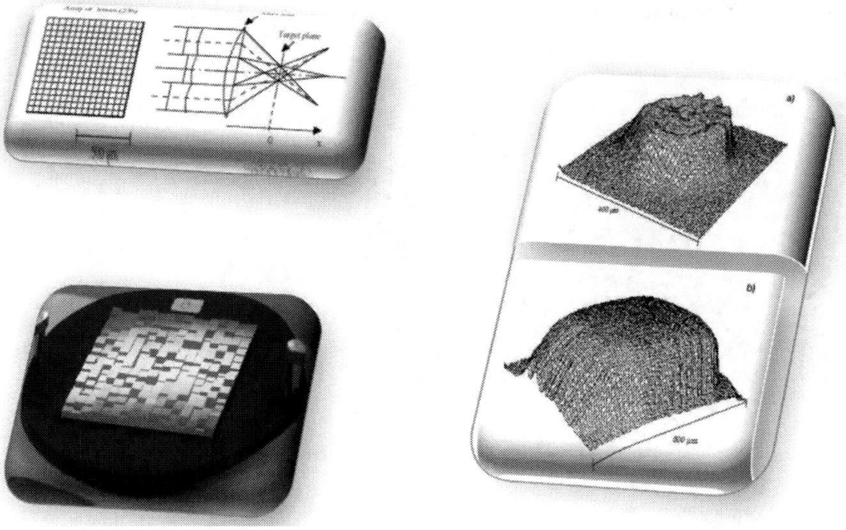

Figure 3. Beam smoothing by ISI plates, showing the different beam waist profiles obtained.

3. Diagnostics

3.1. Radiation and particle diagnostics

A wide set of diagnostics is used for the measurement of the radiation and particles emitted by the plasma. Faraday cups at 10 different angles to the target perpendicular obtain time of flight analysis of the ions emitted by the plasma. Diamond detectors are also used for time of flight analysis, being scarcely affected by electromagnetic noise they offer a better response to fast particles, which are detected in the early phases of the discharge. The time response of diamond detectors depends on the sensor's dimension and structure, Fig. 4 shows the 2 detectors used in ABC with related signals. Ion analysis can also be performed with Thomson spectrometers, which allow ion discrimination. A 4-pinhole X ray camera is producing 4 plasma images on a microchannel plate detector through Be filters which select different energy bands. A time resolved set of X ray images can also be produced on gated X-ray camera Kenntech slix 4.

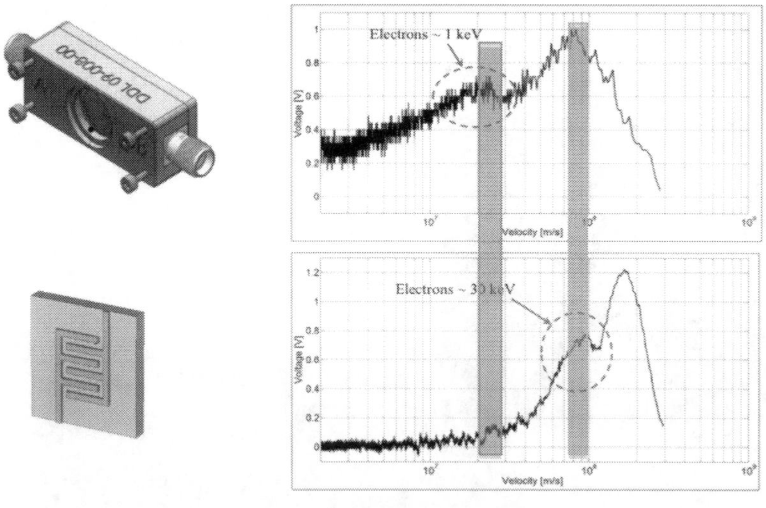

Figure 4. Two diamond detectors used in ABC: (top) High Purity Single Crystal Diamond SCD464650D (Diamond Detectors Ltd); (bottom) Diamonds with Interdigital Structure (Tor Vergata University, Rome).

Optical diagnostics for plasma self-emission include a visible streak camera with 200 ps resolution, and spectroscopy. Shadowgraphy and interferometry of the plasma are performed by use of an optical probe beam originating from the main beam and duplicated in frequency, examples are shown in Fig. 5.

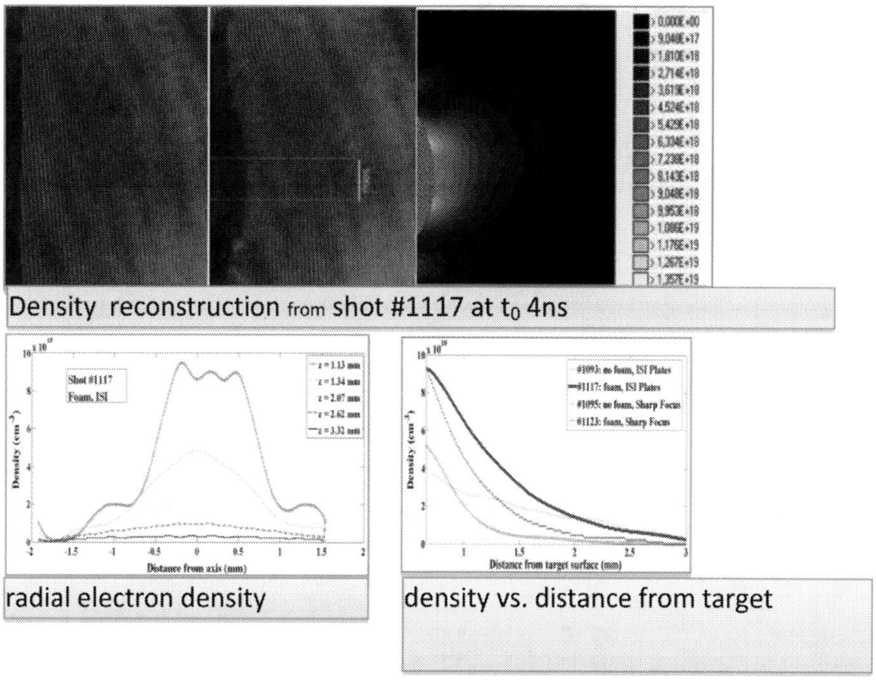

Figure 5. Interferogram and analysis of a laser produced plasma in ABC.

4. Experimental programs

4.1. Study of porous materials as absorbers of laser radiation

Foams of low-density materials were irradiated in ABC to study their behaviour in the ablation process. The use of foams with initial density greater than the critical value for the laser wavelength can be beneficial in compression experiments by distributing the input energy more uniformly, thus reducing the onset of hydrodynamic instabilities. The homogenisation is due to the formation of small plasma cells inside the volume of the porous material. In the most recent experimental campaign the energy absorption of Polystyrene foams with densities in the range 10-60 mg/cm^3 was measured by depositing the foam specimen on thick Aluminum substrates. The high pressure shock wave, originated at the ablation front and reaching the substrate, melts and evaporates a volume of matter $V_{crater} = \frac{\sigma K_p E_L}{\alpha w_0 \rho_0}$, which can be assumed as a measurement of the foam transmission efficiency σK_p. (E_L = Energy of the laser pulse; K_p = laser-plasma energy coupling; σ = efficiency of conversion to shock wave; α = thermal/kinetic ratio in the shock adiabat, w_0 = specific vaporisation heat, ρ_0 = solid metal density).

Accurate measurements of craters volumes have been performed using an optical confocal profilometer LEICA DCM 3D which allowed measurements of their surface with an accuracy of ~ 1.7 µm on the planar dimensions and ~ 45 nm on the depth. Fig. 7 gives an example of the crater reading.

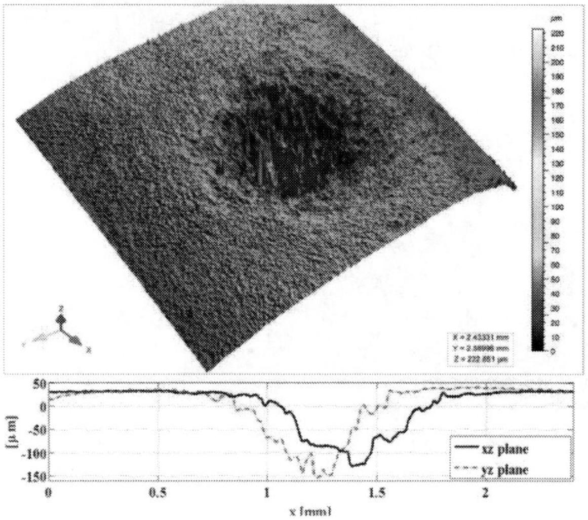

Figure 6. The profile of a crater measured with the confocal microscope (shot #1117).

The dependence of the crater volumes on foam thickness is shown in Fig. 7 for various densities of the absorber.

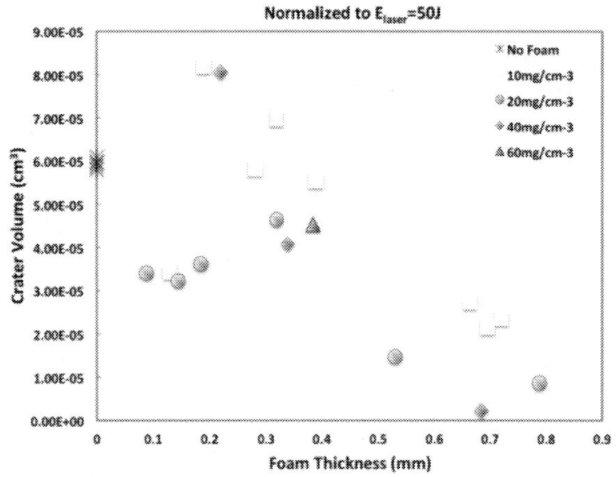

Figure 7. Dependence of crater volumes from foam thickness.

4.2. Study on p+^{11}B fusion reactions in nanoseconds laser plasmas

Studies on the neutron-less p+^{11}B fusion reactions have found interest in nuclear astrophysics and can be important for energy production. Hydrogen and boron are abundant in nature and the fusion energy is released in the form of kinetic energy of α particles. This makes the reaction a candidate for third generation fusion power plants. The possibility of generating p+^{11}B fusions in laser produced plasmas has been demonstrated in experiments with picosecond lasers at intensities of I ≈ 10^{18} W/cm^2. Experiments at lower intensities (I ≈ 10^{14} W/cm^2, 3 ns pulse duration) have been performed with the two-beams ABC laser facility in ENEA Frascati, on Boron-doped polyethylene targets, giving some evidence of the production of α particles. Due to the low number of fusion reactions expected in these conditions it is important that the α particles are detected with minimum incertitude. The most recent experimental activity has been dedicated to further analysis of CR 39 with an increased angular coverage of the measurements. In order to obtain a better discrimination against spurious effects due to other particles, the tracks produced in CR39 by α' s with different energies are compared with template images simulated numerically according to the model presented in [9]. The possibility of producing α's by high-energy Boron atoms impinging on the protons inside the CR39 rather than in the plasma has been proven experimentally as not relevant.

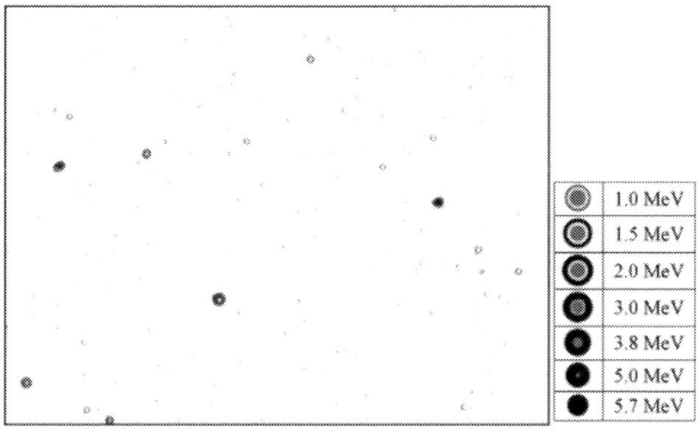

Figure 8. Tracks from α particles on CR-39 after irradiation of Polythene doped with Boron, observed (left) and simulated (right, with enlarged scale).

4.3. Study of microwave fields

A general study of the electromagnetic fields, which develop during the laser- plasma interaction, has been recently started. This study is based on the measurement of signals detected by antennas placed in different positions with respect to the target, both inside and outside the experimental chamber. General purpose antennas as well as specifically designed antennas have been used, the first results obtained in ABC, will be compared with data collected on experiments performed in other laser laboratories.

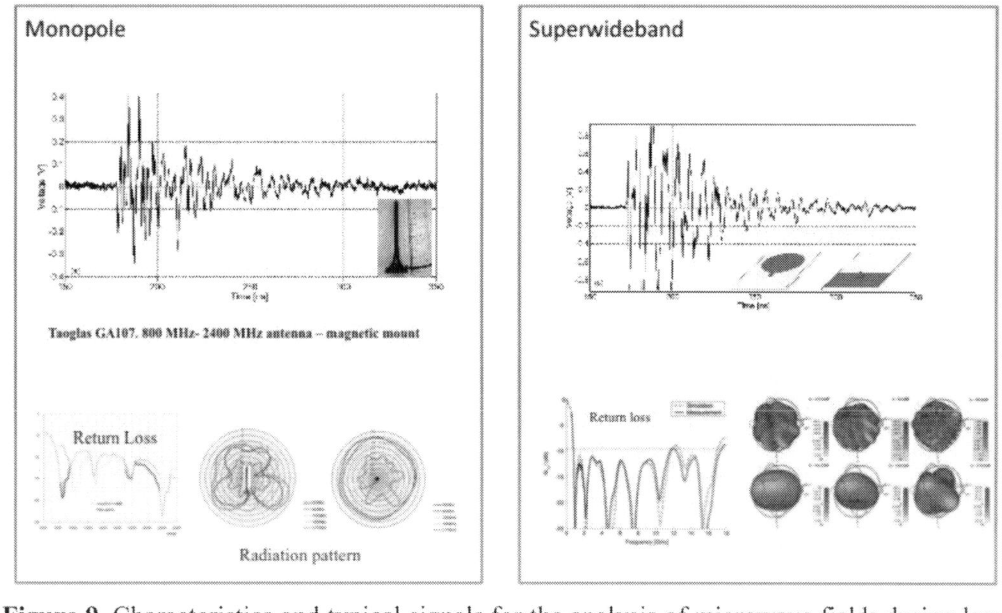

Figure 9. Characteristics and typical signals for the analysis of microwave fields during laser plasma interaction.

Acknowledgements

The authors thank A. Bonasera, M. Calamosca, D. Giulietti, M. Kalal for constant collaboration in the activities reported here.

References

[1] S.Yu. Gus'kov et al, JETP **81**, 296 (1995);
[2] S.Yu. Gus'kov et al, Laser Part. Beams **17**, 287 (1999);
[3] A. Caruso et al, Laser Part. Beams **18**, 25 (2000);

[4] F. Consoli, R. De Angelis et al, NIMA **A 720** 149–152 (2013);
[5] S.Yu. Gus'kov et al, J. Russ. Laser Res., **31**(6), 574-587 (2010)
[6] C. Strangio et al. Quant. Electr. **36**(5) 424-428 (2006)
[7] M. Kalal SPIE Volume 1983, Part II, pp. 686-687
[8] F. Consoli, R. De Angelis et al, Proc. 40th EPS Conf. on Plasma Phys., 2013, Espoo, Finland
[9] D. Nikezic and K.N. Yu, Comput. Phys. Comm. **174**, 160 (2006); Comput. Phys. Comm. **178**, 591 (2008)

AUTHOR INDEX

Aladi, M ... 95
Alifano, P ... 159
Ando', L ... 113
Andreoli, P ... 180
Andrighetto, A ... 104
Anzalone, A ... 54
Badziak, J ... 31
Belloni, F ... 133
Bozzetti, M P ... 169
Branca, C ... 154
Brancato, V ... 149
Buccolieri, A ... 174
Buccolieri, G ... 118, 174
Calcagno, L ... 7, 48, 68
Calderolla, M ... 104
Carini, G ... 154
Carpinelli, M ... 54
Castellano, A ... 174
Castrizio, E D ... 138
Cavallaro, S ... 113, 129
Chodukowski, T ... 31
Cikhardt, J ... 39
Cirrone, G A P ... 54
Consoli, F ... 180
Corradetti, S ... 104
Cristofari, G ... 180
Crupi, C ... 154
Cutroneo, M ... 7, 48, 54, 68, 113, 129, 138, 143, 149, 164
Cuttone, G ... 54
D'Angelo, G ... 154
De Angelis, R ... 180
De Benedittis, A ... 118
De Marco, M ... 39
De Martinis, C ... 54
Di Giorgio, G ... 180
Di Giulio, M ... 174
Di Marco, G ... 154
Di Paolo Emilio, M ... 61
Földes, I B ... 95
Franci, A ... 104
Friscini, A ... 169
Gasior, P ... 89
Genovese, C ... 143
Giove, D ... 54
Giuffreda, E ... 123, 159, 169, 174
Giulietti, D ... 1
Gurin, A ... 99
Jablonski, S ... 31
Klir, D ... 39
Korn, G ... 54
Krása, J ... 14, 21, 39, 129
Krousky, E ... 31, 39
Kubkowska, M ... 89
Kucharik, M ... 31
La Via, F ... 48
Liska, R ... 31
Lorusso, A ... 21
Maggiore, M ... 54
Malinowski, K ... 89
Manti, L ... 54
Manzolaro, M ... 104
Margarone, D ... 54
Mihailescu, A ... 108
Monetti, A ... 104
Musumarra, A ... 54
Musumeci, P ... 48
Nassisi, V ... 14, 21, 26, 39, 75, 118, 123, 159, 169, 174
Nicolosi, P ... 104
Pais, V ... 108
Paladini, F ... 118, 123
Palladino, L ... 61
Parys, P ... 31
Pedullà, E ... 143
Perozziello, F M ... 54
Petrovic, I ... 54
Pfeifer, M ... 39
Pisarczyk, T ... 31
Pisciotta, P ... 54
Prete, G ... 104
Ra¸czka, P ... 31
Rapisarda, E ... 143
Renis, M ... 54
Ristic-Fira, A ... 54
Rohlena, K ... 43
Romano, F ... 54
Romano, F P ... 54
Rosinski, M ... 7, 31, 89
Rossignoli, M ... 104
Roszkowska, A M ... 82
Sadowski, M J ... 89
Scarpa, D ... 104
Schettino, G ... 54
Schillaci, F ... 54
Scolaro, C ... 143, 164
Scuderi, V ... 54
Side, D Delle ... 14, 21, 26, 75, 118, 123, 159, 169, 174
Skladnik-Sadowska, E ... 89
Specchia, V ... 169
Stancalie, V ... 108
Stránský, M ... 43
Talà, A ... 159
Tempera, G ... 143
Tomaselli, A ... 104
Torrisi, A ... 68, 138
Torrisi, L ... 7, 31, 48, 54, 68, 82, 113, 129, 138, 143, 149, 154, 164, 174
Tramontana, A ... 54
Tredici, S M ... 159
Tudisco, S ... 54

AUTHOR INDEX

Ullschmied, J 7, 31, 48, 129
Vasquez, J..104
Velardi, L.... 14, 21, 26, 75, 118, 123, 159, 164, 169, 174
Velyhan, A 113
Visco, A M 149
Wanderlingh, U...154

Institute of Physics
Dirac House, Temple Back
Bristol BS1 6BE UK

CURRAN ASSOCIATES INC
proceedings.com